MOLECULAR BIOLOGY
INTELLIGENCE
UNIT

Molecular Biology of the Parathyroid

Tally Naveh-Many, Ph.D.
Minerva Center for Calcium and Bone Metabolism
Nephrology Services
Hadassah Hebrew University Medical Center
Jerusalem, Israel

LANDES BIOSCIENCE / EUREKAH.COM
GEORGETOWN, TEXAS
U.S.A.

KLUWER ACADEMIC / PLENUM PUBLISHERS
NEW YORK, NEW YORK
U.S.A.

MOLECULAR BIOLOGY OF THE PARATHYROID

Molecular Biology Intelligence Unit

Landes Bioscience / Eurekah.com
Kluwer Academic / Plenum Publishers

Copyright ©2005 Eurekah.com and Kluwer Academic / Plenum Publishers

All rights reserved.
No part of this book may be reproduced or transmitted in any form or by any means, electronic or mechanical, including photocopy, recording, or any information storage and retrieval system, without permission in writing from the publisher, with the exception of any material supplied specifically for the purpose of being entered and executed on a computer system; for exclusive use by the Purchaser of the work.

Printed in the U.S.A.

Kluwer Academic / Plenum Publishers, 233 Spring Street, New York, New York, U.S.A. 10013
http://www.wkap.nl/

Please address all inquiries to the Publishers:
Landes Bioscience / Eurekah.com, 810 South Church Street
Georgetown, Texas, U.S.A. 78626
Phone: 512/ 863 7762; FAX: 512/ 863 0081
www.Eurekah.com
www.landesbioscience.com

Molecular Biology of the Parathyroid, edited by Tally Naveh-Many, Landes / Kluwer dual imprint / Landes series: Molecular Biology Intelligence Unit

ISBN: 0-306-47847-1

While the authors, editors and publisher believe that drug selection and dosage and the specifications and usage of equipment and devices, as set forth in this book, are in accord with current recommendations and practice at the time of publication, they make no warranty, expressed or implied, with respect to material described in this book. In view of the ongoing research, equipment development, changes in governmental regulations and the rapid accumulation of information relating to the biomedical sciences, the reader is urged to carefully review and evaluate the information provided herein.

Library of Congress Cataloging-in-Publication Data

Molecular biology of the parathyroid / [edited by] Tally Naveh-Many.
   p. ; cm. -- (Molecular biology intelligence unit)
 Includes bibliographical references and index.
 ISBN 0-306-47847-1
 1. Parathyroid glands--Molecular aspects. 2. Parathyroid hormone. I. Naveh-Many, Tally. II. Series: Molecular biology intelligence unit (Unnumbered)
 [DNLM: 1. Parathyroid Glands--physiology. 2. Molecular Biology. 3. Parathyroid Glands--physiopathology. 4. Parathyroid Hormone--physiology. WK 300 M718 2005]
 QP188.P3M654 2005
 612.4'4--dc22

2004023419

To Dani, Assaf, Yoav and Amir

CONTENTS

Preface xiii

1. **Development of Parathyroid Glands** 1
   Thomas Günther and Gerard Karsenty
   Physiology of the Parathyroid Glands 1
   Development of Parathyroid Glands in Vertebrates 1
   Genetic Control of Parathyroid Gland Development 3

2. **Parathyroid Hormone, from Gene to Protein** 8
   Osnat Bell, Justin Silver and Tally Naveh-Many
   The Prepro PTH Peptide 8
   Homology of the Mature PTH 9
   The PTH mRNA 10
   Cloning of the PTH cDNAs 11
   Homology of the cDNA Sequences 12
   Structure of the PTH mRNA 18
   The PTH Gene 21
   The 5' Flanking Region 24
   The 3' Flanking Region 24
   Chromosomal Location of the Human PTH Gene 25

3. **Toward an Understanding of Human Parathyroid Hormone Structure and Function** 29
   Lei Jin, Armen H. Tashjian, Jr., and Faming Zhang
   PTH and Its Receptor Family 29
   PTH Structural Determination 30
   Structural Based Design of PTH Analogs 37

4. **The Calcium Sensing Receptor** 44
   Shozo Yano and Edward M. Brown
   Biochemical Characteristics of the CaR 45
   Disorders Presenting with Abnormalities in Calcium
      Metabolism and in the CaR 47
   Signaling Pathways of the CaR 49
   Drugs Acting on the CaR 50

5. **Regulation of Parathyroid Hormone mRNA Stability by Calcium and Phosphate** 57
   Rachel Kilav, Justin Silver and Tally Naveh-Many
   - Regulation of the Parathyroid Gland by Calcium and Phosphate 57
   - Protein Binding and PTH mRNA Stability 58
   - Identification of the PTH mRNA 3'-UTR Binding Proteins and Their Function 61
   - Identification of the Minimal *cis* Acting Protein Binding Element in the PTH mRNA 3'-UTR 62
   - The Structure of the PTH *cis* Acting Element 64

6. **In Silico Analysis of Regulatory Sequences in the Human Parathyroid Hormone Gene** 68
   Alexander Kel, Maurice Scheer and Hubert Mayer
   - Global Homology of PTH Gene between Human and Mouse 71
   - Computer Assisted Search for Potential Cis-Regulatory Elements in PTH Gene 75
   - Phylogenetic Footprint: Identification of TF Binding Sites by Comparison of Regulatory Regions of PTH Gene of Different Organisms 78
   - Discussion 80

7. **Regulation of Parathyroid Hormone Gene Expression by 1,25-Dihydroxyvitamin D** 84
   Tally Naveh-Many and Justin Silver
   - Transcriptional Regulation of the PTH Gene by $1,25(OH)_2D_3$ 84
   - Calreticulin and the Action of $1,25(OH)_2D_3$ on the PTH Gene 89
   - PTH Degradation 90
   - Secondary Hyperparathyroidism and Parathyroid Cell Proliferation 90

8. **Vitamin D Analogs for the Treatment of Secondary Hyperparathyroidism in Chronic Renal Failure** 95
   Alex J. Brown
   - Pathogenesis of Secondary Hyperparathyroidism in Chronic Renal Failure 95
   - Treatment of Secondary Hyperparathyroidism 96
   - Mechanisms for the Selectivity of Vitamin D Analogs 104
   - Future Perspectives 109

9. **Parathyroid Gland Hyperplasia in Renal Failure** 113
   Adriana S. Dusso, Mario Cozzolino and Eduardo Slatopolsky
   Parathyroid Tissue Growth in Normal Conditions
   and in Renal Failure 114
   Dietary Phosphate Regulation of Parathyroid Cell Growth
   in Uremia 116
   Vitamin D Regulation of Uremia- and High Phosphate-Induced
   Parathyroid Cell Growth 120
   Calcium Regulation of Uremia-Induced Parathyroid Growth 123

10. **Molecular Mechanisms in Parathyroid Tumorigenesis** 128
    Eitan Friedman
    Oncogenes Involved in Parathyroid Tumor Development 129
    Tumor Suppressor Genes Involved in Parathyroid
    Tumorigenesis 130
    Other Molecular Pathways Involved in Parathyroid
    Tumorigenesis 132

11. **Molecular Genetic Abnormalities in Sporadic
    Hyperparathyroidism** 140
    Trisha M. Shattuck, Sanjay M. Mallya and Andrew Arnold
    Implications of the Monoclonality of Parathyroid Tumors 141
    Molecular Genetics of Parathyroid Adenomas 142
    Molecular Genetics of Parathyroid Carcinoma 151
    Molecular Genetics of Secondary and Tertiary
    Hyperparathyroidism 152

12. **Genetic Causes of Hypoparathyroidism** 159
    Rachel I. Gafni and Michael A. Levine
    Disorders of Parathyroid Gland Formation 159
    Disorders of Parathyroid Hormone Synthesis or Secretion 167
    Parathyroid Gland Destruction 170
    Resistance to Parathyroid Hormone 171

13. **Skeletal and Reproductive Abnormalities in *Pth*-Null Mice** 179
    *Dengshun Miao, Bin He, Beate Lanske, Xiu-Ying Bai,
    Xin-Kang Tong, Geoffrey N. Hendy, David Goltzman
    and Andrew C. Karaplis*
    Results 180
    Discussion 188
    Materials and Methods 193

Index 197

EDITOR

Tally Naveh-Many
Minerva Center for Calcium and Bone Metabolism
Nephrology Services
Hadassah Hebrew University Medical Center
Jerusalem, Israel
Chapters 2, 5, 7

CONTRIBUTORS

Andrew Arnold
Center for Molecular Medicine
University of Connecticut Health Center
Farmington, Connecticut, U.S.A.
Chapter 11

Xiu-Ying Bai
Division of Endocrinology
Department of Medicine and Lady Davis
   Institute for Medical Research
Sir Mortimer B. Davis-Jewish General
   Hospital
McGill University
Montreal, Canada
Chapter 13

Osnat Bell
Minerva Center for Calcium
   and Bone Metabolism
Nephrology Services
Hadassah Hebrew University
   Medical Center
Jerusalem, Israel
Chapter 2

Alex J. Brown
Renal Division
Washington University School
   of Medicine
St. Louis, Missouri, U.S.A.
Chapter 8

Edward M. Brown
Endocrine-Hypertension Unit
Brigham and Women's Hospital
Boston, Massachusetts, U.S.A.
Chapter 4

Mario Cozzolino
Renal Division
Washington University School
   of Medicine
St. Louis, Missouri, U.S.A.
Chapter 9

Adriana S. Dusso
Renal Division
Washington University School
   of Medicine
St. Louis, Missouri, U.S.A.
Chapter 9

Eitan Friedman
Institute of Genetics
Sheba Medical Center
Tel Hashomer, Israel
Chapter 10

Rachel I. Gafni
Division of Pediatric Endocrinology
University of Maryland Medical Systems
Baltimore, Maryland, U.S.A.
Chapter 12

David Goltzman
Calcium Research Laboratory
   and Department of Medicine
McGill University Health Centre
   and Royal Victoria Hospital
McGill University
Montreal, Canada
Chapter 13

Thomas Günther
Department of Obstetrics
  and Gynecology
Freiburg University Medical Center
Freiburg, Germany
Chapter 1

Bin He
Division of Endocrinology
Department of Medicine and Lady Davis
  Institute for Medical Research
Sir Mortimer B. Davis-Jewish General
  Hospital
McGill University
Montreal, Canada
Chapter 13

Geoffrey N. Hendy
Calcium Research Laboratory
  and Department of Medicine
McGill University Health Centre
  and Royal Victoria Hospital
McGill University
Montreal, Canada
Chapter 13

Lei Jin
Suntory Pharmaceutical Research
  Laboratories LLC
Cambridge, Massachusetts, U.S.A.
Chapter 3

Andrew C. Karaplis
Division of Endocrinology
Department of Medicine and Lady Davis
  Institute for Medical Research
Sir Mortimer B. Davis-Jewish General
  Hospital
McGill University
Montreal, Canada
Chapter 13

Gerard Karsenty
Department of Molecular
  and Human Genetics
Baylor College of Medicine
Houston, Texas, U.S.A.
Chapter 1

Alexander Kel
Department of Research
  and Development
BIOBASE GmbH
Wolfenbüttel, Germany
Chapter 6

Rachel Kilav
Minerva Center for Calcium
  and Bone Metabolism
Nephrology Services
Hadassah Hebrew University
  Medical Center
Jerusalem, Israel
Chapter 5

Beate Lanske
Department of Oral and Developmental
  Biology
Forsyth Institute and Harvard School
  of Dental Medicine
Boston, Massachusetts, U.S.A.
Chapter 13

Michael A. Levine
Department of Pediatric Endocrinology
The Children's Hospital
  at The Cleveland Clinic
Cleveland Clinic Lerner College
  of Medicine of Case Western
  Reserve University
Cleveland, Ohio, U.S.A.
Chapter 12

Dengshun Miao
Calcium Research Laboratory
  and Department of Medicine
McGill University Health Centre
  and Royal Victoria Hospital
McGill University
Montreal, Canada
Chapter 13

Sanjay M. Mallya
Center for Molecular Medicine
University of Connecticut School
 of Medicine
Farmington, Connecticut, U.S.A.
Chapter 11

Hubert Mayer
Department of Gene Regulation
Gesellschaft für Biotechnologische
 Forschung
Braunschweig, Germany
Chapter 6

Maurice Scheer
Department of Research
 and Development
BIOBASE GmbH
Wolfenbüttel, Germany
Chapter 6

Trisha M. Shattuck
Center for Molecular Medicine
University of Connecticut School
 of Medicine
Farmington, Connecticut, U.S.A.
Chapter 11

Justin Silver
Minerva Center for Calcium
 and Bone Metabolism
Nephrology Services
Hadassah Hebrew University
 Medical Center
Jerusalem, Israel
Chapters 2, 5, 7

Eduardo Slatopolsky
Renal Division
Washington University School
 of Medicine
St. Louis, Missouri, U.S.A.
Chapter 9

Armen H. Tashjian, Jr.
Department of Cancer Cell Biology
Harvard School of Public Health
 and Department of Biological
 Chemistry and Molecular
 Pharmacology
Harvard Medical School
Boston, Massachusetts, U.S.A.
Chapter 3

Xin-Kang Tong
Division of Endocrinology
Department of Medicine and Lady Davis
 Institute for Medical Research
Sir Mortimer B. Davis-Jewish General
 Hospital
McGill University
Montreal, Canada
Chapter 13

Shozo Yano
Department of Nephrology
Ichinomiya Municipal Hospital
Ichinomiya, Aichi, Japan
Chapter 4

Faming Zhang
Lilly Research Laboratories
Eli Lilly & Company
Indianapolis, Indiana, U.S.A.
Chapter 3

PREFACE

Maintaining extracellular calcium concentrations within a narrow range is critical for the survival of most vertebrates. PTH, together with vitamin D, responds to hypocalcemia to increase extracellular calcium levels, by acting on bone, kidney and intestine. The recent introduction of PTH as a major therapeutic agent in osteoporosis has directed renewed interest in this important hormone and in the physiology of the parathyroid gland. The parathyroid is unique in that low serum calcium stimulates PTH secretion. As hypocalcemia persists, there is also an increase in PTH synthesis. Chronic hypocalcemia leads to hypertrophy and hyperplasia of the parathyroid gland together with increased production of the hormone. Phosphate is also a key modulator of PTH secretion, gene expression and parathyroid cell proliferation.

Understanding the biology of the parathyroid as well as the mechanisms of associated diseases has taken great strides in recent years. This book summarizes the molecular mechanisms involved in the function of the parathyroid gland. The first chapter reviews the development of the parathyroid gland and the genes involved in this process as identified using genetically manipulated mice. Then the biosynthetic pathway of PTH from gene expression to its intracellular processing and the sequences in the gene controlling its transcription as well as those regulating mRNA processing, stability and translation are described. Studies on the structure of PTH with correlations to its function are presented and provide a starting point for understanding the recognition of the PTH ligand by its receptor the PTH/PTHrP or PTH1 receptor. The calcium sensing receptor regulates PTH secretion, gene expression and parathyroid cell proliferation. A chapter on the calcium receptor focuses on the signalling pathways that it activates and the associated disorders that involve the calcium receptor gene and lead to excess or decreased PTH secretion. Calcium and phosphate regulate PTH gene expression post-transcriptionally. The mechanisms of this regulation and the *cis* and *trans* acting factors that are involved in determining PTH mRNA stability are described. Vitamin D's active metabolite, $1,25(OH)_2$-vitamin D_3, regulates PTH gene transcription. The regulatory sequences in the human PTH gene and the studies on the regulation of PTH gene transcription by $1,25(OH)_2$-vitamin D_3 as well as the subsequent use of vitamin D analogs for the treatment of secondary hyperparathyroidism are all reviewed. Patients with chronic renal failure develop excessive activity of the parathyroid gland that causes severe bone disease. The known factors involved in its pathogenesis are $1,25(OH)_2$-vitamin D_3, a low serum calcium and a high serum phosphate. Insights into the mechanisms implicated in secondary hyperparathyroidism of renal failure are now being revealed and are discussed. Additional chapters are devoted to the pathophysiology of abnormalities of the parathyroid. The genetic alterations involved in parathyroid tumorigenesis are summarized. In addition, the genetic causes of sporadic hyperparathyroidism and hypoparathyroidism are reviewed. The genetic mutations leading to diseases of hyper- or hypoactivity of the parathyroid have elucidated a host of interacting transcription factors that have a central role in normal physiology. Finally, the last chapter focuses on the characteristics of PTH-null mice and the skeletal and reproductive abnormalities that they present.

Together the chapters of this book offer a state of the art description of the major aspects of the molecular biology of the parathyroid gland, PTH production and secretion. The book is designed for students and teachers as well as scientists and investigators who wish to acquire an overview of the changing nature of the PTH field. I would like to express my deep appreciation to all the authors who have contributed to this book for their comprehensive and stimulating chapters and for making the book what it is. I am especially grateful to Justin Silver for his help and support that have made this book possible. I also thank Landes Bioscience for giving me the opportunity to edit this book.

Tally Naveh-Many, Ph.D.

CHAPTER 1

Development of Parathyroid Glands

Thomas Günther and Gerard Karsenty

Summary

The parathyroid glands (PG) are the main source for circulating parathyroid hormone (PTH), a hormone that is essential for the regulation of calcium and phosphate metabolism. The PGs develop during embryogenesis from the pharyngeal pouches with contributions from endodermal and neural crest cells. A few genes have been attributed to the formation, migration and differentiation of the PG anlage. In studies mostly done in genetically manipulated mice it could be demonstrated that *Rae28*, *Hoxa3*, *Pax1*, *Pax9* and *Gcm2* are essential for proper PG formation. Recently, candidate genes involved in the DiGeorge syndrome have been identified as well.

Physiology of the Parathyroid Glands

The parathyroids are small glands located in the cervical region in close proximity to the thyroids. The main function of the PGs is the secretion of PTH. It is on top of a complex hormonal cascade regulating serum calcium concentration (Fig. 1). The latter is remarkably constant in diverse organisms under various physiological conditions. This tight regulation is important since calcium is essential for many functions such as muscle contraction, neuronal excitability, blood coagulation, mineralization of bone and others. A reduction of the serum calcium concentration to less than 50% will lead to tetany and subsequently to death. The importance of a strict regulation of the serum calcium is also reflected by the rapid secretion of PTH within seconds, new synthesis of the hormone within minutes and new transcription within hours following a decrease in serum calcium concentration which is detected through the calcium sensing receptor expressed in the PGs. The overall role of PTH is to increase calcium concentration. It fulfils this function through three different means. First it prevents calcium elimination in the urine, second it favors the hydroxylation in one of the 25 hydroxycholecalciferol and as a results it favors indirectly intestinal calcium absorption. Lastly PTH favors through still poorly understood mechanisms bone resorption and as a result increases the extracellular calcium concentration (Fig. 1).

Development of Parathyroid Glands in Vertebrates

The PGs derive from the pharyngeal pouches which are transient structures during embryonic development. They are evolutionary homologous to gill slits in fish. The foregut endoderm and cells originating from the neural crest of rhombomere 6 and 7 contribute to the anlage of the PGs. The neural crest originates at the apposition of neuroectoderm and ectoderm during the formation of the neural tube. Therefore neural crest cells have to migrate

Molecular Biology of the Parathyroid, edited by Tally Naveh-Many. ©2005 Eurekah.com and Kluwer Academic / Plenum Publishers.

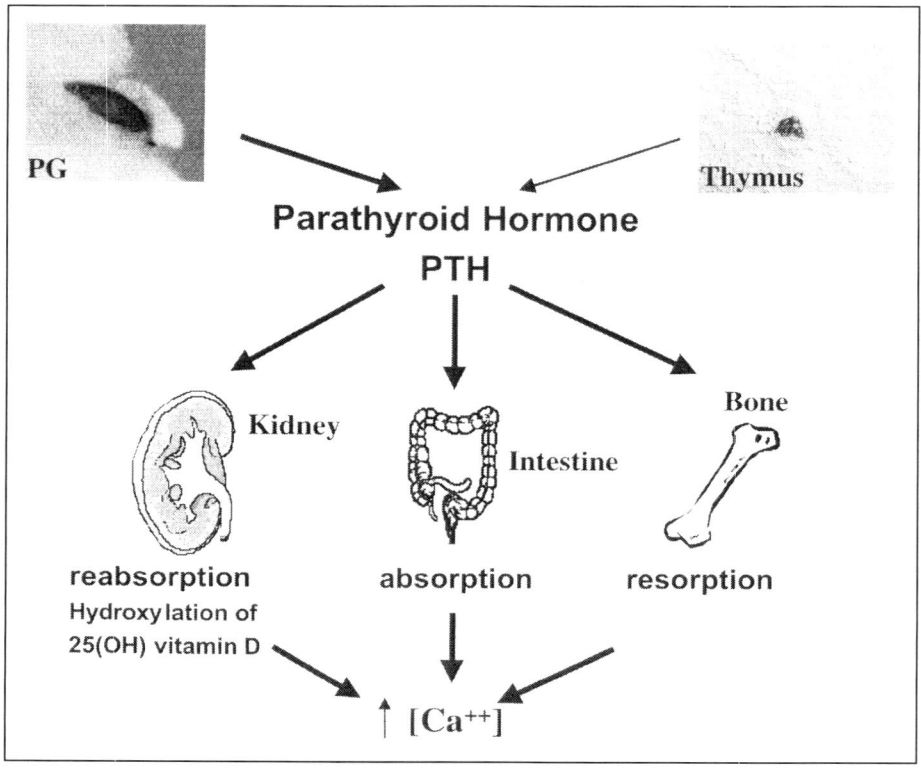

Figure 1. Regulation of calcium homeostasis. Parathyroid hormone is on top of a hormonal cascade regulating serum calcium concentration. PTH secretion leads to an increase of serum calcium through renal reabsorption and intestinal absorption, the latter is caused by the induction of the synthesis of the active form of vitamin D in the kidney. Bone is the main reservoir for calcium containing more than 99% of the body content. Calcium is released through bone resorption. The main source for circulating PTH are the parathyroid glands (PG) while *Pth*-expressing cells in the thymus can function as a backup in mice.

towards the foregut endoderm first before they can add to the anlage of the PGs. Neural crest of rhombomere 6 migrates towards the third branchial arch while the fourth branchial arch is primarily invaded by neural crest cells from rhombomere 7 (Fig. 2).

Mice only have one pair of PGs deriving from the third pharyngeal pouch homologous to the inferior PGs in men while the superior ones derive from the fourth pharyngeal pouch. The anlage of the PGs in mice first becomes visible between embryonic day 11 (E11) and E11.5 histologically in a very limited area in the dorsal region of the cranial wall of the third endodermal pouch while the caudal portion of the very same pouch develops into the thymus which is involved in the maturation of the immune system (Fig. 2).[1] Both domains are demarkated by the complementary expression of *Gcm2* and *Foxn1* (the latter mutated in *nude* mice, lacking a functional thymus), respectively already two days before the anlagen are morphologically visible.[2] In contrast to thymus development, induction of the ectoderm is not necessary for the formation of the PGs.[3] In mammals both structures start to migrate shortly thereafter towards the caudal end before at around E14 they seperate. While the thymus moves on further in the direction of the heart the PGs become incorporated to the thyroid gland between E14 and E15.

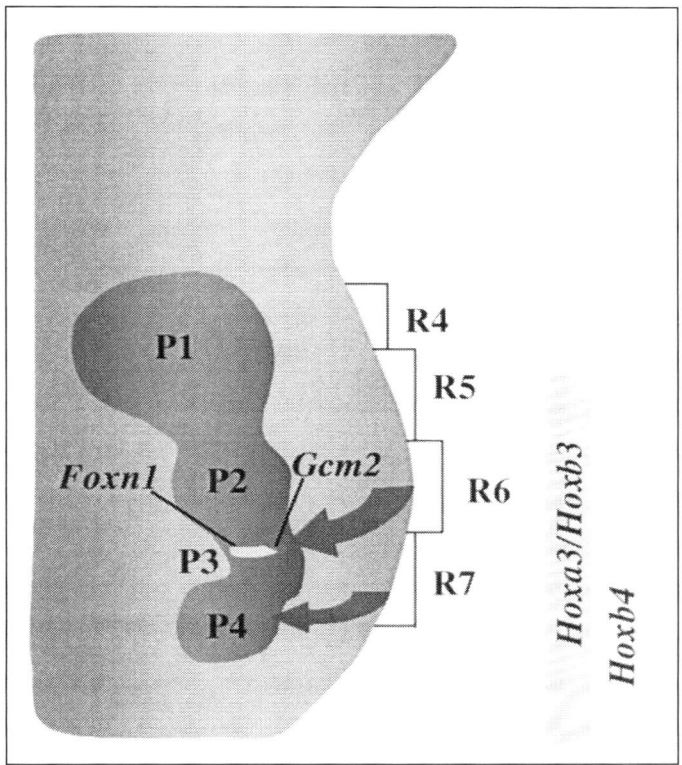

Figure 2. Specification of the parathyroid gland anlage. The parathyroid glands develop from the third pahryngeal pouch (in humans from P3 and P4). Neural crest cells evaginating from rhombomere six and seven (R6, R7) of the hindbrain and pharyngeal endoderm contribute the primordium of PGs and thymus. Both anlagen are demarcated by the expression of *Gcm2* and *Foxn1*, respectively, already two days before the anlagen become histological visible. The identity of the neural crest is determined by genes of the Hox cluster. The anterior expression borders of *Hoxa/b3* and *Hoxb4* are depicted.

Pth is expressed already in the anlage of the PGs at E11.5[4] and contributes to fetal serum calcium regulation to some extent although placental transport involving parathyroid hormone related protein (PTHrP) is more important.[5] The parathyroid gland is not the only source of PTH. The protein is also synthesized by a few cells in the hypothalamus[6] and in the thymus.[4] It has been shown in mice that the thymic *Pth*-expressing cells actually contribute to the circulating hormone keeping the level of serum calcium even in the absence of PGs at a concentration compatible with life.[4]

Genetic Control of Parathyroid Gland Development

Three different steps can be used to separate the formation of the PGs mechanistically. They include (I) formation of the PGs, (II) migration towards their final destination and (III) the differentiation towards PTH producing cells (Fig. 3). Mouse mutants that highlight the role of the few genes known to be involved in these different processes have been generated in the last decade.

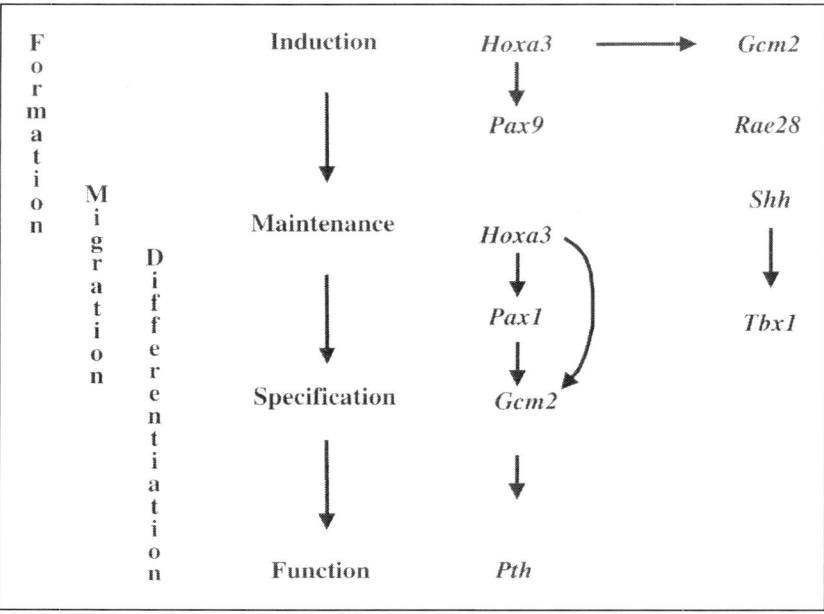

Figure 3. Schematic representation of parathyroid gland development. Parathyroid gland development can be mechanistically seperated into formation of the anlage, caudal migration towards their final location within the thyroid glands and differentiation into PTH-secreting cells. The genetic interactions between factors involved in induction, maintenance, specification and function are shown.

Both, neural crest cells and the pharyngeal endoderm contribute to the anlage of the PGs. Neural crest cells possibly already maintain information about their localization along the anterior-posterior axis before they start to migrate ventrally. They derive this information from a group of evolutionary conserved transcription factors containing a homebox, the Hox genes, organized in four paralogous genomic clusters (Hoxa, b, c and d). Hox genes are expressed in the neural crest prior to, during and after migration into the pharyngeal arches and endodermal epithelia express Hox genes as well.

I. *Rae28* is the mouse homologue of the Drosophila polyhomeotic gene which is required for the proper expression of hometic genes along the anterior-posterior axis. Similar, absence of *Rae28* causes an anterior shift of anterior expression boundaries of several genes of the Hox cluster including *Hoxa3*, *Hoxb3* and *Hoxb4*. Mice deficient for *Rae28* are characterized by malformations of tissues partly derived from neural crest like altered localization of PGs as well as PG and thymic hypoplasia and cardiac anomalies.[7] How the altered hox expression pattern influences PG formation still needs to be evaluated.

The first reported malformation of PGs caused by a deletion through homologous recombination in mouse embryonic stem cells were represented by *Hoxa3*-deficient animals. Among other defects knockout mice are devoid of PGs and thymus and exhibit thyroid hypoplasia.[8] This coincides very well with *Hoxa3* expression in the third and fourth pharyngeal arches and in the pharyngeal endoderm. The Hoxa3 signal does neither effect the number of neural crest cells nor their migration pattern. Mutant cells rather lost their capacity to induce differentiation of surrounding tissues.[10]

Absence of the paired box containing transcription factor *Pax9* in targeted mice also displays absence of PGs and thymus. *Pax9* is expressed in the pharyngeal endoderm. The epithelial buds separating from the third pharyngeal pouch did not form in the mutant mice. This phenotype could be traced back to delayed development of the third pouch already at E11.5 and coincides with the expression of *Pax9* in the pharyngeal endoderm.[9]

II. PGs develop normally in mice deficient for the paralogous *Hoxb3* and *Hoxd3*. However further removal of a single *Hoxa3* allele leads to the inability of the normally formed anlge of the PGs to migrate to their position next to the thyroid gland.[10] Therefore, development and migration of the PGs are separable events which is consistent with the fact that in other vertebrates like fish and birds PGs do not migrate from location of their origination.

III. Glial cell missing2 (Gcm2) is the homoloug of the Drosophila GCM transcription factor. Unlike its glia cell fate determining function in fruit flies implies, mouse *Gcm2* exclusively characterizes parathyroid cells and starts to be expressed around E10 in the pharyngeal endoderm.[11] The pattern rapidly becomes restricted to the cranial portion of the third pharyngeal pouch.[2] Mice deficient for *Gcm2* revealed that *PTH* is never expressed in the PG anlage although parathyroid like cells characterized by *Pax9* expression are still present at E14.5.[4] This clearly points out that Gcm2 is essential for the specification of precursors to become *Pth*-expressing cells rather than for the induction of the precursors itself (Fig. 3). Interestingly, Pth-positive cells still could be detected in the thymus of mutant mice indicating that at least 2 pathways for the specification of *Pth*-expressing cells exist (Fig. 1). *Gcm1* expressed in the thymus is the most likely candidate to compensate for Gcm2 function. It will be compelling to determine if a ‚backup mechanism' for the parathyroid gland also exists in man. In this direction it is very interesting to note that the first human homozygous mutation for *GCM2* has been identified in hypoparathyroidic patients.[12]

It has been discovered just recently that newborn *Pax1*-deficient mice exhibit severely reduced PGs.[13] The reduction in size could be traced back to the beginning of PG development at E11.5. The hypoplasia of the anlage was even more severe in *Hoxa3+/-Pax1-/-* embryous and PGs were absent at late gestational stages.[13] Interestingly, *Gcm2* expression although properly initiated at E10.5 was reduced at E11.5 in *Pax1*-deficient embryos while the reduction was even more severe in the compound mutant. *Hoxa3*-deficient embryos exhibit no *Gcm2* signal at all.[13] Therefore, *Hoxa3* is necessary for *Gcm2* induction while both *Hoxa3* and *Pax1* are substantial for the proper maintenance of Gcm2 expression. *Pax1* expression in the PG primordium on the other hand is reduced in *Hoxa3*-deficient mice.[3,13] This would place *Hoxa3* genetically upstream of *Pax1* and both upstream of *Gcm2* which in turn is required for *PTH* expression in PGs (Fig. 3).

A long time known conglomerate of congenital malformations in humans including dysplasia or absence of the PGs and thymus as well as malformations of the heart outflow is the DiGeorge syndrome. The organs affected derive in part from neural crest so that mutations in one or several genes influencing these cells have been suspected to be the cause for the disease. It could be shown that most patients are hemizygous for a megabase deletion on chromosome 22q11. Recently, two groups came up with a good candidate gene for several of the features in DiGeorge syndrome including PG defects simultaneously.[14,15] Both laboratories generated hemizygous megabase deletions comprising more than a dozen genes on the synthenic mouse chromosome 16 that reflected the human malformations including PG abnormalities. *TBX1* was among them and it could be shown that the gene is expressed in the pharyngeal endoderm and mesoderm-derived core but not in neural crest-derived mesenchyme.[14-16] *Tbx1* expression in the pharyngeal arches is possibly induced through the morphogen Sonic hedgehog.[16] Mice heterozygous for a *Tbx1* deletion by homologous recombination reflected the pharyngeal arch artery malformations while homozygous-deficient mice exhibited PG hypoplasia.[14,15,17]

DiGeorge syndrome patients resemble hemizygous deletions. This suggests that other genes of this region may contribute to the PG phenotype. Indeed, Guris and colleagues[18] could demonstrate that mice homozygous for a targeted null mutation for *Crkol* dysplay cardiovascular, PG and thymus defects. The migration and early proliferation of neural crest cells was not altered pointing out that Crkol influences the function of neural crest during later stages. *CRKL* (homolog human gene name) also maps within the common deletion region for the DiGeorg syndrome.

Deletions on chromosome 10p also cause DiGeorge like malformations. The locus includs a subregion that encodes for the hypoparathyroidism, sensorineural deafness, renal anomaly (HDR) syndrome. Van Esch and her colleagues[19] could demonstrate that two heterozygous patients exhibit loss of function mutations in *GATA3*. The transcription factor is indeed expressed in the affected organs during human and mouse embryonic development. Surprisingly though, heterozygous knockout mice have been reported to be normal while homoyzgous mice die around E12.[20]

The understanding of the contribution from several gene products to the development of PGs from these critical regions still awaits further analysis.

Concluding Remark

Clinical studies indicate that multiple mutations can account for the malfunction of serum calcium regulation through PTH in humans. These include the synthesis of PTH, sensing of the calcium content in the blood stream as well as the development PGs and proper specification of PTH translating cells. It is astonishing how rather little is known so far on the molecular level in comparison to the formation of other organs. Surely, the genome sequencing projects for mice and man and the use of microarrays to compare different cDNA pools will shed new light on this issue in the near future.

References

1. Cordier AC and Haumont SM. Development of thymus, parathyroids, and ultimo- branchial bodies in NMRI and nude mice. Am J Anat 1980; 157:227-263.
2. Gordon J, Bennett AR, Blackburn CC. Gcm2 and Foxn1 mark early parathyroid- and thymus-specific domains in the developing third pharyngeal pouch. Mech Dev 2001; 103:141-143.
3. Manley NR and Capecchi MR The role of Hoxa-3 in mouse thymus and thyroid development. Development 1995; 121:1989-2003.
4. Günther T, Chen ZF, Kim J et al. Genetic ablation of parathyroid glands reveals another source of parathyroid hormone. Nature 2000; 406:199-203.
5. Kovacs CS, Manley NR, Moseley JM et al. Fetal parathyroids are not required to maintain placental calcium transport. J Clin Invest 2001; 107:1007-1015.
6. Pang PK, Kaneko T, Harvey S. Immunocytochemical distribution of PTH immunoreactivity in vertebrate brains. Am J Physiol 1988; 255:R643-647.
7. Takihara Y, Tomotsune D, Shirai M et al. Targeted disruption of the mouse homologue of the Drosophila polyhomeotic gene leads to altered anteroposterior patterning and neural crest defects. Development 1997; 124:3673-3682.
8. Chisaka O and Capecchi MR Regionally restricted developmental defects resulting from targeted disruption of the mouse homeobox gene hox-1.5. Nature 1991; 350:473-479.
9. Peters H, Neubüser A, Kratochwil K et al. Pax9-deficient mice lack pharyngeal pouch derivatives and teeth and exhibit craniofacial and limb abnormalities. Genes Dev 1998; 12:2735-2747.
10. Manley NR and Capecchi MR Hox group 3 paralogs regulate the development and migration of the thymus, thyroid, and parathyroid glands. Dev Biol 1998; 195:1-15.
11. Kim J, Jones BW, Zock C et al. Isolation and characterization of mammalian homologs of the Drosophila gene glial cells missing. Proc Natl Acad Sci USA 1998; 95:12364-12369.

12. Ding C, Buckingham B, Levine MA Familial isolated hypoparathyroidism caused by a mutation in the gene for the transcription factor GCMB. J Clin Invest 2001; 108:1212-1220.
13. Su D, Ellis S, Napier A et al. Hoxa3 and Pax1 regulate epithelial cell death and proliferation during thymus and parathyroid organogenesis. Dev Biol 2001; 236:316-329.
14. Merscher S, Funke B, Epstein J A et al. TBX1 is responsible for cardiovascular defects in velo-cardio-facial/DiGeorge syndrome. Cell 2001; 104: 619-629.
15. Lindsay EA, Vitelli F, Su H et al. Tbx1 haploinsufficieny in the DiGeorge syndrome region causes aortic arch defects in mice. Nature 2001; 410: 97-101.
16. Garg V, Yamagishi C, Hu T et al. Tbx1, a DiGeorge syndrome candidate gene, is regulated by sonic hedgehog during pharyngeal arch development. Dev Biol 2001; 235:62-73.
17. Jerome LA and Papaioannou VE. DiGeorge syndrome phenotype in mice mutant for the T-box gene, Tbx1. Nat Genet 2001; 27:286-291.
18. Guris DL, Fantes J, Tara D et al. Mice lacking the homologue of the human 22q11.2 gene CRKL phenocopy neurocristopathies of DiGeorge syndrome. Nat Genet 200; 27: 293-298.
19. Van Esch H, Groenen P, Nesbit MA et al. GATA3 haplo-insufficiency causes human HDR syndrome. Nature 2000; 406:419-422.
20. Pandolfi PP, Roth ME, Karis A et al. Targeted disruption of the GATA3 gene causes severe abnormalities in the nervous system and in fetal liver haematopoiesis. Nat Genet 1995; 11:40-44.

CHAPTER 2

Parathyroid Hormone, from Gene to Protein

Osnat Bell, Justin Silver and Tally Naveh-Many

Abstract

The biosynthetic pathway of parathyroid hormone (PTH) has been studied from gene expression to PTH intracellular processing.[1] The processing of PTH has been described and involves the synthesis of an initial translational product, preProPTH, and two proteolytic cleavages that in turn produce ProPTH and PTH. The genes and cDNAs from ten different species have been cloned, sequenced and characterized. This chapter will summarize the molecular biology of PTH, from the gene to the mRNA, the initial translational product, preProPTH and the processed mature secreted form of PTH. It will describe the sequences of the PTH gene and mRNA in different species and the specific elements in the PTH mRNA that determine mRNA processing, stability and translation.

The Prepro PTH Peptide

The primary form of PTH, which is stored and secreted, contains 84 amino acids.[2] PTH is initially synthesized as a precursor, preProPTH. Two proteolytic cleavages produce the ProPTH and the secreted form of PTH. The proPTH sequence contains six extra amino acids at the N-terminus.[3,4] Conversion of ProPTH to PTH occurs about 15 to 20 min after biosynthesis at about the time ProPTH reached the Golgi apparatus.[5]

The Structure of the Pre-Peptide

Evidence that the translational product of PTH mRNA was larger than ProPTH was initially obtained by translation of a crude preparation of bovine parathyroid RNA in the wheat germ cell-free system.[6] The primary translational product migrated slower than ProPTH when analyzed by electrophoresis on either acidic-urea or sodium dodecyl sulfate-containing acrylamide gels. At that time, a similar phenomenon had been observed only for myeloma light chains.[7] In further studies, preProPTH was shown to be synthesized in cell-free systems of reticulocyte lysates.[8] Translation of human parathyroid RNA also produced an analogous preProPTH.[9]

The observation that the carboxyl terminal peptides of bovine PTH and preProPTH were identical indicated that the extra amino acids in preProPTH were at the amino terminus. This was confirmed by incorporating selected radioactive amino acids into preProPTH and determining the location of the radioactivity by automated Edman degradation.[10] By analyzing overlap of these radioactive amino acids with those in ProPTH, the length of the bovine pre-peptide was shown to be 25 amino acids. The entire sequence of the bovine pre-peptide was determined eventually by this microsequencing technique[11] and was later confirmed by structural studies of both the bovine PTH cDNA and gene.[12-14] The sequence of human pre-peptide was also partially determined by this microsequencing technique.[9] The complete amino acid sequence was derived from the human PTH cDNA sequence[15] and later confirmed by the determination of the structure of the human gene.[16] The amino acid sequence of the rat pre-peptide was derived from the sequence of the rat PTH gene[17] and partially by analysis of cloned rat PTH cDNA.[18]

The amino acid sequences of the pre-peptides show that the human and bovine pre-peptides are 80% homologous while the rat sequence is 64% homologous to the bovine and human.[1] This is somewhat lower than the homology of 89 and 77% in the Pro and PTH regions for bovine/human and rat/bovine-human, respectively (Fig. 1). The fact that the pre-peptide is less conserved than the rest of the molecule is consistent with pre-peptides or signal peptides of many eukaryotic proteins.[19] General structural features of the signal peptides are a central hydrophobic core and, in many cases, charged amino acids at the N-terminal and C-terminal ends of the central core. These features are largely retained in the pre-peptides of the three preProPTH molecules. Only conservative changes are present within the central core of uncharged amino acids from amino acids 10 to 21.[1]

Conversion of PrePro to ProPTH

The removal of the pre-peptide to produce ProPTH is mediated by an enzyme associated with microsomes.[8] In reticulocyte and wheat germ systems that contain little or no microsomal membranes, the primary transcriptional product of PTH mRNA is preProPTH.[6,8] Addition of microsomal membranes from dog pancreas or chicken oviduct results in the synthesis of ProPTH.[8,20]

The first evidence that pre or signal peptides function by binding to a limited number sites in the microsomal membrane was obtained by studies on a synthetic prePro-peptide of bovine preProPTH.[21] The identification of the signal recognition particle as a signal peptide receptor, later on, confirmed this mechanism for most secreted and membrane proteins.[22]

The pre peptide of preProPTH is rapidly degraded after its proteolytic cleavage from preProPTH. In studies of PTH biosynthesis in intact cells, no labeled pre-peptide could be detected.[23] The proteolytic removal of the pre-peptide probably occurs before completion of the ProPTH nascent chain, since preProPTH is difficult to detect in intact cells.

Homology of the Mature PTH

The mature PTH has been determined or predicted by the cDNAs in several species. The sequence of PTH of mouse, rat, man, non-human primates, horse, dog, cat, cow, pig, and chicken is shown in Figure 1. The resulting phylogenetic tree obtained from alignment of the protein sequences is shown in Figure 3A.

A comparison of the amino acid sequences of PTH from several species revealed high conservation of the protein amongst all species apart from gallus (Fig. 1). In addition, three relatively conserved regions could be observed.[17] The first two regions comprise the biologically active region of PTH and would be expected to be conserved. The addition or loss of a single amino acid at the amino terminus greatly reduces biological activity, and the region is involved in binding of PTH to the receptor. In addition there is a region of conservation at the C-terminal region that is itself of interest, particularly since this region may have a separate biological effect at least on osteoclasts.[24] Analyses of the silent changes that occur between the nucleotide sequences suggest that the conservation in the C-terminal region may be related to pre-translational events. Analysis by Perler et al[25] described replacement changes that result in changes in amino acids and silent changes that do not alter the encoded amino acid.

```
                                    pre          pro
              1                      ↓            ↓                              50
murine    MMSANTVAKV  MIIMLAVCLL  TQTDGKPVRK  RAVSEIQLMH  NLGKHLASME
rat       MMSASTMAKV  MILMLAVCLL  TQADGKPVKK  RAVSEIQLMH  NLGKHLASVE
human     MIPAKDMAKV  MIVMLAICFL  TKSDGKSVKK  RSVSEIQLMH  NLGKHLNSME
macaca    MIPAKDMAKV  MIVMLAICFL  TKSDGKSVKK  RSVSEIQLMH  NLGKHLNSME
Equine                                     K  RSVSEIQLMH  NLGKHLNSVE
canine    MMSAKDMVKV  MIVMFAICFL  AKSDGKPVKK  RSVSEIQFMH  NLGKHLSSVE
feline    MMSAKDMVKV  MVVMFAICFL  AKSDGKPVKK  RSVSEIQFMH  NLGKHLSSVE
bovine    MMSAKDMVKV  MIVMLAICFL  ARSDGKSVKK  RAVSEIQFMH  NLGKHLSSME
porcine   MMSAKDTVKV  MVVMLAICFL  ARSDGKPIKK  RSVSEIQLMH  NLGKHLSSLE
gallus    MTSTKNLAKA  IVILYAICFF  TNSDGRPMMK  RSVSEMQLMH  NLGEHRHTVE

              51                                                                100
murine    RMQWLRRKLQ  DMHNFVSLGV  QMAARDGSHQ  KPTKKEENVL  VD........
rat       RMQWLRRKLQ  DVHNFVSLGV  QMAAREGSYQ  RPTKKEENVL  VD........
human     RVEWLRKKLQ  DVHNFVALGA  PLAPRDAGSQ  RPRKKEDNVL  VE........
macaca    RVEWLRKKLQ  DVHNFIALGA  PLAPRDAGSQ  RPRKKEDNIL  VE........
Equine    RVEWLRKKLQ  DVHNFIALGA  PIFHRDGGSQ  RPRKKEDNVL  IE........
canine    RVEWLRKKLQ  DVHNFVALGA  PIAHRDGSSQ  RPLKKEDNVL  VE........
feline    RVEWLRRKLQ  DVHNFVALGA  PIAHRDGGSQ  RPRKKEDNVP  AE........
bovine    RVEWLRKKLQ  DVHNFVALGA  SIAYRDGSSQ  RPRKKEDNVL  VE........
porcine   RVEWLRKKLQ  DVHNFVALGA  SIVHRDGSSQ  RPRKKEDNVL  VE........
gallus    RQDWLQMKLQ  DVHS......  ..ALEDARTQ  RPRNKEDIVL  GEIRNRRLLP

              101                      128
murine    .....GNPKS  LGEGDKADVD  VLVKSKSQ
rat       .....GNSKS  LGEGDKADVD  VLVKAKSQ
human     .....SHEKS  LGEADKADVN  VLTKAKSQ
macaca    .....SHEKS  LGEADKADVD  VLTKAKSQ
Equine    .....SHQXS  LGEADKADVD  VLSKTKSQ
canine    .....SYQKS  LGEADKADVD  VLTKAKSQ
feline    .....NHQKS  LGEADKADVD  VLIKAKSQ
bovine    .....SHQKS  LGEADKADVD  VLIKAKPQ
porcine   .....SHQKS  LGEADKAAVD  VLIKAKPQ
gallus    EHLRAAVQKK  SIDLDKAYMN  VLFKTKP~
```

Figure 1. Alignment of the amino acid sequences of PTH from the 10 different species. Alignments were obtained using the default setting of PileUp program (Accelrys Inc. Madison WI). Comparison of the amino acid sequences of PTH for mouse (mus), rat, human, non human primates (macaca), horse (equine), dog (canine), cat (feline), cow (bovine), pig (porcine) and chicken (gallus). Gaps indicated by dashes were introduced to maximize the homology to the gallus sequence. The N terminal sequence of the equus PTH is not available. The arrows indicate the protolytic cleavage sites required for the conversion of preProPTH to ProPTH and PTH.

The PTH mRNA

Bovine preProPTH mRNA was initially more extensively characterized than the mRNAs from the other species. Preparations of bovine parathyroid RNA were obtained that contained about 50% PTH mRNA as estimated by gel electrophoresis and RNA excess hybridization to radioactive cDNA.[26] The size of the mRNA was estimated to be about 750 nucleotides by sucrose gradient centrifugation. About two thirds of the translatably active mRNA was retained by oligo(dT) cellulose, and the sizes of the poly(A) extension was broadly distributed around an average size of 60 adenylate residues, though this may be an under estimation of the actual size. While not directly determined, PTH mRNA probably contains a 7-methylguanosine cap since the translation of PTH mRNA was inhibited by 7-methylguanosine-5'-phosphate. The human and bovine

PTH mRNAs appear to be heterogeneous at the 5' terminus (see section on genes). The sizes of the rat and human PTH mRNAs have been determined by Northern blot analysis to be about 800 and 850 nucleotides, respectively.[15,17] Therefore, PTH mRNAs are typical eukaryotic mRNAs that contain a 7-methyguanosine cap at the 5' terminus and a polyadenylic acid (poly A) stretch at the 3' terminus. The PTH mRNAs are twice as long as necessary to code for the primary translational product, due to 5' and 3' untranslated regions at both ends of the mRNA.

Cloning of the PTH cDNAs

To date the sequence of the full cDNA of rat,[17] man,[15] dog,[27] cat (un published), cow,[13] pig,[18] and chicken[28] and the partial sequence of horse[29] and non human primates[30] have been determined. The cDNA of mouse PTH was determined from the genomic PTH sequence.[31] Table 1 shows the Gene Bank accession number for the PTH sequences of the different species and the length of the cDNAs of each of the mRNAs as they appear in the NCBI and Gene Bank databases. In addition, the hypothalamus PTH cDNA was sequenced after the PTH mRNA had been detected in neuronal tissue.[32]

The first PTH cDNAs identified were the DNAs complementary to bovine[12,13] and human[15] PTH mRNA that had been cloned into the Pst 1 site of pBR322 by the homopolymer extension technique. The rat PTH cDNA[18] was cloned by the Okayama and Berg method. The bovine mRNA was isolated from normal parathyroid glands, and the human mRNA was isolated from parathyroid adenomas. The sequence of the rat mRNA has been derived partially from the rat cDNA and from the sequence of the cloned gene.[17]

Kronenberg et al[12] initially determined the sequence of a bovine cDNA clone, pPTHm1, which contained about 60% of the PTH mRNA, including the entire region coding for pre-ProPTH. Restriction analysis of near full-length double-stranded cDNA, synthesized enzymatically from partially purified bovine PTH mRNA, indicated that about 200 nucleotides from the 3' untranslated region were missing in the clone.[33] Analysis of several additional bovine PTH cDNA clones and the sequencing of cDNA of the 5' terminus of PTH mRNA, which was synthesized by extension of a primer with reverse transcriptase, provided the full bovine DNA sequence.[34]

Nucleotide sequences of the parathyroid (PTH) gene of 12 species of non-human primates belonging to suborder Anthropoidea were characterized.[30] The deduced amino acid sequences of exons II and III of the PTH gene of the 12 species of non- human primates was compared to the human PTH and revealed no amino acid substitution in the mature PTH among orangutans, chimpanzees, and humans. The results indicated that the PTH gene is highly conserved among primates, especially between great apes and humans.[30]

The 5' end of the bovine mRNA sequence, which was determined by sequencing DNA complementary to the 5' end of PTH mRNA produced by primed reverse transcription,[34] produced multiple 5' termini of the mRNA. The heterogeneity at the beginning of the 5' end of the mRNA was confirmed by S1 nuclease mapping.[14] The longest reverse transcribed cDNA was isolated and sequenced. Surprisingly, this cDNA contained a canonical TATA sequence at the beginning, which was in the proper position to direct the transcription of the shorter mRNAs. This result suggested that a second TATA sequence would be present 5' to the one detected in the cDNA and would direct the synthesis of the longer mRNAs. The predicted second TATA sequence was discovered when the gene was sequenced. The 5' end of the rat PTH mRNA was also analyzed by S1 nuclease mapping and was less heterogeneous than the bovine mRNA.[17] The single species of rat PTH mRNA corresponded to the larger of the bovine mRNAs. The size of the human mRNA, based on the cDNA sequence, is about 100 nucleotides longer than the bovine and rat mRNAs (Table 1). Northern blot analysis of the mRNAs was consistent with these predicted sizes.[17] The 3' untranslated region (UTR) of the avian PTH mRNA is 1236 nt long, much larger than any of the PTH mRNA 3'-UTRs (Table 1). In general the difference in size in

Table 1. List of the known sequences for the PTH gene and the sizes of the mRNA, 5'-UTR, coding region and 3'-UTR

	NCBI Accession Number	mRNA	5'UTR	CDS	3'UTR
Mus musculus	Af066074: gene, exon 1 Af066075: gene, exons 2 and 3 and complete mRNA (deduced)	714	127	348	239
Rat	K01267: gene, exon 1 K01268: gene, exon 2 and 3 X05721: mRNA, complete	704	118	348	238
Canis familiaris	U15662: mRNA, complete	692	88	348	256
Felis catus	Af309967: mRNA, complete	737	63	348	326
Human	J00300: gene, 3' end J00301: gene, coding region and 3' flank V00597: mRNA, complete	772	74	348	350
Macaca fascicularis	Af130257: gene, complete cds	398*		348	50*
Bovine	K01938: gene, complete cds and flank M25082: mRNA, complete	699	127	348	224
Equus caballus	Af134233: gene, partial cds	311*		267*	44*
Porcine	X05722: mRNA, complete	698	96	348	254
Gallus gallus	M36522: mRNA, complete	1723	127	360	1236

The NCBI accession number of the different sequences and the size of the mRNAs are indicated. The asteryxes show sequences that have been partially sequenced.

the PTH mRNA of the different species primarily results from the difference in the size of the 3'-UTR (Table 1). The significance of this finding has not been studied.

The overall nucleotide compositions of the cDNAs are similar. All the sequences are A-T rich. The 3' noncoding region has a particularly large portion of A and T, ranging from 68 to 74%, making it an AU rich element (ARE). The rat sequence differs from the other sequences in that the 5' noncoding region is only 50% A and T compared to 63 to 65% for the human and bovine.

Homology of the cDNA Sequences

Alignment of the PTH cDNAs of the nine preProPTH sequences is shown in Figure 2. Gaps have been introduced in the 5' and 3' untranslated regions to maximize homology. For

```
            1                                                            50
murine      ~~~ctgcata  tgaaactcag  acttgaagaa  ctgcagtcca  gttcatcagc
rat         ~~~ctgcata  tgaaactcag  gcttgaagaa  ctgcagtcca  gttcatcagc
canine      ~~~~~~~~~~  ~~~~~~~~~~  ~~~~~~~~~~  ~~~~~~~~~c  ggcacgagca
feline      ~~~~~~~~~~  ~~~~~~~~~~  ~~~~~~~~~~  ~~~~~~~~~~  ~~~~~~~~~~
human       ~~~~~~~~~~  ~~~~~~~~~~  ~~~~~~~~~~  ~~~~~~~~~~  ~~~~~~~~~~
macaca      ~~~~~~~~~~  ~~~~~~~~~~  ~~~~~~~~~~  ~~~~~~~~~~  ~~~~~~~~~~
bovine      atatataaaa  gtcacattga  agggtctaca  gctcaattta  tcagccttct
equine      ~~~~~~~~~~  ~~~~~~~~~~  ~~~~~~~~~~  ~~~~~~~~~~  ~~~~~~~~~~
porcine     ~~~~~~~~~~  ~~~~~~~~~~  ~~~~~~~~~~  ~~~~aattca  tcagccttct

            51                                                           100
murine      tgtctggttt  actccagctt  actacagcat  cagtttgtgc  atccccgaag
rat         tgtctggctt  actccagctt  aatacagggt  cact......  ...cctgaag
canine      caagtttact  caacttcgaa  aaagcatcag  ctgccgatac  acctgaa...
feline      ~~~~~~~~~~  ~~gcacgagg  aaagtatcag  ctgtcaagac  acctgaa...
human       ~~~~~~tgtc  tttagtttac  tcagcatcag  ctactaacat  acctgaacga
macaca      ~~~~~~~~~~  ~~~~~~~~~~  ~~~~~~~~~~  ~~~~~~~~~~  ~~~~~~~~~~
bovine      caggtttact  caactttgag  aaagcatcag  ctgctaatac  atttgaaaga
equine      ~~~~~~~~~~  ~~~~~~~~~~  ~~~~~~~~~~  ~~~~~~~~~~  ~~~~~~~~~~
porcine     cgggtttact  caactttgag  aaagcatcag  ctgctaacac  acctgaaaga

            101                         ▼                               150
murine      gatccccttt  gagagtcatt  gtatgtaaag  atgatgtctg  caaacaccgt
rat         gatcctctct  gagagtcatt  gtatgtgaag  atgatgtctg  caagcaccat
canine      agatcttgtc  acaagacatt  gtgtgtgaag  atgatgtctg  caaaagacat
feline      agatcttgtc  aca..acctt  gtgtgtgaag  atgatgtctg  cgaaagacat
human       agatcttgtt  ctaagacatt  gtatgtgaag  atgatacctg  caaaagacat
macaca      ~~~~~~~~~~  ~~~~~~~~~~  ~~~~~~~~~~  atgatacctg  caaaagacat
bovine      agattgtatc  ctaagac...  gtgtgttaat  atgatgtctg  caaaagacat
equine      ~~~~~~~~~~  ~~~~~~~~~~  ~~~~~~~~~~  ~~~~~~~~~~  ~~~~~~~~~~
porcine     agatcgtgtc  ctaagacgtt  gtgtgtgaag  atgatgtctg  caaaagacac

            151                                                          200
murine      ggctaaagtg  atgatcatca  tgctggcagt  ctgtcttctt  acccaaacgg
rat         ggctaaggtg  atgatcctca  tgctggcagt  ttgtctcctt  acccaggcag
canine      ggttaaagta  atgattgtca  tgtttgcaat  ttgttttctt  gcaaagtcag
feline      ggttaaagtc  atggttgtca  tgtttgcaat  ttgctttctt  gcaaaatcgg
human       ggctaaagtt  atgattgtca  tgttgcaat   ttgttttctt  acaaaatcag
macaca      ggctaaagta  atgattgtca  tgttggcaat  ttgctttctt  acaaaatcag
bovine      ggttaaggta  atgattgtca  tgcttgccat  ctgtttcctt  gcaagatcag
equine      ~~~~~~~~~~  ~~~~~~~~~~  ~~~~~~~~~~  ~~~~~~~~~~  ~~~~~~~~~~
porcine     agttaaagta  atggttgtca  tgcttgcaat  ttgttttctt  gcaagatcag

            201    ▼              ▼                                      250
murine      atgggaaacc  cgtgaggaag  agagctgtca  gtgaaataca  gcttatgcac
rat         atgggaaacc  cgttaagaag  agagctgtca  gtgaaataca  gcttatgcac
canine      atgggaaacc  tgttaagaag  agatctgtga  gtgaaataca  gtttatgcat
feline      atgggaaacc  tgttaagaag  aggtctgtga  gtgaaataca  gtttatgcat
human       atgggaaatc  tgttaagaag  agatctgtga  gtgaaataca  gcttatgcat
macaca      atgggaaatc  tgttaagaag  agatctgtga  gtgaaataca  gcttatgcat
bovine      atgggaagtc  tgttaagaag  agagctgtga  gtgaaataca  gtttatgcat
equine      ~~~~~~~~~~  ~accaggaag  agatctgtga  gtgaaataca  gcttatgcat
porcine     atgggaagcc  tattaagaag  agatctgtga  gtgaaataca  gcttatgcat
```

Figure 2. Part 1, see legend page 16.

```
             251                                                            300
murine       aacctgggca aacacctggc ctccatggag aggatgcaat ggctgagaag
rat          aacctgggca aacacctggc ctctgtggag aggatgcaat ggctgagaaa
canine       aacctgggca aacatctgag ctccatggag agggtggaat ggctacggaa
feline       aacctgggca agcatctgag ctccgtggag agggtagaat ggctgcggag
human        aacctgggaa aacatctgaa ctcgatggag agagtagaat ggctgcgtaa
macaca       aacctgggaa aacatctgaa ctcgatggag agagtagaat ggctgcgtaa
bovine       aacctgggca aacatctgag ctccatggaa agagtggaat ggctgcggaa
equine       aacctgggca aacatctgaa ctcagtggaa agggtggaat ggctgcggaa
porcine      aacctgggca aacacctgag ctctctggag agagtggaat ggctgcgaaa
             301                                                            350
murine       gaagctgcaa gatatgcaca attttgttag tcttggagtc caaatggctg
rat          aaagctgcaa gatgtacaca attttgttag tcttggagtc caaatggctg
canine       gaagctccag gatgtacaca actttgttgc ccttggagct ccaatagctc
feline       gaaactacag gatgtacaca actttgtcgc cctcggagct ccaatagctc
human        gaagctgcag gatgtgcaca attttgttgc ccttggagct cctctagctc
macaca       gaagctgcag gatgtgcaca attttattgc ccttggagct cctctagctc
bovine       aaagctacag gatgtgcaca actttgttgc ccttggagct tctatagctt
equine       gaagctgcag gatgtgcaca attttattgc cctcggagct cctatatttc
porcine      gaagctgcag gatgtgcaca actttgttgc cctcggagct tctatagttc
             351                                                            400
murine       ccagagatgg cagtcaccag aagcccacca agaaggagga aaatgtcctt
rat          ccagagaagg cagttaccag aggcccacca agaaggagga aaatgtcctt
canine       acagagatgg tagttcccag aggccctaa  aaaaggaaga caatgtccta
feline       acagagatgg tggttcccag aggccccgaa aaaaggaaga caatgtcccg
human        ccagagatgc tggttcccag aggccccgaa aaaaggaaga caatgtcttg
macaca       ccagagatgc tggttcccag aggccccgaa aaaaggaaga caatatcttg
bovine       acagagatgg tagttcccag agacctcgaa aaaaggaaga caatgtcctg
equine       acagagatgg tggttcccag aggcctcgaa aaaaggaaga caatgtgctg
porcine      acagagatgg tggttcccag agaccccgaa aaaaggaaga caatgtcctg
             401                                                            450
murine       gttgatggca atccaaaaag tcttggtgag ggagacaaag ctgatgtgga
rat          gttgatggca attcaaaaag tcttggcgag ggggacaaag ctgatgtgga
canine       gttgagagct atcaaaaaag tcttggagaa gccgacaaag ctgatgtgga
feline       gctgagaacc atcaaaaaag tcttggagaa gcagacaaag ctgatgtgga
human        gttgagagcc atgaaaaaag tcttggagag gcagacaaag ctgatgtgaa
macaca       gtagagagcc atgaaaaaag tcttggagag gcagacaaag ctgatgtgga
bovine       gttgagagcc atcagaaaag tcttggagaa gcagacaaag ctgatgtgga
equine       attgagagcc atcaaraaag tcttggagaa gcagacaaag ctgatgtgga
porcine      gttgagagcc atcaaaaaag tctcggagaa gcagataaag ctgctgtgga
             451                                                            500
murine       tgtattagtt aaatcaaaat ctcagtaaat gctgatttat tctagacagt
rat          tgtattagtt aaggctaaat ctcagtaaat gctgacgtat tctagaccgt
canine       tgtattaact aaagctaaat cccagtgacg ataca..... ......tcag
feline       tgtgttaatc aaagctaaat cccagtgaag acaga..... ......gcag
human        tgtattaact aaagctaaat cccagtgaaa atgaaaacag atattgtcag
macaca       tgtattaact aaagctaaat cccaatgaaa atgaaaatag atatggtcag
bovine       tgtattaatt aaagctaaac cccagtgaa. ...aacagat atgatc..ag
equine       tgtgttaagt aaaactaaat cccagtgar. ...aacagat aggatc..ag
porcine      tgtattaatt aaagctaaac cccagtgaa. ...aacacat atgatcagag
```

Figure 2. Part 2, see legend page 16.

```
          501                                                      550
murine    gcagggcact gacatatgct gctacctttt caagct.tat gaagatcacc
rat       gctgagcaat aacatatgct gctatccttt caagctccac gaagatcacc
canine    ggcadtgctg tagacagcat agggcaacaa cattacaagc tgctaacatt
feline    agcactgcta tacacaggat agggcaacaa aattacatgc tgctaacatt
human     agttctgctc tagacagtgt agggcaacaa tacatgctgc taattcaaag
macaca    agttctgctc tagacagtgt agggcaac~~ ~~~~~~~~~~ ~~~~~~~~~~
bovine    atcactgttc tagacagcat agggc.aaca atattacatg ctgctaatgt
equine    agcatcgctc tagacagcat asggc.aaa~ ~~~~~~~~~~ ~~~~~~~~~~
porcine   agcactgctc tagacagcat aaggcaaaca atatttcatg ctgctaatgt
          551                                                      600
murine    aagtgctaat acttctactg taatgaaact ttggaatttt tttgattaca
rat       aagtgctaat tcttctactg taataaaagt ttgaaa.... tttgattcca
canine    ttcaagctct taagattaat aaatgccaaa atttacatgt aatccattgt
feline    ctcaagcttt gaagatcacc aaatgccaat atttacgtct aatccatggc
human     ctctattaag atttccaagt gccaatattt ctgatataac aaactacatg
macaca    ~~~~~~~~~~ ~~~~~~~~~~ ~~~~~~~~~~ ~~~~~~~~~~ ~~~~~~~~~~
bovine    gttcaccttc tattaagtgc cagtagttct atgaccaacc tttattgcta
equine    ~~~~~~~~~~ ~~~~~~~~~~ ~~~~~~~~~~ ~~~~~~~~~~ ~~~~~~~~~~
porcine   tttcaatctc tattaagatt aagtgccaat atttctaata ttactaaact
          601                                                      650
murine    ttttgctca tttaaggtct ctttcaatga ttccatttca atatgctctt
rat       cttttgctca tttaaggtct cttccaatga ttccatttca atatattctt
canine    tagccatgat agctgaaatt ttaattgatt gttttgattc tagtttaatt
feline    tagccacgat agctgaaatt ctaattgatt gttttgattc tacttttatt
human     taatccatca ctagccatga taactgcaat tttaattgat tattctgatt
macaca    ~~~~~~~~~~ ~~~~~~~~~~ ~~~~~~~~~~ ~~~~~~~~~~ ~~~~~~~~~~
bovine    gctgtgatac ctacaatttt aattgagtat tttgattcta ctttattcat
equine    ~~~~~~~~~~ ~~~~~~~~~~ ~~~~~~~~~~ ~~~~~~~~~~ ~~~~~~~~~~
porcine   tgatgggtaa tcattgctag ccatgattgc tgaaatttta attgatcatt
          651                                                      700
murine    cttttaaag tactactcat ttccacttct ctccttaaat ataaataaag
rat       cttttaaag tattacacat ttccacttct ctccttaaat ataaataaag
canine    catttaagag ctctttaat tgttctattt ctattgttta ttctttttaa
feline    catgtaaggc ctctttaat tattccattt ctgttgttta ttctttttaa
human     ccacttttat tcatttgagt tattttaatt atcttttcta ttgttattc
macaca    ~~~~~~~~~~ ~~~~~~~~~~ ~~~~~~~~~~ ~~~~~~~~~~ ~~~~~~~~~~
bovine    ctaagagctc ttttaataat tctatttcta ttgattccaa ataaatgaag
equine    ~~~~~~~~~~ ~~~~~~~~~~ ~~~~~~~~~~ ~~~~~~~~~~ ~~~~~~~~~~
porcine   ttgattctac ttttactcat ttaagagctt cttttaacaa ttctatttct
          701                                                      750
murine    ctttaatgct catgaatc~~ ~~~~~~~~~~ ~~~~~~~~~~ ~~~~~~~~~~
rat       ~tttaatgat catgaaccaa a~~~~~~~~~ ~~~~~~~~~~ ~~~~~~~~~~
canine    agtatgtttt tgcataattt ataaaagaat aaaattgcac ttttt~~~~~
feline    agtatgttat tgcataattt ataaaagaat aaaattgcac tttgtaacct
human     tttttaaagt atgttattgc ataatttata aagaataaa attcgacttt
macaca    ~~~~~~~~~~ ~~~~~~~~~~ ~~~~~~~~~~ ~~~~~~~~~~ ~~~~~~~~~~
bovine    ttaagtatt~ ~~~~~~~~~~ ~~~~~~~~~~ ~~~~~~~~~~ ~~~~~~~~~~
equine    ~~~~~~~~~~ ~~~~~~~~~~ ~~~~~~~~~~ ~~~~~~~~~~ ~~~~~~~~~~
porcine   attgattcta aataaatgaa gtatttcttc cttgtt~~~~ ~~~~~~~~~~
```

Figure 2. Part 3, see legend page 16.

```
            751                                                        800
murine      ~~~~~~~~~~  ~~~~~~~~~~  ~~~~~~~~~~  ~~~~~~~~~~  ~~~~~~~~~~
rat         ~~~~~~~~~~  ~~~~~~~~~~  ~~~~~~~~~~  ~~~~~~~~~~  ~~~~~~~~~~
canine      ~~~~~~~~~~  ~~~~~~~~~~  ~~~~~~~~~~  ~~~~~~~~~~  ~~~~~~~~~~
feline      ctctcccatc  gtacactgca  aaataaaaat  ttaatgatca  taattttaaa
human       taaacctctc  ttctacctta  aaatgtaaaa  caaaaatgta  atgatcataa
macaca      ~~~~~~~~~~  ~~~~~~~~~~  ~~~~~~~~~~  ~~~~~~~~~~  ~~~~~~~~~~
bovine      ~~~~~~~~~~  ~~~~~~~~~~  ~~~~~~~~~~  ~~~~~~~~~~  ~~~~~~~~~~
equine      ~~~~~~~~~~  ~~~~~~~~~~  ~~~~~~~~~~  ~~~~~~~~~~  ~~~~~~~~~~
porcine     ~~~~~~~~~~  ~~~~~~~~~~  ~~~~~~~~~~  ~~~~~~~~~~  ~~~~~~~~~~

            801                     828
murine      ~~~~~~~~~~  ~~~~~~~~~~  ~~~~~~~~
rat         ~~~~~~~~~~  ~~~~~~~~~~  ~~~~~~~~
canine      ~~~~~~~~~~  ~~~~~~~~~~  ~~~~~~~~
feline      aaaaaaaaaa  aaaaa~~~~~  ~~~~~~~~
human       gtctaaataa  atgaagtatt  tctcactc
macaca      ~~~~~~~~~~  ~~~~~~~~~~  ~~~~~~~~
bovine      ~~~~~~~~~~  ~~~~~~~~~~  ~~~~~~~~
equine      ~~~~~~~~~~  ~~~~~~~~~~  ~~~~~~~~
porcine     ~~~~~~~~~~  ~~~~~~~~~~  ~~~~~~~~
```

Figure 2. Alignment of the known nucleotide sequences of PTH mRNA for different species. Alignments were obtained using the default setting of PileUp program. Alignment of the nucleotide sequences for mouse (murine), rat, dog (canine), cat (feline), human, non-human primate (macaca), cow (bovine), horse (equine) and pig. The gallus sequence was not included in the alignment because of the large differences in sequence and size from all other published species. Gaps indicated by dashes were introduced to maximize the homology in the 5' and 3' -UTRs. The arrows indicate the positions of the two introns in the gene. The closed triangles indicate the protolytic cleavage sites required for the conversion of preProPTH to ProPTH and PTH. The nt in the dark gray box show the coding sequence, and the sequences 5' and 3' to this region are the 5'-UTR and 3'-UTR respectively. The nt that are surrounded by the square comprise the proximal PTH mRNA 3'-UTR protein binding element and the nt that are on a light gray background are the distal *cis* acting functional element. Nucleotides that are not identical to the bovine (proximal element) and the rat (distal element) sequence are shown in bold.

simplicity the gallus cDNA was not included in the alignment of the pre ProPTH mRNAs in Figure 2. This PTH mRNA is significantly longer than the other cloned cDNAs (Table 1) and is the least preserved compared to the other species, even in the coding sequence (Table 2 and Fig. 1).

Comparison of the sequences show that human and macaca; canis and felis; rat and mouse; and bovine and pig are the most similar to each other (Table 2). The lowest homology is seen when the sequence of gallus PTH mRNA is compared to each of the other sequences, even in the translated coding region of the mRNA that is, as expected, the most conserved region. The coding sequences of the other species are the most preserved as expected. The 5'-UTR is relatively well conserved with homologies about 15% less than the coding region. The 3'-UTR is the least conserved region (Table 2).

Interestingly, a 26 nt *cis* acting functional protein binding element at the distal region of the 3' UTR is highly conserved in the PTH mRNA 3'-UTRs of rat, mouse, man, dog and cat (Table 3, distal element). In the 26 nt element, the identity amongst species varies between 73 and 89%. In particular, there is a stretch of 14 nt within the element that is present in all five species. We have previously characterized this distal protein binding element in the rat PTH mRNA 3'-UTR as a *cis*-acting sequence that determines the stability of the PTH mRNA and its regulation by calcium and phosphate (P).

Table 2. Similarity (ratio) of the nucleotide sequences for the PTH mRNAs of different species

	mRNA	5'-UTR	CDS	3'-UTR
Bovine/Porcine	7.864	8.229	9.253	6.643
Bovine/Canine	7.341	6.364	9.224	6.049
Bovine/Human	6.682	5.541	8.879	5.571
Bovine/Rat	5.863	4.610	7.730	4.107
Human/Macaca			9.613	9.6
Human/Canine	7.542	5.824	9.052	6.605
Human/Rat	5.804	4.743	7.816	4.660
Rat/Murine	8.743	8.5	9.253	8.118
Rat/Canine	5.925	4.511	7.989	4.176
Canine/Feline	8.503	7.206	9.195	8.504
Bovine/Gallus	4.565	4.283	5.330	3.705
Human/Gallus	4.519	4.554	5.474	3.809
Rat/Gallus	4.207	3.847	5.043	3.765
Canine/Gallus	4.444	4.250	5.388	3.691

The comparisons between each two sequences were performed using the default setting of the GAP program (Accelrys Inc, Madison WI). This program considers all possible alignments and gap positions between two sequences and creates a global alignment and evaluates its significance. The average alignment score, plus or minus the standard deviation, of all randomized alignments is reported in the output file as the 'quality' score. Ratio is the quality divided by the number of bases in the shorter segment of each two sequences.

In addition, a 22 nt protein binding element in the 3' UTR (Table 3 proximal element) was also identified in bovine and porcine, as well as human, non-human primates, equus, canis and felis (Table 3 proximal element), but not in rat and mouse. The functionality of the proximal element remains to be determined. The conserved sequences within the 3'-UTR suggest that the binding elements represent a functional unit that has been evolutionarily conserved (see 'conserved elements in the 3' UTR).

The 3'-UTR in the human and feline sequences are more than 100 nucleotides longer than the other 3' UTR sequences, with the exception of the gallus PTH mRNA. Large gaps have to be introduced to maximize homology to the human 3'-UTR (Fig. 2). Hendy et al[15] suggested that the extra sequence in the 3' region of the human cDNA, corresponding to the large gap in the bovine sequence, might have been the result of a gene duplication since it contained some homology to the region around the polyadenylation signal, including a second consensus polyadenylation signal. Interestingly, in the rat sequence, large gaps also must be introduced in this region, but they do not coincide exactly with that of the bovine sequence. Phylogenetic trees obtained from alignment of the protein and mRNA sequences are shown in Figure 3. The same phylogenetic tree is obtained from the amino acid sequences and from the coding regions of the mRNA (Fig. 3A). Phylogenetic comparison based on nt similarity of the full PTH mRNAs or the 3'-UTRs is shown in Figure 3B. This map does not include macaca and equine PTH sequences where there are only partial sequences of the cDNA available. The gallus is very different from all the other species indicating a separate evolutionary branch. Interestingly, based on amino acid sequence and the coding region of the mRNA, the bovine and porcine were grouped closest to canis and felis but not by the full-length mRNA or 3'-UTR sequences. This mainly represents the large

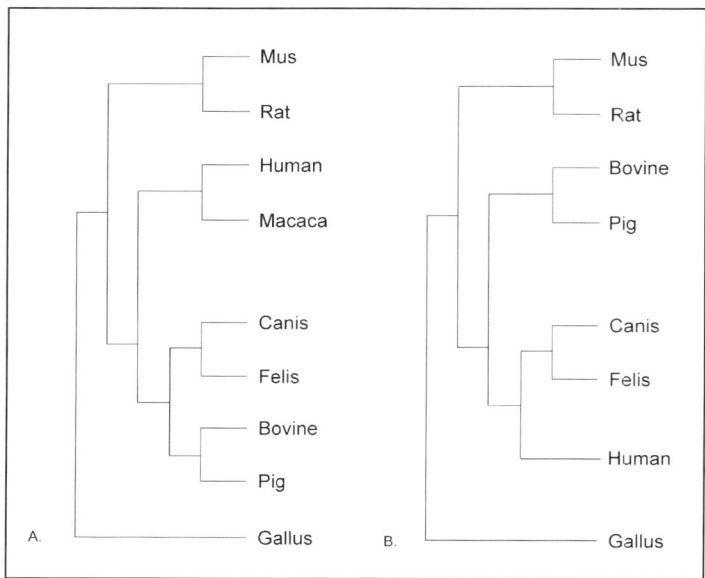

Figure 3. Phylogenetic tree obtained from alignment of the amino acid sequences and nucleotide sequences of PTH and PTH mRNA. The phylogenetic trees were obtained using the default setting of PileUp program. A) Phylogenetic tree based on amino acid similarities, or the nt sequence of the coding regions of the PTH mRNAs for mouse (mus), rat, dog (canis), cat (felis), human, non-human primate (macaca), cow (bovine), pig (porcine) and chicken (gallus) PTH. The horse PTH was not included in this study because only partial amino acid sequence is available. B) Phylogenetic tree according to nt sequence of the full-length PTH mRNAs for mouse (mus), rat, dog (canine), cat (feline), human, cow (bovine), pig (porcine) and chicken (gallus). The horse and non-human primate (macaca) PTH mRNA were not included in this study because only partial sequences of these RNAs are available. The same Phylogenetic tree is also obtained when the 3'-UTR sequences are analized separately. Interestingly, based on amino acid sequence, the bovine and pig were grouped closest to canis and felis but not by RNA sequence. This corresponds to the presence of the distal and proximal protein-binding elements in the 3'-UTRs.

differences in the 3'-UTRs and correlates with the conservation of protein-binding elements (Table 3). The mouse and rat species are separate because they only have the distal PTH mRNA 3'-UTR element. The human, canis, felis, bovine and porcine are grouped together, all containing the proximal element. But in this group, the bovine and porcine represent a separate branch expressing only the proximal element, and the human, felis and canine are a distinct branch, which corresponds with their expression of both the proximal and distal elements.

Structure of the PTH mRNA

The 5' Untranslated Region

The 5' untranslated sequence of the longer forms of the human and bovine mRNAs and rat PTH mRNA contains about 120 nucleotides, and the shorter bovine and human cDNAs contain about 100 nucleotides in the 5' noncoding region. The average length of the 5' UTR in eukaryotic mRNAs is 80-120 nucleotides.[35] As a result, the m^7G cap at the 5' terminus of the mRNA is a considerable distance from the initiator codon. In the bovine sequence, a

Table 3. The sequences of the 26 nt proximal cis acting element and the 22 nt distal element of the PTH mRNA 3'-UTR in different species

	Proximal Element	Distal Functional Element
Rat	–	ATATATTCTTCTTTTTAAAGTA
Mus musculus	–	ATAT**GC**TCTTCTTTTTAAAGTA
Bovine	TGTTCTAGACAGCATAGGGCAA	–
Porcine	TG**C**TCTAGACAGCATA**A**GGCAA	–
Equus caballus	**C**G**C**TCTAGACAGCATA*GGCAA	?
Canis familiaris	TG**CT**GTAGACAGCATAGGGCAA	AT**TGT**T**A**TTCTTTTTAAAGTA
Felis catus	TG**CTATAC**ACAG**G**ATAGGGCAA	**G**T**TGT**T**A**TTCTTTTTAAAGTA
Human	TG**C**TCTAGACAG**TG**TAGGGCAA	AT**TGT**T**A**TTCTTTTTAAAGTA
Macaca fascicularis	TG**C**TCTAGACAG**TG**TAGGGCAA	?
Gallus gallus	–	–

The sequences that are not available are indicated by a ?; species lacking a particular element are indicated by a -. The * in the equine proximal element indicates an unidentified nt in the gene bank. The nt in the proximal and distal elements that are different from the bovine and rat sequence respectively are shown in bold.

possible hairpin loop may bring the 5' end closer to the initiator codon. However, in both the human and rat sequences, deletions of 11 and 16 nucleotides respectively, largely eliminate the sequences involved in the stem of the loop. Thus, there seems to be little functional significance related to the bovine secondary structure. In the rat PTH mRNA 5' terminus the first 19 nt of the mRNA may form a stable stem loop structure that could affect PTH mRNA translation, but its function has not been determined. (T. Naveh-Many, unpublished data). However the most outstanding conclusion from a comparative analysis of the sequences is that no region in the 5' untranslated region is conserved that has any known functional significance.

The Coding Region

The actual initiator ATG codons for the human and bovine PTH mRNAs have been identified by sequencing in vitro translation products of the mRNAs.[36,10] In the bovine sequence, the first ATG codon is the initiator codon, in accord with many other eukaryotic mRNAs.[37,38] The human and rat sequences have ATG triplets prior to the probable initiator ATG, which are present ten nucleotides before the initiation codon and are immediately followed by a termination codon. In the rat, another ATG is present 115 nucleotides before the initiator codon. The designation of the third ATG codon of the rat sequence as the initiator codon is based on indirect evidence, primarily by comparison with the bovine and human cDNAs. Regardless, the presence of termination codons in phase with the earlier ATG prohibits the synthesis of a long protein initiated at these codons, as is the case in some other genes with premature ATG codons.[37,38] However in some systems, small peptides that are translational products of upstream ATGs have been shown to have regulatory functions.[39] Whether this is the case in the rat PTH mRNA is not known. The most stringent requirement for optimal initiation of synthesis is for a purine at the –3 position. Since non of the premature ATG codons in the rat and human PTH mRNAs has a purine at the –3 position, they are likely to be weak initiators. In contrast, the probable initiator ATG codon has an A at the –3 position in each sequence.

The 3' Untranslated Region

As noted above, the 3' untranslated region is the most variable region of the cDNAs requiring significant gaps to maximize homology. The termination codon in all species except for the rat and mouse is TGA and is followed closely by a second in-phase termination codon. In the rat and mouse TAA is the termination codon and no following termination codon is present (Fig. 2).

Conservation of Protein Binding Elements in the PTH mRNA 3'-UTR

We have defined the *cis* sequence in the rat PTH mRNA 3'-UTR that determines the stability of the PTH mRNA and its regulation by calcium and phosphate (P). PTH gene expression is regulated post-transcriptional by Ca^{2+} and P, with dietary induced hypocalcemia increasing and dietary induced hypophosphatemia decreasing PTH mRNA levels. This regulation of PTH mRNA stability correlates with differences in binding of *trans* acting cytosolic proteins to a *cis* acting instability element in the PTH mRNA 3'-UTR. There is no PT cell line and therefore to study PTH mRNA stability we performed in vitro degradation assays. We did this by incubating the labeled PTH transcript with cytosolic PT proteins from rats on the different diets and measuring the amount of intact transcript remaining with time. PT proteins from low Ca^{2+} rats stabilized and low P PT proteins destabilized the PTH transcript compared to PT proteins of control rat. This rapid degradation by low P was dependent upon the presence of the terminal 60 nt protein binding region of the PTH mRNA.[40] We have defined the *cis* sequence in the rat PTH mRNA 3'-UTR that determines the stability of the PTH transcript and to which the *trans* acting PT proteins bind. A minimum sequence of 26 nt was sufficient for RNA-protein binding (Table 3, distal element). One of the *trans* acting proteins that binds and prevents degradation of the PTH mRNA was identified by affinity purification. This protein is AU rich element binding protein 1 (AUF1) that is also involved in half life of other mRNAs.[41,42]

To study the functionality of the *cis* sequence in the context of another RNA, a 63 bp PTH cDNA sequence consisting of the 26 nt and flanking regions was fused to the growth hormone (GH) cDNA. Since there is no parathyroid (PT) cell line an in vitro degradation assay was used to determine the effect of PT cytosolic proteins from rats fed the different diets on the stability of RNA transcripts for GH and the chimeric GH-PTH 63 nt.[43,44] The GH transcript was more stable than PTH RNA and was not affected by PT proteins from the different diets. The chimeric GH PTH 63 nt transcript, like the full-length PTH transcript was stabilized by PT proteins from rats fed a low calcium diet and destabilized by proteins from rats fed a low phosphate diet. Therefore, the 63 nt protein binding region of the PTH mRNA 3'-UTR is both necessary and sufficient to regulate RNA stability and to confer responsiveness to changes in PT proteins by calcium and phosphate.[43] The regulation of PTH mRNA stability by calcium and phosphate is discussed in detail in the chapter by Levin et al.

Sequence analysis of the PTH mRNA 3'-UTR of different species revealed a preservation of the 26 nt core protein-binding element in rat, mouse, human, cat and canine 3'-UTRs (Table 3). The *cis* acting element identified is at the 3' distal end in all species that express it and is therefore designated the distal functional *cis* element. The conservation of the sequence suggests that the binding element represents a functional unit that has been evolutionarily conserved. Protein binding experiments by UV cross linking and RNA electrophoretic mobility shift assays showed that there is specific binding of rat and human parathyroid extracts to an in vitro transcribed probe for the rat and human PTH mRNA 3'-UTR 26 nt elements.

In contrast, the 26 nt distal *cis* element was not present in the 3'-UTR of bovine, porcine and gallus PTH mRNA. To determine the protein binding pattern of the bovine PTH mRNA, binding experiments were performed with bovine parathyroid gland extracts and RNA probes for different regions of the bovine PTH mRNA. Binding and competition experiments revealed a 22 nt minimal protein binding element in the bovine PTH mRNA 3'-UTR that was sufficient for protein binding. The 22 nt element is at the 5' portion of the 3'-UTR (Fig. 2) and is the proximal *cis* element. Interestingly this element was also present in the 3'-UTRs of man, dog, cat, non-human primates, horse and porcine PTH mRNA. Therefore the PTH mRNA 3'-UTRs of man, dog and cat have both sequences, the distal functional *cis* element of 26 nt that has been characterized in rat PTH mRNA and the 22 nt proximal protein binding element initially characterized in bovine PTH mRNA (Table 3). The bovine and porcine mRNAs only have the 22 nt element and the gallus PTH mRNA has neither of the elements. It is not known if the 26 nt element is present in the horse and macaca because there is only partial sequencing of these mRNAs. Though the 22 nt sequence is a protein binding element, its functionality remains to be determined.

The Polyadenylation Signal

Another region that is well conserved in the PTH mRNA 3'-UTR is the AATAAA polyadenylation signal. In the bovine sequence, only a single AATAAA has been detected in the 3' noncoding region, whereas in the human and rat sequences two potential polyadenylation sites are found. The second AATAAA region in the human sequence is about 60 bases upstream from the first and has been suggested to have resulted from a gene duplication;[15] however, other than the AATAAA regions, there is little homology surrounding the two sites. Sequences analogous to the human upstream AATAAA are missing in both the bovine and rat sequences. No cDNAs were detected in which the upstream AATAAA was utilized as a polyadenylation signal; however the probability that these sites function as a polyadenylation signal cannot be ruled out. The rat sequence also has a second AATAAA site about 115 nucleotides earlier than the functional one. A single rat PTH mRNA was detected by Northern blot analysis,[17] suggesting that only one polyadenylation site is used, and the size of the mRNA was consistent with the second AATAAA being the site. There is no direct evidence for the location of the 3' end by analysis of the rat PTH mRNA or cDNA.

The PTH Gene

The genes for human,[16] bovine,[14] rat[17] and mouse[31] PTH have each been cloned and characterized from genomic libraries in lambda phage. The human gene was isolated from a total human fetal DNA library prepared in λ phage Charon 4A. The library was screened initially by filter hybridization with human cloned cDNAs as a probe and later by the recombination selection method. The structure of the human gene was determined by the analysis of two overlapping clones. For the bovine gene, Southern analysis of the total bovine DNA showed that the PTH gene was present on an 8000 bp EcoR1 fragment. To clone the gene, bovine liver DNA was digested with EcoR1 and fragments in the range of 5,000 to 10,000 bp were isolated by sucrose gradient centrifugation. A partial library was then constructed by ligating the EcoR1 fragments to λ phage Charon 31 arms, which had been isolated after digestion with EcoR1. Several independent clones were isolated by plaque filter hybridization using cloned bovine PTH cDNA as probes. The rat gene was isolated from a λ phage Charon 4A rat liver DNA library produced by partial EcoR1 digestion of the rat DNA. Two independent positive plaques were obtained. The insert of each of the two phages contained the entire rat PTH gene. The sequence of the mouse gene was determined from a mouse genomic library.[31] One recombinant clone contained 14 kb of DNA, encompassing the entire PTH gene. The transcriptional unit spans 3.2 kb of genomic DNA, analogous to the human PTH gene.

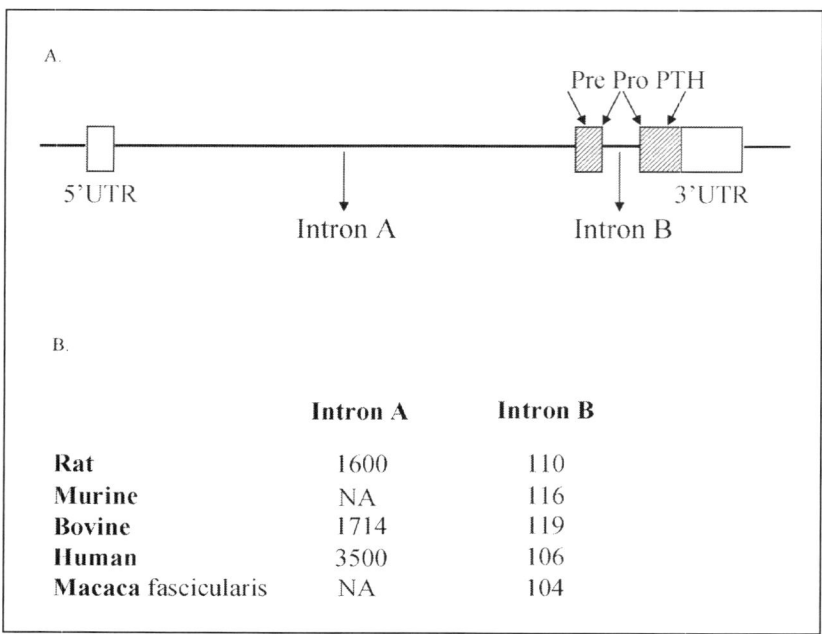

Figure 4. Schematic representation of the rat and bovine PTH gene structure and the length of the introns. A) The PTH gene including exons 1-3 and introns A and B. Exons are indicated by the rectangles and the shaded areas indicate regions in the gene that code for preProPTH. B) The known sizes, in bp, of introns A and B in rat, mouse (murine), bovine, human and non-human primate (macaca fascicularis). The full length sequence of intron A is not available (NA) for mouse and non-human primate.

Structure of the Gene

The overall structure of the bovine or rat PTH gene is shown schematically in Figure 4A. In the human gene the larger intron A is aproximatly twice as long as this intron in bovine and rat (Fig. 4B). All sequenced genes contain two introns. The exact location of the bovine and human gene introns was determined by comparing the sequence of the gene to the previously determined cDNA structure. The location of intron A of the rat was determined by comparing the gene sequence with the sequence of cDNA to the 5' end of the mRNA. The cDNA was synthesized with reverse transcriptase using a synthetic pentadecamer as primer. Intron B in the rat gene was determined indirectly by homology of the sequence to the human and bovine cDNA sequences.

The locations of the introns are identical in each case as has been found with most other genes.[45] Intron A splits the 5' untranslated sequence five nucleotides before the initiator methionine codon (Fig. 2). Intron B splits the fourth codon of the region that codes for the pro sequence of preProPTH. The three exons that result, thus, are roughly divided into three functional domains. Exon I, 95 to 121 nucleotides long, contains the 5' untranslated region. ExonII, has 91 nucleotides and codes for the pre sequence, or signal peptide and exon III, 375 to 486 nucleotides long, codes for PTH as well as the 3' untranslated region. The structure of the PTH gene is thus consistent with the proposal that exons represent functional domains of the mRNA.[45]

The length of the introns in the species where the sequence is available is shown in Figure 4B. Although the introns are at the same location, the size of the large intron A in human is about twice as large as those in the rat and bovine (Fig. 4B). It is of interest that the human gene is considerably longer in both intron A and the 3' untranslated region of the cDNA compared to the bovine, rat and mouse. Knowledge of the structures of other PTH genes from other species will be necessary in order to determine whether the extra sequence was inserted or is less susceptible to deletion in the human gene.

Both introns have the characteristic splice site elements. They have the GT-AG nucleotides at the 5' and 3' ends of the intron and the pyrimidine tract at the 3' end of the intron. The large intron A, has about 75% homology between the bovine and human PTH genes in over 200 bp of the intron, similar to the homology in the other non-translated regions of the genes. The rat intron A is only 55-57% homologous to the other species. The sequences of introns are generally only conserved at the *cis* elements essential for splicing and the relatively large homology for the PTH genes suggests that there may be some constrains on the basis of changes some distance from the intron /exon border.

The second exon, containing 106 and 121 nt in the human and bovine pre-mRNA, is much smaller and more homologous in size among the genes than intron A. The sequence of intron B is well conserved with homology of 74 and 68% of bovine/human and human/rat, respectively, but is relatively poorly conserved between the rat and bovine genes, with a homology of 49%.[1]

In each of the species, only a single PTH gene appears to be present. Extensive Southern blot analysis of bovine DNA with cloned PTH cDNA as probe produced single hybridizing bands for restriction enzymes that do not cut within the probe sequence.[13] The restriction map determined from the Southern analysis of bovine DNA was consistent with that of the cloned gene. With the exception of a single nucleotide in the 3' untranslated region, the sequence of the cloned cDNA was identical to the sequence of the exons in the gene. Less extensive Southern blot analysis of the human[16] and rat[17] genes also were consistent with a single gene per haploid genome. Furthermore, in the human studies, probes from the 5' and 3' ends of the cDNA both hybridized to the same sized fragment, and the strength of the signal from the genomic DNA was about the same as one gene-equivalent of the gene cloned in λ phage. Thus, the PTH gene is a single gene. The genes for PTH and PTHrP (PTH-related protein) are located in similar positions on sibling chromosomes 11 and 12. It is therefore likely that they arose from a common precursor by chromosomal duplication.

Initiation Site for RNA Transcription

As noted above in the discussion on the cDNA, the 5' termini of bovine PTH mRNA are heterogeneous. The large mRNAs contain a TATA sequence in the appropriate location to direct the synthesis of the smaller mRNAs. It was postulated that a second TATA would be found in the gene sequence 5' of the first one.[13] In both the human and bovine gene sequences, a second TATA sequence is present in the 5' flanking region about 25 base pairs from the first one in the appropriate position to direct the synthesis of the larger mRNAs. The heterogeneity of the 5' end of the bovine PTH mRNA, originally detected by reverse transcription of the mRNA,[13] was confirmed by S1 nuclease mapping.[14] The initiation sites for human PTH mRNA have not been determined directly, but were proposed[17] on the basis of analogy with the bovine sequence and the consensus TATA sequences. The presence of multiple functional TATA sequences has been reported for several other eukaryotic genes. The rat mRNA appears to be relatively homogeneous at the 5' terminus on the basis of both primed reverse transcriptions near the 5' end of the mRNA and S1 nuclease mapping.[17] The single initiation site for the rat mRNA can be explained by the changes in the rat sequence which alter the second downstream TATA sequence. The sequence, TATATATAAAA, in the human and bovine genes, is changed to TGCATATGAAA in the rat gene,[1] which is no longer a consensus TATA sequence. While this change seems the most likely explanation for the difference in length at the 5' termini between the mRNAs, there are other changes that also occur in this region of the gene and may play a role.[17]

The smaller bovine PTH mRNAs are also heterogeneous with initiation occurring over a range of about eight nucleotides at the 5' terminus. The second TATA sequence in the bovine sequence is unusual since the sequence TA is repeated five times, and thus the TATA-like sequence is spread over 12 base pairs. This may result in a less rigorous delineation of the appropriate start site.

The conclusion that the 5' end of bovine mRNA is heterogeneous has not been conclusively proven. Both the S1 nuclease mapping and the primed reverse transcriptase techniques require that the mRNA is intact and not degraded. Since in the studies described above, it was not demonstrated that all the mRNA had a 5' methylguanosine cap and thus was intact, the possibility that heterogeneity was introduced during isolation of the mRNA cannot be excluded. However, the additional indirect evidence provided by the presence of two TATA sequences considerably strengthens the theory that two regions are utilized for initiation of transcripts.

The 5' Flanking Region

The three PTH genes of human, bovine and rat show homology in the first 200 bp upstream of the RNA initiation site of the 5' flanking region.[1] The homology in this region is similar to that in the 5' untranslated region of the mRNAs.[1] There are few stretches of sequence in the 5' flanking region that are completely conserved in all three sequences except for the TATA sequences. A C-rich sequence, GCACCGCCC, about 75 bp to the 5' side of the upstream TATA sequence is present in all three sequences, and an AT-rich region of about 25 bp immediately prior to this C-rich region is strongly conserved. A sequence, CAGAGAA, about 25 bp to the 5' side of the TATA sequence, is also present in all three sequences. No CAAT sequence is present 5' of the TATA sequences. In the bovine gene, an extraordinary stretch of almost 150 nucleotides, located from 250 to 400 nucleotides before the transcript initiator, consists primarily of alternating AT.[1] A similar region is not present in the rat gene, suggesting it is not critical for the function of the gene. There are defined functional response elements in the 5'-flanking region that regulate PTH gene transcription, such as the vitamin D response element (VDRE) and the cyclic AMP response element (CRE) that are discussed in detail in the chapters by Kel et al and Silver et al.

The 3' Flanking Region

In the 3' flanking region, again there is also considerable homology between the bovine and human sequences. A small inverted repeat region, that could form a hairpin loop in the transcript, is followed by a stretch of 7 Ts. There is no direct evidence that this region serves as a transcriptional stop signal in the PTH genes. A difference in the stem in the human compared to bovine is matched by a second change in the human that maintains the base pairing in the stem. A similar sequence is not present in the approximately 110 bp of 3' flanking sequence reported for the rat sequence. The rat in fact has little homology with either of the other two sequences beyond the polyadenylation signal.[1] This is surprising in view of the homology retained between the rat PTH gene and the other genes in the 5' flanking and intron regions. Perhaps the polyadenylation signal for the rat sequence is derived from a different region of the gene, which was moved into its present position by a deletion of sequence or translocation. Large gaps must be introduced into the bovine and rat sequences just prior to the polyadenylation signal supporting the idea that this may be a relatively unstable region of the gene.[1]

Overall, the PTH genes are typical eukaryotic genes that contain the consensus sequences for initiation of RNA synthesis, RNA splicing, and polyadenylation. The PTH genes appear to be represented only once in the haploid genome. Perhaps the most striking characteristic of the DNA in the region of the genes is its stability. In addition, regions that diverge rapidly in other genes are relatively stable in the PTH genes, particularly between the human and bovine sequences and to a lesser extent with the rat sequence. Thus, considerable homology is observed between 5' and 3' flanking and untranslated regions, internal regions of introns, and potential sites for silent changes in the coding region. Since these regions that do not change the amino acid sequence have been estimated to diverge at a rate of 1% 10^6 years, relatively low homologies would be expected from these sequences that diverged about 60 to 80 x 10^6 years ago.[1] Whether this conservation of sequence occurs because the genes happen to be present in a region of the chromosome that is usually stable or reflects some functional constraints inherent in the PTH gene, remains to be elucidated.

The rat and mouse sequences are considerably less homologous to the human and bovine sequences than these sequences are to each other. This observation is difficult to explain, since evolutionarily each of the sequences is about equidistant from another. Potentially, differences in the physiology or nutrition of calcium in the rat and mouse compared to the other two species may have resulted in increased acceptance of mutations in the rat PTH gene.

Chromosomal Location of the Human PTH Gene

The location of the human PTH gene on chromosome 11 has been determined independently by two groups. The assignments were made by screening panels of human-mouse[46] or human-mouse and human-Chinese hamster cell[47] hybrids with a human cDNA clone or a cloned fragment of human genomic DNA. The PTH gene was further localized to the short arm of the chromosome 11 by analysis of human-mouse hybrids with various translocations.[46] The short arm of chromosome 11 contains several other polymorphic genes including the β globin gene cluster, insulin, and the human oncogene Harvey *ras* (C-Ha-ras-1).[48,49] The polymorphisms in these genes and PTH were used to determine whether the genes are genetically linked and their order on the chromosome. In addition to these genes, the gene for calcitonin has also been mapped to the short arm of chromosome 11. Thus, the short arm contains genes for both of the polypeptide hormones that regulate calcium metabolism. Whether this is a mere coincidence or is somehow related to the evolution or regulation of these calcium regulating genes remains a matter of speculation. The porcine PTH gene was localized to chromosome 14q25-q28 by in situ hybridization[50] and the equus gene is on chromosome 11p15.3.[29]

Summary

The PTH genes and cDNAs have been isolated and characterized in 10 species. The gene contains two introns, which are in the same position in each species, and dissect the gene into 3 exons that code, respectively, for the 5' untranslated region, the signal peptide, and PTH plus the 3' untranslated region. The mRNAs contain a 7-methyl quanosine cap at the 5' terminus and a polyadenylation signal at the 3' terminus. They are about twice as long as necessary to code for preProPTH. The 5' termini of the bovine and human mRNAs are heterogeneous at the 5' terminus, the basis of which is two TATA sequences in the 5' flanking regions of the gene. In contrast, the mouse and rat gene contain a single TATA sequence and the mRNA has a single 5' terminus. The initial translational product of the mRNA is preProPTH, and the pre-peptide of 25 amino acids and the pro sequence of 6 amino acids are removed by two proteolitic cleavages.

The mRNAs are very homologous in the region that codes for preProPTH. But substantial homology is also retained in the mRNA untranslated regions and flanking regions and introns, where sequences are available. The gallus PTH mRNA is the most distant sequence of the PTH mRNAs. In the PTH mRNA the 3'-UTR is the region less conserved amongst species. However two protein binding elements in the 3'-UTR were identified and show high homology. One of these elements is the distal 26 nt *cis* acting functional element that has been shown to mediate the regulation of PTH mRNA stability in response to changes in serum calcium and phosphate. This element is expressed in the 3'-UTR of rat, man, dog, cat and mouse. An additional proximal element of 22 nt is present in the 3'-UTR of bovine, pig, macaca, horse and also in man, cat, dog. This element binds cytosolic proteins but its function has not been demonstrated. The conservation of such elements in the 3'-UTR suggests that they represent an evolutionary conserved function. PTH is central to normal calcium homeostasis and bone strength and the PTH peptide is highly conserved amongst species apart from Gallus. This conservation is evident in the coding sequence but also, to a less extent in the 5'- and 3'-UTRs.

Acknowledgements

This chapter quotes widely, with the author's permission, from the outstanding review by Byron Kemper where there is a detailed analysis of the bovine, rat and human PTH genes that were published at that time (ref. 1). We are extremely grateful to him both for his contribution and his generosity.

References

1. Kemper B. Molecular biology of parathyroid hormone. CRC Crit Rev Biochem 1986; 19:353-379.
2. Brewer HBJr, Ronan R. Bovine parathyroid hormone: Amino acid sequence. Proc Natl Acad Sci USA 1970; 67:1862-1869.
3. Cohn DV, MacGregor RR, Chu LL et al. Calcemic fraction-A: Biosynthetic peptide precursor of parathyroid hormone. Proc Natl Acad Sci USA 1972; 69:1521-1525.
4. Kemper B, Habener JF, Potts JT Jr et al. Proparathyroid hormone: Identification of a biosynthetic precursor to parathyroid hormone. Proc Natl Acad Sci USA 1972; 69:643-647.
5. Habener JF, Amherdt M, Ravazzola M et al. Parathyroid hormone biosynthesis: Correlation of conversion of biosynthetic precursors with intracellular protein migration as determined by electron microscope autoradiography. J Cell Biol 1979; 80:715-731.
6. Kemper B, Habener JF, Mulligan RC et al. Pre-proparathyroid hormone: A direct translation product of parathyroid messenger RNA. Proc Natl Acad Sci USA 1974; 71:3731-3735.
7. Milstein C, Brownlee GG, Harrison TM et al. A possible precursor of immunoglobulin light chains. Nat New Biol 1972; 239:117-120.
8. Dorner A, Kemper B. Conversion of preproparathyroid hormone to proparathyroid hormone by dog pancreatic microsomes. Biochemistry 1978; 17:5550-5555.
9. Habener JF, Kamper B, Potts JT Jr et al. Preproparathyroid hormone identified by cell-free translation of messenger RNA from hyperplastic human parathyroid tissue. J Clin Invest 1975; 56:1328-1333.
10. Kemper B, Habener JF, Ernst MD et al. Pre-proparathyroid hormone: analysis of radioactive tryptic peptides and amino acid sequence. Biochemistry 13-1-1976; 15:15-19.
11. Habener JF, Rosenblatt M, Kemper B et al. Pre-proparathyroid hormone; amino acid sequence, chemical synthesis, and some biological studies of the precursor region. Proc Natl Acad Sci U S A 1978; 75:2616-2620.
12. Kronenberg HM, McDevitt BE, Majzoub JA et al. Cloning and nucleotide sequence of DNA coding for bovine preproparathyroid hormone. Proc Natl Acad Sci USA 1979; 76:4981-4985.
13. Weaver CA, Gordon DF, Kemper B. Nucleotide sequence of bovine parathyroid hormone messenger RNA. Mol Cell Endocrinol 1982; 28:411-424.
14. Weaver CA, Gordon DF, Kissil MS et al. Isolation and complete nucleotide sequence of the gene for bovine parathyroid hormone. Gene 1984; 28:319-329.
15. Hendy GN, Kronenberg HM, Potts JTJ et al. Nucleotide sequence of cloned cDNAs encoding human preproparathyroid hormone. Proc Natl Acad Sci USA 1981; 78:7365-7369.

16. Vasicek TJ, McDevitt BE, Freeman MW et al. Nucleotide sequence of the human parathyroid hormone gene. Proc Natl Acad Sci USA 1983; 80:2127-2131.
17. Heinrich G, Kronenberg HM, Potts JT Jr et al. Gene encoding parathyroid hormone. Nucleotide sequence of the rat gene and deduced amino acid sequence of rat preproparathyroid hormone. J Biol Chem 1984; 259:3320-3329.
18. Schmelzer HJ, Gross G, Widera G et al. Nucleotide sequence of a full-length cDNA clone encoding preproparathyroid hormone from pig and rat. Nucleic Acids Res 1987; 15:6740-6741.
19. Watson ME. Compilation of published signal sequences. Nucleic Acids Res 11-7-1984; 12:5145-5164.
20. Mead DA, Skorupa ES, Kemper B. Single stranded DNA SP6 promoter plasmids for engineering mutant RNAs and proteins: synthesis of a 'stretched' preproparathyroid hormone. Nucleic Acids Res 1985; 13:1103-1118.
21. Majzoub JA, Rosenblatt M, Fennick B et al. Synthetic pre-poroparathyroid hormone leader sequence inhibits cell-free processing of placental, parathyroid, and pituitary pre-hormones. J Biol Chem 1980; 255:11478-11483.
22. Jackson RC, Blobel G. Post-translational cleavage of presecretory proteins with an extract of rough microsomes from dog pancreas containing signal peptidase activity. Proc Natl Acad Sci USA 1977; 74:5598-5602.
23. Habener JF, Rosenblatt M, Dee PC et al. Cellular processing of pre-proparathyroid hormone involves rapid hydrolysis of the leader sequence. J Biol Chem 1979; 254:10596-10599.
24. Divieti P, John MR, Juppner H et al. Human PTH-(7-84) inhibits bone resorption in vitro via actions independent of the type 1 PTH/PTHrP receptor. Endocrinol 2002; 143:171-176.
25. Perler F, Efstratiadis A, Lomedico P et al. The evolution of genes: the chicken preproinsulin gene. Cell 1980; 20:555-566.
26. Stolarsky L, Kemper B. Characterization and partial purification of parathyroid hormone messenger RNA. J Biol Chem 1978; 253:7194-7201.
27. Rosol TJ, Steinmeyer CL, McCauley LK et al. Sequences of the cDNAs encoding cannine parathyroid hormone-related protein and parathyroid hormone. Gene 1995; 160:241-243.
28. Khosla S, Demay M, Pines M et al. Nucleotide sequence of cloned cDNAs encoding chicken preproparathyroid hormone. J Bone Miner Res 1988; 3:689-698.
29. Caetano AR, Shiue YL, Lyons LA et al. A comparative gene map of the horse (Equus caballus). Genome Res 1999; 9:1239-1249.
30. Malaivijitnond S, Takenaka O, Anukulthanakorn K et al. The nucleotide sequences of the parathyroid gene in primates (suborder Anthropoidea). Gen Comp Endocrinol 2002; 125:67-78.
31. He B, Tong TK, Hiou-Tim FF et al. The murine gene encoding parathyroid hormone: genomic organization, nucleotide sequence and transcriptional regulation. J Mol Endocrinol 2002; 29:193-203.
32. Nutley MT, Parimi SA, Harvey S. Sequence analysis of hypothalamic parathyroid hormone messenger ribonucleic acid. Endocrinology 1995; 136:5600-5607.
33. Gordon DF, Kemper B. Synthesis, restriction analysis, and molecular cloning of near full length DNA complementary to bovine parathyroid hormone mRNA. Nucleic Acids Res 1980; 8:5669-5683.
34. Weaver CA, Gordon DF, Kemper B. Introduction by molecular cloning of artifactual inverted sequences at the 5' terminus of the sense strand of bovine parathyroid hormone cDNA. Proc Natl Acad Sci USA 1981; 78:4073-4077.
35. Suzuki Y, Ishihara D, Sasaki M et al. Statistical analysis of the 5' untranslated region of human mRNA using "Oligo-Capped" cDNA libraries. Genomics 2000; 64:286-297.
36. Habener JF, Kemper B, Potts JT Jr et al. Parathyroid mRNA directs the synthesis of preproparathyroid hormone and proparathyroid hormone in the Krebs ascites cell-free system. Biochem Biophys Res Commun 1975; 67:1114-1121.
37. Kozak M. An analysis of vertebrate mRNA sequences: intimations of translational control. J Cell Biol 1991; 115:887-903.
38. Kozak M. Structural features in eukaryotic mRNAs that modulate the initiation of translation. J Biol Chem 1991; 266:19867-19870.
39. Kwon HS, Lee DK, Lee JJ et al. Posttranscriptional regulation of human ADH5/FDH and Myf6 gene expression by upstream AUG codons. Arch Biochem Biophys 2001; 386:163-171.

40. Moallem E, Silver J, Kilav R et al. RNA protein binding and post-transcriptional regulation of PTH gene expression by calcium and phosphate. J Biol Chem 1998; 273:5253-5259.
41. Sela-Brown A, Silver J, Brewer G et al. Identification of AUF1 as a parathyroid hormone mRNA 3'-untranslated region binding protein that determines parathyroid hormone mRNA stability. J Biol Chem 2000; 275:7424-7429.
42. Guhaniyogi J, Brewer G. Regulation of mRNA stability in mammalian cells. Gene 2001; 265:11-23.
43. Kilav R, Silver J, Naveh-Many T. A conserved cis-acting element in the parathyroid hormone 3'-untranslated region is sufficient for regulation of RNA stability by calcium and phosphate. J Biol Chem 2001; 276:8727-8733.
44. Naveh-Many T, Bell O, Silver J et al. Cis and trans acting factors in the regulation of parathyroid hormone (PTH) mRNA stability by calcium and phosphate. FEBS Lett 2002; 529:60-64.
45. Gilbert W. Why genes in pieces? Nature 1978; 271:501.
46. Naylor SL, Sakaguchi AY, Szoka P et al. Human parathyroid hormone gene (PTH) is on short arm of chromosome 11. Somatic Cell Genet 1983; 9:609-616.
47. Mayer H, Breyel E, Bostock C et al. Assignment of the human parathyroid hormone gene to chromosome 11. Hum Genet 1983; 64:283-285.
48. Gerhard DS, Kidd KK, Kidd JR et al. Identification of a recent recombination event within the human beta-globin gene cluster. Proc Natl Acad Sci USA 1984; 81:7875-7879.
49. Fearon ER, Antonarakis SE, Meyers DA et al. c-Ha-ras-1 oncogene lies between beta-globin and insulin loci on human chromosome 11p. Am J Hum Genet 1984; 36:329-337.
50. Ji Q, Zhang X, Chen Y. [Localization of PTH gene on pig chromosome 14q25-q28 by in situ hybridization]. Yi Chuan Xue Bao 1997; 24:218-222.horse and non-human primate (macaca) PTH mRNA were not included in this study because only partial sequences of these RNAs are available. The same Phylogenetic tree is also obtained when the 3'-UTR sequences are analized separately. Interestingly, based on amino acid sequence, the bovine and pig were grouped closest to canis and felis but not by RNA sequence. This corresponds to the presence of the distal and proximal protein-binding elements in the 3'-UTRs.Macaca fascicularis Af130257: gene, complete cds mRNA (partial, deduced).

CHAPTER 3

Toward an Understanding of Human Parathyroid Hormone Structure and Function

Lei Jin, Armen H. Tashjian, Jr., and Faming Zhang

PTH and Its Receptor Family

Parathyroid hormone (PTH) is synthesized as a 115 amino acid precursor and secreted as an 84 amino acid polypeptide that regulates extracellular calcium homeostasis via actions directly on kidney and bone and indirectly on the intestine by facilitating calcium absorption.[1] PTH and a related molecule, parathyroid hormone-related protein (PTHrP), act on cells via a common G protein-coupled, seven-transmembrane helix receptor (PTH/PTHrP or PTH1 receptor).[2] PTH has both anabolic and catabolic effects on the skeleton. Persistent elevation of plasma PTH causes predominately increased bone resorption, whereas intermittently administered PTH results in enhanced bone formation.[3,4] The mechanism by which PTH exhibits its dual effects is not fully known. PTH interacts with the PTH1 receptor to stimulate adenylyl cyclase (AC)(5) and phospholipase C (PC) pathways.[6]

The naturally occurring hPTH(1-37) fragment as well as hPTH(1-34) maintains the full spectrum of bone-relevant activities of the intact 1-84 hormone. Studies, both in vitro and in vivo, have shown that hPTH(1-34) has the same biological activities as the intact hormone in eliciting cAMP responses and in stimulating bone formation.[7] Thus hPTH(1-34) has all the structural elements necessary for binding and activation of the PTH1 receptor. Once daily subcutaneous administration of hPTH(1-34) stimulates bone formation and increases bone mass in patients with osteoporosis[4] and in ovariectomized monkeys.[8] Consequently, hPTH(1-34) represents a novel class of therapeutics for the treatment of osteoporosis.[9-11] Truncation and mutagenesis studies on PTH(1-34) have revealed that the N-terminal region of the peptide is critical for full activation of receptor signaling, while the N-terminal truncated peptide PTH (3-34) is a partial agonist, and the further shortened peptide PTH (7-34) becomes a low affinity antagonist.[12,13] Residues 17 to 31, near the C-terminus of PTH (1-34), are required for high affinity receptor binding.[14]

PTHrP is over-expressed in certain tumors and causes the syndrome of malignancy-associated humoral hypercalcemia.[15] Under physiological conditions, PTHrP is produced locally in a wide variety of tissues and is involved in cell growth, differentiation, and development of the skeleton. There are six identical amino acids in the first 13 amino acids in the known PTH and PTHrP sequences (Fig. 1). Like PTH, PTHrP binds to the same G protein-coupled receptor, and its N-terminal fragment PTHrP(1-34) has many functions that mimic those of

Molecular Biology of the Parathyroid, edited by Tally Naveh-Many. ©2005 Eurekah.com and Kluwer Academic / Plenum Publishers.

										10						20						30						
PTH_HUMAN	S	V	S	E	I	Q	L	M	H	N	L	G	K	H	L	N	S	M	E	R	V	E	W	L	R	K	K	L	Q	D	V	H	N	F
PTH_BOVIN	A	V	S	E	I	Q	F	M	H	N	L	G	K	H	L	S	S	M	E	R	V	E	W	L	R	K	K	L	Q	D	V	H	N	F
PTH_CANFA	S	V	S	E	I	Q	F	M	H	N	L	G	K	H	L	S	S	M	E	R	V	E	W	L	R	K	K	L	Q	D	V	H	N	F
PTH_PIG	S	V	S	E	I	Q	L	M	H	N	L	G	K	H	L	S	S	L	E	R	V	E	W	L	R	K	K	L	Q	D	V	H	N	F
PTH_RAT	A	V	S	E	I	Q	L	M	H	N	L	G	K	H	L	A	S	V	E	R	M	Q	W	L	R	K	K	L	Q	D	V	H	N	F
PTH_CHICK	S	V	S	E	M	Q	L	M	H	N	L	G	E	H	R	H	T	V	E	R	Q	D	W	L	Q	M	K	L	Q	D	V	H	S	A
PTHrP_HUMAN	A	V	S	E	H	Q	L	L	H	D	K	G	K	S	I	Q	D	L	R	R	R	F	F	L	H	H	L	I	A	E	I	H	T	A
PTHrP_CANFA	A	V	S	E	H	Q	L	L	H	D	K	G	K	S	I	Q	D	L	R	R	R	F	F	L	H	H	L	I	A	E	I	H	T	A
PTHrP_RAT	A	V	S	E	H	Q	L	L	H	D	K	G	K	S	I	Q	D	L	R	R	R	F	F	L	H	H	L	I	A	E	I	H	T	A
PTHrP_MOUSE	A	V	S	E	H	Q	L	L	H	D	K	G	K	S	I	Q	D	L	R	R	R	F	F	L	H	H	L	I	A	E	I	H	T	A
PTHrP_CHICK	A	V	S	E	H	Q	L	L	H	D	K	G	K	S	I	Q	D	L	R	R	R	I	F	L	Q	N	L	I	E	G	V	N	T	A
TIP39_BOVIN (SL)	A	L	A	D	D	A	A	F	R	E	R	A	R	L	L	A	A	L	E	R	R	H	W	L	N	S	Y	M	H	K	L	L	V	L	(DAP)

Figure 1. Sequence alignment of known species of PTH(1-34), PTHrP(1-34) and TIP39. The invariant residues are shaded in orange. The conserved residues are shaded in yellow in PTH(1-34), and in blue in PTHrP(1-34). TIP39 shares less sequence identity with PTH and PTHrP.

full-length PTHrP, PTH(1-34) and PTH(1-84).[15,16] In addition, NMR studies indicate that hPTH(1-34) and hPTHrP(1-34) have similar three-dimensional structures.[17,18]

The recently identified PTH2 receptor natural ligand, tuberoinfundibular peptide 39 (TIP39), has been characterized structurally as a PTH homolog.[19] The human PTH2 receptor shares 70% sequence similarity with the human PTH1 receptor. The PTH1 receptor is activated by PTH and PTHrP but not by TIP39, while the rat PTH2 receptor is activated by PTH and TIP39 and responds very weakly to PTHrP.[20] Further studies demonstrated that TIP39 and its truncated analogs bind to the PTH1 receptor but its N-terminal domain failed to stimulate cAMP accumulation; therefore, they function as antagonists at the PTH1 receptor (21). NMR studies indicate that TIP39, similar to hPTH(1-34), contains two stable α-helices at the N and C termini separated by a flexible hinge region. TIP39 is in a somewhat extended conformation; the N-terminal helix shares a high structural and sequence homology with hPTH(1-34). The differences are the lengths and amphipathic character of the helices as well as the location of the flexible hinge region.[19]

Comparison of PTH, PTHrP and TIP39 as well as their binding differences with the PTH1 and PTH2 receptors provides a starting point to study the structure-function relationships between the ligands and receptors. Studies such as mutagenesis in both ligand and receptors, creation of receptor chimeras, and cross-linking between ligand and receptors have been performed extensively in the PTH family.[2] Biophysical techniques such as circular dichroism (CD), nuclear magnetic resonance (NMR), x-ray crystallography and molecular modeling provide information on PTH structure and its possible interactions with receptors for cell signaling.

PTH Structural Determination

Various methods have been used to determine the structure of PTH, including dark-field electron microscopy, fluorescence spectroscopy, CD, NMR spectroscopy[22-29] and X-ray crystallography.[30] Results from these diverse approaches have not yet yielded a consistent structure for this peptide. In part, this uncertainty arises from the flexible nature of most small peptides in solution, as well as different experimental conditions such as peptide concentration, solvent conditions, pH, temperature and differences in methods used for data interpretation. There is general agreement that PTH(1-34) and PTHrP(1-34) have an N-terminal helix and a C-terminal helix, which vary in length and stability, depending on the specific experimental conditions, and are connected by a highly flexible mid-region. The C-terminal helix is more stable than the N-terminal helix. In aqueous solution, PTH(1-34) and PTHrP(1-34) form fewer and less stable secondary structural elements than under membrane-mimicking conditions such as in

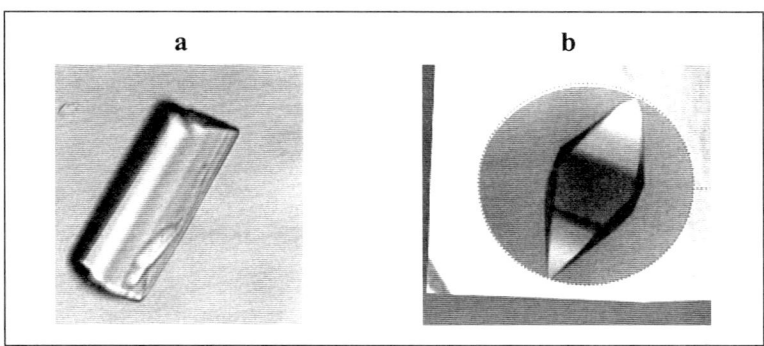

Figure 2. Crystals of hPTH(1-34). **a**) Crystal grown from 20% isopropanol, 0.2M NaCitrate, 0.1M TrisHCl, at pH 8.0. **b**) Crystal grown from 2.5M Ammonium Sulfate, 5% isopropanol, 0.1M Na Acetate, at pH 4.5. The high-resolution structure was derived from the **b** crystal form.

dodecylphosphocholine micelles,[26] palmitoyloleoylphosphatidylserine vesicles,[29] or in the presence of a secondary structure-inducing solvent such as trifluoroethanol (TFE).[24-26,28,29] Here we shall describe our x-ray crystallographic studies on hPTH(1-34) and give a brief review of recent NMR and molecular modeling results.

X-Ray Crystallography

Human PTH(1-34) (Forteo, Eli Lilly and Company, Indianapolis, IN, U.S.A.) was expressed in *E. coli* and subsequently purified and refolded by reverse phase and cation exchange chromatography. For phase determination, selenomethionine hPTH(1-34) was synthesized on an ABI-430A peptide synthesizer using Boc seleno-L-methionine. hPTH(1-34) could be readily crystallized in many different forms under slightly different solvent conditions. The best crystals were grown at 20°C by the hanging drop vapor diffusion method by mixing 20 mg ml^{-1} of hPTH (1-34) in 20% glycerol, at 1:1 ratio (v/v), with a solution containing 2.5 M ammonium sulfate, 5% isopropanol and 0.1 M sodium acetate buffer, pH 4.5. Crystals appeared overnight and continuously grew to 0.6 x 0.2 x 0.1 mm^3 in a week (Fig. 2a,b). Ultra-high 0.9 Å resolution data were collected by a Mar CCD detector at the Industrial Macromolecule Crystallography Association beam line ID-17 in Argonne National Laboratories. The crystals belong to haxagonal space group P6$_5$ with unit cell dimensions of a= 30.18 Å and c=110.44 Å. Three data sets were collected for a selenomethionine hPTH (1-34) crystal at wavelengths of 0.9795, 0.97936 and 0.9840 Å for multiwavelength anomalous dispersion (MAD) phasing calculations.[30]

The structure was solved by the program SOLVE with the MAD data. The polypeptide chain was fitted to the electron density using the program O.[31] The model was refined to 2.0 Å resolution using the MAD data, and to 0.9 Å resolution using the native data in X-PLOR98[32] by simulated annealing. The model was then further refined in SHELX 97[33] by the conjugate-gradient algorithm with riding hydrogens. Six-parameter anisotropic temperature factors for all non-H atoms were included in and after anisotropic refinement. Sequential model building processes were performed in QUANTA (Molecular Simulations, Inc) against 2F$_o$-F$_c$ and F$_o$-F$_c$ maps. The final R factor for all data was 13.7%, R-free was 14%. The final structure contains 660 non-hydrogen peptide atoms and 104 water molecules. All residues are in the most favorable conformation in Ramachandran plot.

The crystal structure of hPTH(1-34) is a slightly bent helix (Fig. 3a). The bend is located between residues 12 and 21 with a bending angle of 15° between the N-terminal helix (resi-

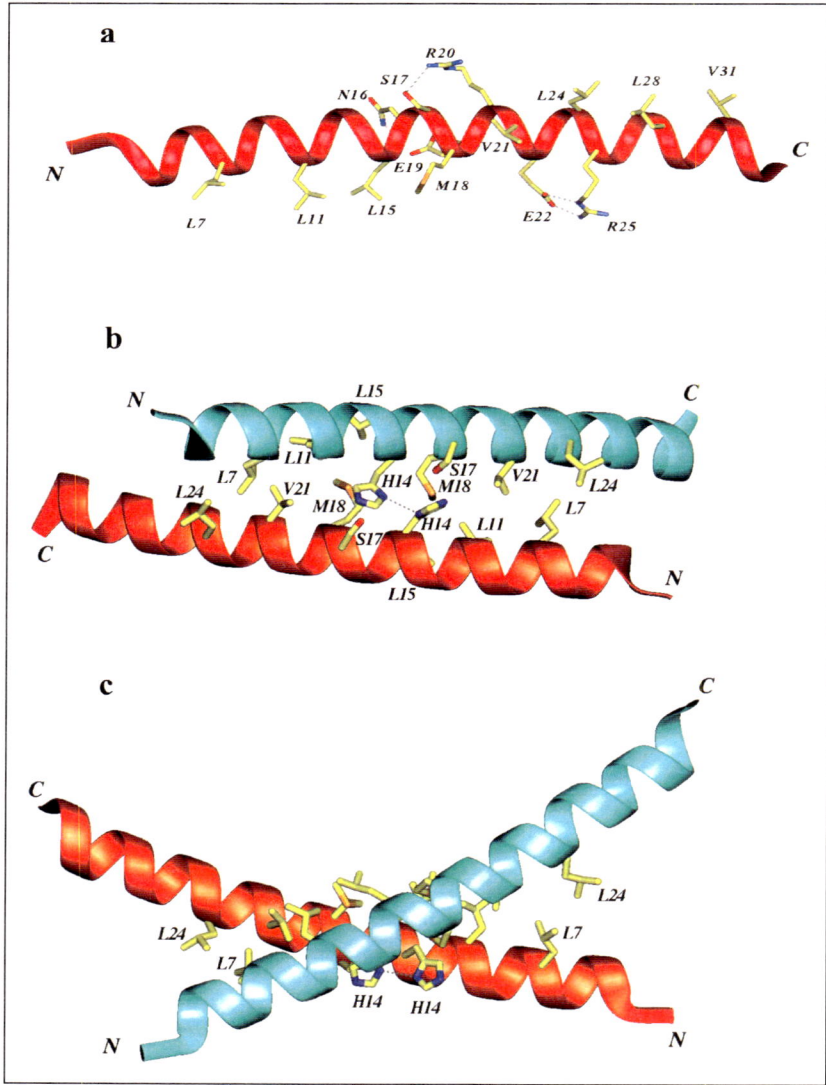

Figure 3. Overall structure of hPTH(1-34). **a)** hPTH(1-34) monomer is a slightly bent helix presented as a red ribbon. The hydrophobic residue side chains of the amphiphilic helices (Leu7, Leu11, Leu15, Met18, Val21, Leu24, Leu28 and Val31) are shown. **b, c)** hPTH(1-34) dimer is presented as blue and orange ribbons and the residues forming the dimer interface are highlighted.

dues 3-11) and the C-terminal helix (residues 21-33). Hydrogen bonds between the side chains of Asn16 and Glu19, Ser17 and Arg20, and a salt bridge between Glu22 and Arg25 were observed. Although hPTH(1-34) is a continuous helix, residues 6-20 and residues 21-33 form two amphiphilic helices with their hydrophobic sides facing in different directions. Thus, the hydrophobic residues of hPTH(1-34) form a twisted belt from the N-terminus to the C-terminus with the crossing-point near residue Arg20 (Fig. 3a). Gly12 is a conserved residue in all the known PTH and PTHrP species (Fig. 1). Despite the flexible nature of glycine, Gly12 is in a strict helical conformation in the crystal structure. Substitution of Gly12 with Ala, a helix promotor, in [Tyr34]hPTH(1-34)NH$_2$ was well tolerated; while substitution with Pro, a helix breaker, decreased receptor binding affinity by 840 fold and adenylyl cyclase-stimulating activity by 3500 fold.[34] Together, these findings indicate that the helical conformation around Gly12 may play an essential role for full biological activity.

hPTH(1-34) crystallizes as a dimer in the hexagonal space group P6$_5$. His14 and Ser17 from both molecules are located at the crossing-point of the X-shaped dimer (Fig. 3b,c). The Nδ of His14 from one molecule forms a hydrogen bond with Nδ of His14 from another molecule, while Ser17 from one molecule packs against the imidazole ring of His14 from the other molecule. Within the dimeric interface, hydrophobic interactions extend from the crossing-point toward the N- and C-termini. Residues Leu7, Leu11 and Leu15 of the N-terminal amphiphilic helix from one molecule are in van de Waals contact with residues Leu24, Val21 and Met18 of the C-terminal amphiphilic helix of the other molecule. There is no evidence of well-ordered dimerization of PTH under physiological conditions; therefore, the dimer formation is most likely the result of crystal packing artifacts under specific solvent condition. The monomer structure was used for molecular modeling.

NMR Studies

NMR and x-ray crystallography are in many respects complementary. X-ray crystallography deals with the structure of proteins in the crystalline state, while NMR determines structure in solution. X-ray structures are determined at different levels of resolution. At high resolution, most atomic positions can be determined to a high degree of accuracy. NMR has the advantage to determine structures in near-physiological solution.

In NMR, a COSY (correlation spectroscopy) experiment gives peaks between hydrogen atoms that are covalently connected through one or two other atoms. An NOE (nuclear overhauser effect) spectrum, on the other hand, gives peaks between pairs of hydrogen atoms that are close together in space even if they are quite distant in primary sequence. To determine the three-dimensional structure of proteins, combined multidimensional NMR experiments including double-quantum-filtered correlation spectroscopy (DQF-COSY), total correlation spectroscopy (TOCSY), NOE spectroscopy (NOESY) have been used to obtain a list of distance constraints between atoms in the molecule, from these data a set of three-dimensional structures that satisfy these constraints are calculated.

NMR studies of hPTH(1-84) defined three helices between Ser3 to Asn10, Ser17 to Lys27, and Asp30 to Leu37. In the C-terminus, a less well-defined helix between Asn57 to Ser62 and series of loose turns were detected. In contrast to the hPTH(1-34) structure, the intact hormone shows a number of long-range NOEs, specially between helix 1 and helix 2.[35]

Several of the NMR studies have been interpreted to show a "U or V-shaped" tertiary structure with the N- and C-terminal helices interacting with each other to form a hydrophobic core.[24,36] However the majority of the NMR analyses of PTH and PTHrP do not provide evidence of long-range interactions between the two helices. NMR studies of hPTH(1-34) in near-physiological solution revealed a relatively extended two-helical component structure with the absence of any tertiary interactions between the two helices proposed in the U-shaped model.[18]

Extensive NMR studies have been carried out on PTH and PTHrP in different solvent environments.[17] In general, NMR studies show that PTH(1-34) and PTHrP(1-34) form an N-terminal helix and a C-terminal helix connected by a highly flexible region in solution. Figure 4 shows the superposition of the crystal structure with NMR structures of hPTH(1-34) with PDB code 1HPY[37] by superimposing the Cα atoms of the C-terminal helices (residues 18-28). Superposition of the crystal structure with other NMR structures, such as hPTH(1-37)

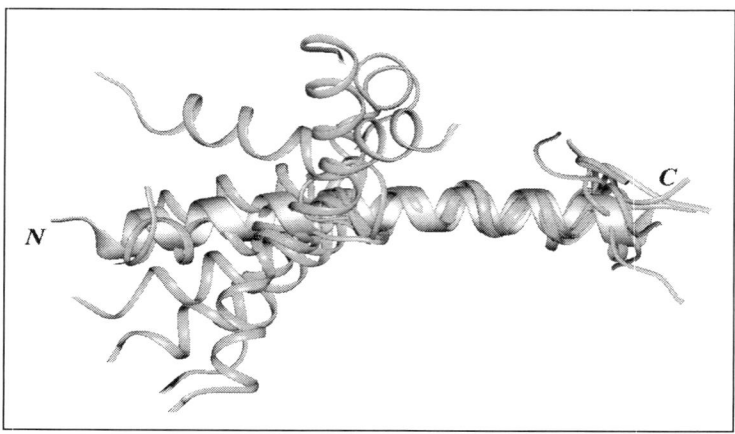

Figure 4. Superposition of the crystal structure and NMR structures of hPTH(1-34). Representative NMR conformations were selected. The Cα atoms of the C-terminal helix (residues 18-28) were superimposed. The crystal structure of hPTH(1-34) is presented as a thick ribbon, NMR structures are presented as thin ribbons. The crystal structure of hPTH(1-34) is in extended helical conformation, which is different from the NMR structures that possess N-terminal and a C-terminal helices connected by a flexible hinge.

(PDB code: 1HPH)[23] and hPTHrP(1-34) (PDB code: 1BZG)[18] yielded similar images as in Figure 4. The highly flexible region in the NMR structures (residues 10-20) is found to form a regular helix in the crystal structure. Evidence from several NMR studies on PTH(1-34) and PTHrP(1-34) concluded that the helical content increases with increasing TFE concentration or conditions that mimic the membrane environment. In 70% TFE, the N- and C-terminal helices (residues 3-13 and 15-29) of PTH(1-34) were very regular with a short discontinuity at residue 14,[38] NMR structures of PTHrP(1-34) in 50% TFE also showed two well-defined helices (residues 3-12 and 17-33).[17] Our crystal structure is similar to the NMR structures determined in high concentrations of TFE or under membrane-mimicking conditions.[17,38] This similarity is not surprising because hPTH(1-34) in the crystal has very limited solvent exposure. The solvent content of the hPTH(1-34) crystal is less than 30% with extensive hydrophobic protein-protein interactions. In fact, the crystal structure might represent the conformation of PTH (1-34) when it is close to its membrane receptor as proposed for the NMR structures under high concentrations of TFE or membrane-mimicking conditions.[29]

Molecular Modeling

Currently, there is no satisfactory method to directly determine the PTH-PTH receptor complex structure. Molecular modeling provides an alternative method to explore the ligand-receptor binding mode. PTH1 and PTH2 receptors belong to a GPCR family from which members have not yet been crystallized. The crystal structure of rhodopsin has provided the first three-dimensional model in atomic resolution for the GPCR families.[39] Although construction of realistic models of certain GPCRs like PTH receptors with large extra-cellular domain remains challenging,[40] the crystal structure of rhodopsin does provide a general starting point for modeling GPCRs, specifically in the seven-transmembrane (TM) helix region. We undertook a molecular modeling with the program QUANTA using the Protein Design tools. The seven transmembrane helical domains of the PTH1 receptor were first determined by several programs provided by the ExPASy Molecular Biology Server. The crystal structure of bacteriorhodopsin at 1.9 Å (PDB code: 1QHJ)[41] was used as a template for the topological orientation and arrangement of the transmembrane helices. Sequences of the TM helices for the PTH1 receptor and bacteriorhodopsin were aligned, and then homology modeling was carried out to create the TM helices of the PTH1 receptor. The conformations of the intracellular and extracellular loops were constructed in QUANTA using the fragment database-searching algorithm. For the N-terminal receptor region 168-198, the NMR structure determined in a lipid environment[42] (PDB code: 1BL1) was incorporated in the model. This was accomplished by aligning the membrane embedded helix (residues 190-196) with the beginning of transmembrane helix 1. The full-length PTH1 receptor contains residues 1-593, our model is only partial, containing residues 168-469.

One hPTH(1-34) monomer, derived from the crystal structure, was docked onto the receptor using two constraints based on cross-linking studies.[43,44] Energy minimization was applied to the complex of hPTH(1-34) and residues 168-198 of the receptor using the default setting in QUANTA until no significant changes were observed. The hPTHrP(1-34) model was produced by homology modeling using the crystal structure of hPTH(1-34) as a template. The model of hPTHrP(1-34) binding to the PTH1 receptor was created similarly to the hPTH(1-34)-receptor complex.

Previous studies on PTH- or PTHrP-receptor interactions have suggested that the juxtamembrane region of the transmembrane helices and extracellular loops (especially the third loop) of the PTH1 receptor interact with the amino-terminus of PTH or PTHrP agonists (the so-called "activation domain") to induce second messenger signaling;[2,45] the amino-terminal extracellular region of the receptor interacts with the carboxyl-terminal region (residues 15-34) of either PTH or PTHrP during ligand binding.[2,13] Results from photoaffinity cross-linking by *p*-benzoylphenylalanine and site-directed mutagenesis identified two contact points in the PTH(1-34):PTH1 receptor complex, Ser1 of hPTH(1-34) to Met425 of the receptor[43] and Lys13 of hPTH(1-34) to Arg186 of the receptor.[44] Our model of hPTH(1-34) bound to the PTH1 receptor was created by incorporating these restraints.

In our model (Fig. 5a), the N-terminal region of hPTH(1-34), responsible for its agonist activity, binds to a pocket consisting of the extracellular portions of TM3, TM4 and TM6 and the second and third extracellular loops of the receptor. The middle region of hPTH(1-34) is sandwiched between the first extracellular loop and the N-terminal extracellular region of the receptor adjacent to TM1. The C-terminal region of hPTH(1-34) forms extensive interactions with the putative binding domain of the PTH1 receptor (Fig. 5b). This interface consists of hydrophobic interactions (residues Leu24, Trp23 and Leu28 of hPTH(1-34), and Phe173, Leu174 of the receptor), and hydrophilic interactions between Arg20 of hPTH(1-34) and Glu180, Glu177 of the receptor, as well as Lys27 of hPTH(1-34) and Glu169 of the receptor. Recent photoaffinity cross-linking studies reveal that residues 23, 27, 28 of native PTHrP are indeed near regions of the amino-terminal extracellular domain of the PTH1 receptor.[46]

Site-directed mutagenesis in the C-terminal region of hPTH(1-34) have suggested that Leu24 and Leu28 are intolerant to mutation.[47] When Leu24 and Leu28 are substituted by Glu, the receptor binding affinities were decreased by 4000 and 1600 fold respectively. A less dramatic reduction of receptor binding affinity (40 fold) is observed when Val31 is replaced by Glu. In contrast, replacement of Asp30 by Lys has no effect on receptor binding. In our model, Leu24 and Leu28 of hPTH(1-34) are located at the center of the hydrophobic interface while Val31 is located at the end of the hydrophobic patch. Asp30 is exposed to solvent; therefore, the lysine mutant at this position would not be likely to change the receptor binding affinity. The hydrophilic interaction between Lys27 of hPTH(1-34) and Glu169 of the PTH1 receptor may be less important for binding than other interactions because a variety of mutations were tolerated at Lys27.[47] Several other models have been proposed in the literature for the binding

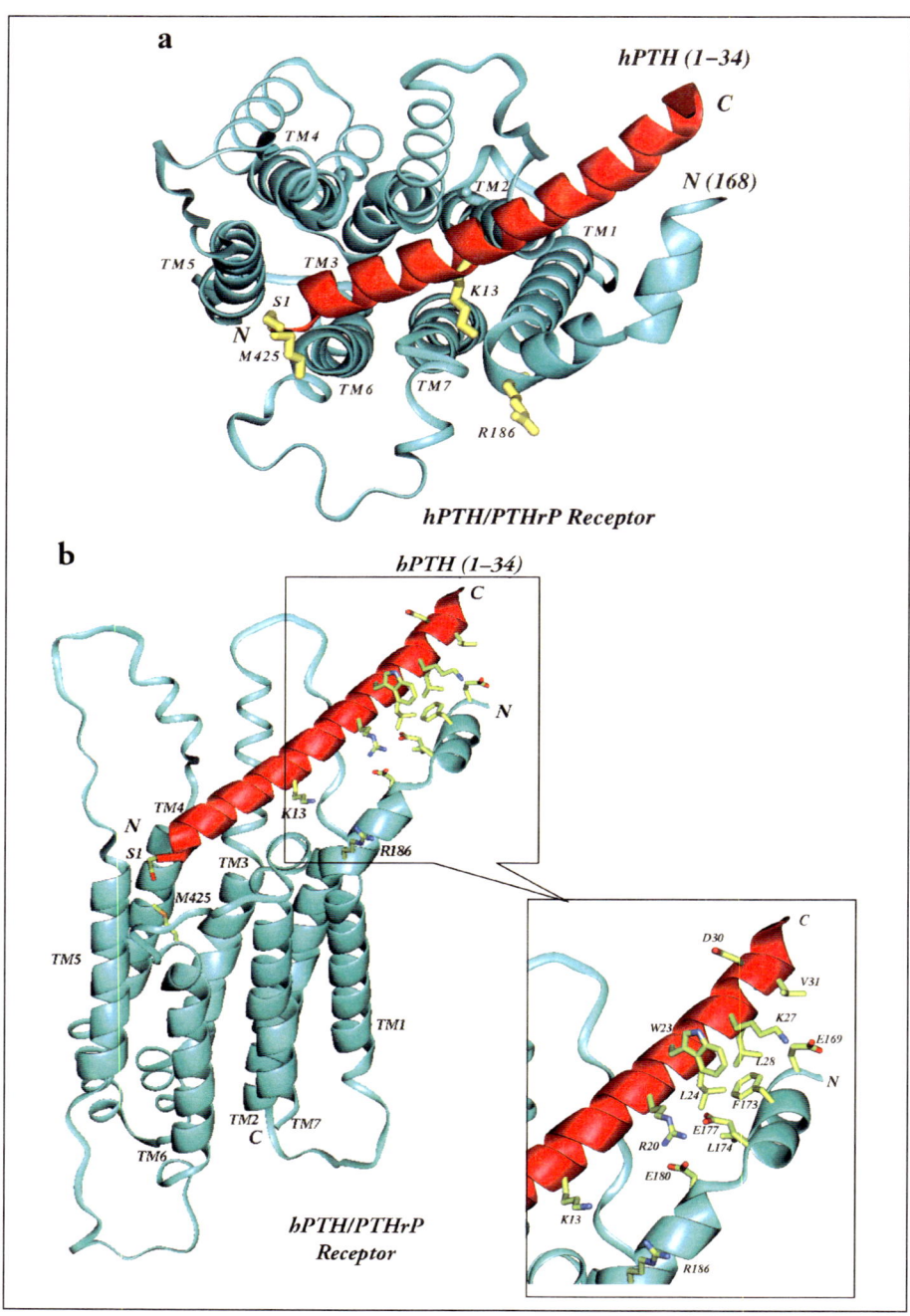

Figure 5. Model of hPTH(1-34) binding to the PTH1 receptor. The crystal structure of hPTH (1-34) is in red and the receptor is in blue. Residues at the ligand-receptor interface are highlighted in yellow. **a**) A top view of the model looking down the seven transmembrane helices. **b**) A side view of the model rotated 90° from the view in **a**. Residues forming the interface between the C-terminus of hPTH(1-34) and the receptor are highlighted.

of hPTH(1-34) to the PTH1 receptor, utilizing the NMR structure of hPTH(1-34) with N- and C-terminal helices connected by a flexible loop.[43,48] The most predictive model will likely be based on a combination of all the mutagenesis, cross-linking and structural experimental data available.

In the PTHrP:PTH1 receptor model (not shown here), residues Arg20, Phe23, Leu24, Ile28 and Ile31 of hPTHrP(1-34) form similar interaction with receptor as the corresponding residues of hPTH (1-34) (Fig. 5b). Residue Leu27 in hPTHrP, which is lysine in hPTH, is included in the extensive hydrophobic interface. hPTH and hPTHrP(1-34) share eight identical amino acids in the region 1-13, but only three identical amino acids in the region 14-34 (Fig. 1). However, the C-termini of both peptides form similar amphiphilic helices that are proposed to be responsible for high affinity receptor binding.[28,29] When residues 22-31 were substituted with a model amphiphilic sequence (ELLEKLLEKL) in the PTHrP analog RS-66271, it demonstrated high in vivo bone anabolic efficacy.[49] CD and NMR studies confirmed that RS-66271 exists in a helical conformation from residues 16 to 32.[50] Our models for the interaction of PTH and PTHrP to the PTH1 receptor support the hypothesis that the amphiphilic helices at the C-terminal regions of the PTH and PTHrP(1-34) are responsible for high affinity peptide-receptor interaction.[28,29]

Structural Based Design of PTH Analogs

Stabilizing α-Helical Conformation

Lactam bridges were introduced at different locations along the peptide to connect the side chains at *i* and *i*+4 positions in an effort to stabilize a helical conformation and identify the bioactive conformation of PTH and PTHrP. Structural and functional studies have suggested that increasing helical content by such conformational constraints may increase biological potency, but this result is highly sensitive to the constrained positions. Condon, S.M. et al,[51] reported that adenylyl cyclase- stimulating activity in ROS 17/2.8 cells was increased when a lactam bridge was introduced between residues 14 and 18 or 18 and 22 of hPTH (1-31)NH$_2$, but decreased when the lactam bridge was introduced between residues 10 and 14. In PTHrP, when lactamization was introduced between residues 13 and 17, adenylyl cyclase-stimulating activity was also increased.[18] However, a lactam bridge introduced between residues 26 and 30 resulted in 400 times lower binding affinity and 30 times lower adenylyl cyclase-stimulating activity.[18] Interestingly, the lactam-containing structures of hPTH(1-31) and hPTHrP(1-34) by NMR were both in extended helical conformations, similar to our crystal structure of hPTH(1-34). In the crystal structure of hPTH(1-34), the three well tolerated lactam bridges (residues 13-17, 14-18 and 18-22) are located on either the convex or concave sides of the arc formed by the slightly bent helix and are in the mid-region of the molecule (Fig. 6a). Thus, it appears that enhancing helical structure in this flexible region of the peptide increases the biological activity of PTH and PTHrP. The poorly tolerated bridges (residues 10-14 and 26-30) are located on the sides of the hPTH(1-34) helical arc. In these cases, the decreased biological activities may be caused by twisting the helical arc sideways or interfering directly with the ligand-receptor interaction. Thus, rigidity in the middle region of hPTH(1-34), as well as the bending direction of the helix appears to have significant functional effects. Therefore, the extended helical conformation observed in the crystal structure may well represent an active receptor-binding conformation of hPTH(1-34). This led us to propose that hPTH(1-34) could be in a flexible conformation in solution as would occur in the extracellular space, but a regularized, slightly bent helical conformation is likely to be induced when the peptide approaches the hydrophobic membrane before receptor binding.

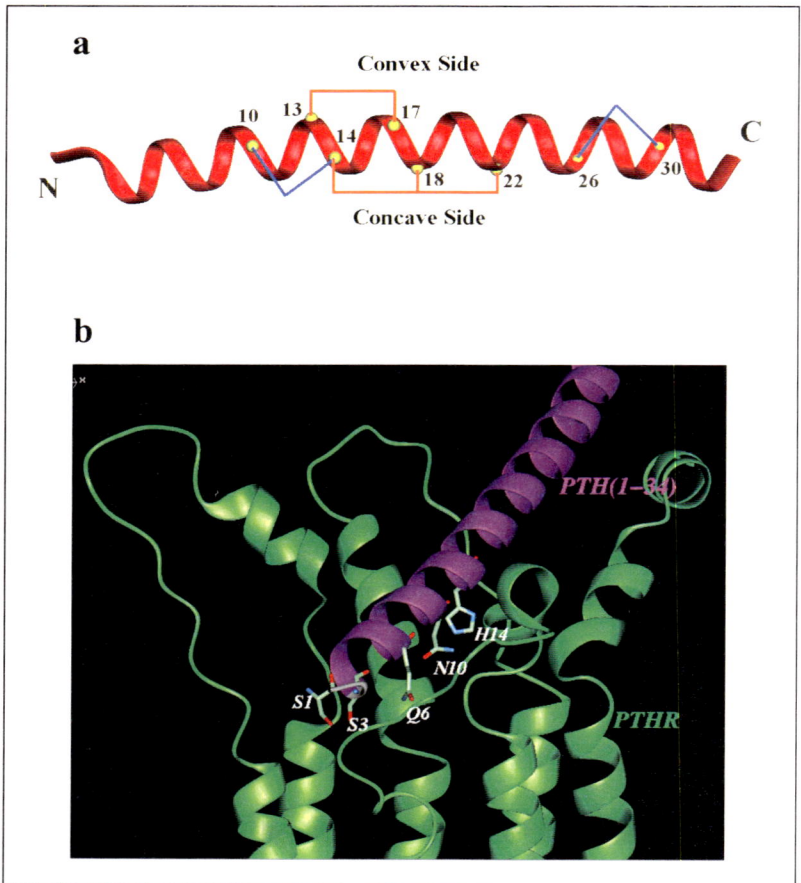

Figure 6. **a**) Positions of lactam bridges introduced in hPTH(1-31) or hPTHrP(1-34). The Cα atoms of the residues that were connected by lactam bridges are shown in yellow balls. The lactam bridges (between residues 13 and 17, 14 and 18, and 18 and 22), which increased the biological activity, are connected by orange lines. They are located in the mid region of the molecule and on either the convex or concave sides of the helical arc. The lactam bridges (between residues 10 and 14, 26 and 30), which significantly decreased biological activity, are connected by blue lines and are located on the sides of the helical arc. **b**) N-terminal (1-14) binding site in the model. Side chains of residues 1, 3, 6, 10,14 all point in the same direction down into the transmembrane domain of the receptor.

Substitution of Key Amino Acids

Extensive amino acid scanning has been done in PTH(1-34), especially in the 1-14 activation domain. In hPTH(1-34), substitution of Gly1 for Ser reduced phospholipase C activity but did not affect adenylyl cyclase activity, while removal of either Gly1 or the alpha-amino group eliminated phospholipase C activity completely.[52,53] Randomly mutated codons for amino acids 1-4 in hPTH(1-34) demonstrated that Val2 and Glu4 are important determinants of receptor binding and activation.[13] Studies with a series hybrid analogs containing both PTH and PTHrP sequences demonstrated that residue 5 (Ile in PTH and His in PTHrP) is the specificity "switch" between the PTH1 and PTH2 receptors,[54] switching residues at 5 and 23 in PTHrP to PTH residues yielded a ligand that avidly bound the PTH2 receptor and fully stimulate cAMP formation in contrast to the totally inactive native PTHrP on the PTH2 receptor.[55] There are two methionine residues in hPTH(1-34) that can be oxidized.[56] Oxidation of Met 8 was reported produce a partial agonist with greatly reduced potency; oxidation of Met 18 full agonist with slightly reduced potency. Thus, Met 8 is important in receptor binding and activation.[57]

Systematic site-directed mutagenesis at multiple sites has been used to create more than 50 hPTH(1-34) analogs. A highly active combination variant has been identified with six substitutions (K13S, E19S, V21Q, E22S, K27Q, D30N) and is 15 times more active for adenylyl cyclase-stimulation.[58] A variety of different unnatural amino acids have also been introduced into the PTH sequence to search for a more potent or signaling pathway specific agonist. Combinations of cyclization and substitution with either natural or unnatural amino acids may yield agonists with greatly enhanced potency.

Minimizing Active PTH Length

It is well established that the hPTH(1-37) and hPTH(1-34) fragments maintain the full spectrum of bone-relevant activities of the intact 1-84 hormone. A slightly shorter peptide, hPTH(1-31)NH2, appears to have nearly equipotent anabolic effects on the skeletons to hPTH(1-34), and predominantly stimulates adenylyl cyclase activity.[59-61] The reported NMR structure of hPTH(1-31)NH2 shows a V-shaped two helical structure.[36] Deletion of less structured residues His32, Asn33, and Phe34 as well as capping and specific hydrophobic interactions between the end of the N-terminal helix and the beginning of the C-terminal helix stabilize the helical conformation. Lactam cyclization stablized α-helical conformation. A lactam bridged analog (Leu27-cyclo(Glu22-Lys26)hPTH(1-31)NH2 demonstrated six times more AC-stimulating activity and slightly enhanced bone anabolic effects in ovariectomized rats.[59]

To probe the role of residue 19 in the N-terminal domain's binding, a series of hPTH (1-20) analogues have also been made and show biological activity.[62] Glu19 is conserved in all PTH sequences; in all PTHrP sequences residue 19 is Arg. NMR studies indicate that PTH (1-20) is an extended α-helix from residue 4 to 19. Interestingly, mutation of Glu19 to Arg19 resulted in a more stably extended α-helix with enhanced biological activity.[63] Arg 20 is conserved in both PTH and PTHrP families. Substitutions of Arg 20 with a series of unnatural amino acids decrease the AC-stimulating activity. Hydrogen bonding, hydrophobicity of the Arg side chain, and stabilization of α-helix in this region are all important for the interaction of the peptide with receptor.[64]

Gardella and colleagues further shortened the active PTH analogs to 14 and 11 residues. PTH(1-14) and PTH(1-11) can stimulate the cAMP-signaling response, although they exhibit extremely weak binding affinities to the PTH1 receptor. Specific substitutions with natural amino acids at positions 3, 11, 14 greatly enhance the potency in cell-based signaling assays.[65] Substitutions of residue 11 in PTH(1-11)NH2 with unnatural amino acids to increase the length and polarization of the side chain resulted in full cAMP agonists.[66] Substitutions of residues 1 and 3 with the sterically hindered and helix-promoting amino acid α-aminoisobutyric acid (Aib) in PTH(1-14) and PTH(1-11) analogs resulted in peptides that are highly active in bone cells.[67] These studies indicate that the α-helical conformation is important for receptor activation and cell signaling and only 11 amino acids are sufficient. NMR data of hPTH(1-14) shows a typical helical conformation from 3-11 in aqueous solution.[68] Functional studies further demonstrated that residues Gln6 and Asn10 play a direct role for intramolecular side chain interactions with the receptor.[69] The findings are consistent with our model that shows residues 6, 10 along with residues 1, 3, 14 in the same face of the helix are involved in ligand-receptor interation (Fig. 6b).

hPTH(1-34) was approved as an agent for the treatment of osteoporosis with the unique pharmacologic activity to build new bone. However, the requirement for subcutaneous injection may limit its use in some patients. The ultimate goal is to create a potent PTH analog or small molecule mimic that can be delivered by the oral or pulmonary routes. The X-ray and NMR structures of hPTH (1-34), combined with the accumulated biochemical data have allowed modeling of the interactions of hPTH and hPTHrP with the PTH1 receptor. These findings provide a conceptual starting point for unraveling the ligand-receptor recognition mechanism and, consequently, to guide structure-based design of novel PTH analogs and mimics.

Acknowledgments

We wish to thank Drs. H. U. Bryant, W. W. Chin, R. D. DiMarchi, C. A. Frolik, J. M. Hock, A. Hunt, B. H. Mitlak, and J. D. Termine for many helpful discussions, Mrs. V. Lawson for assistance.

References

1. Potts JT Jr, Bringhurst FR, Gardella TJ et al. Endocrinology; Parathyroid Hormone: Physiology, Chemistry, Biosynthesis, Secretion, Metabolism and Mode of Action. Philadelphia: W.B. Saunders Company, 1995:920-966.
2. Gardella TJ JŸppner H. Interaction of PTH and PTHrP with their receptors. Rev Endocine & Metab Disorders 2000; 1:317-329.
3. Canalis E, Hock JM Raisz LG. The Parathyriods. New York: Raven Press, 1994:65-82.
4. Neer RM, Arnaud CD, Zanchetta JR et al. Effect of parathyroid hormone (1-34) on fractures and bone mineral density in postmenopausal women with osteoporosis. N Engl J Med 2001; 344:1434-1441.
5. Chase LR, Aurbach GD. The effect of parathyroid hormone on the concentration of adenosine 3',5'-monophosphate in skeletal tissue in vitro. J Biol Chem 1970; 245:1520-1526.
6. Civitelli R, Reid IR, Westbrook S et al. Parathyroid hormone depresses cytosolic pH and DNA synthesis in osteoblast-like cells. Am J Physiol 1988; 255:E660-667.
7. Mosekilde L, Sogaard CH, Danielsen CC et al. The anabolic effects of human parathyroid hormone (hPTH) on rat vertebral body mass are also reflected in the quality of bone, assessed by biomechanical testing: A comparison study between hPTH-(1-34) and hPTH-(1-84). Endocrinology 1991; 129:421-428.
8. Brommage R, Hotchkiss CE, Lees CJ et al. Daily treatment with human recombinant parathyroid hormone-(1-34), LY333334, for 1 year increases bone mass in ovariectomized monkeys. J Clin Endocrinol Metab 1999; 84:3757-3763.
9. Lindsay R, Nieves J, Formica C et al. Randomised controlled study of effect of parathyroid hormone on vertebral-bone mass and fracture incidence among postmenopausal women on oestrogen with osteoporosis. Lancet 1997; 350:550-555.
10. Mitlak BH. Parathyroid hormone as a therapeutic agent. Curr Opin Pharmacol 2002; 2:694-699.
11. Rubin MR, Bilezikian JP. The potential of parathyroid hormone as a therapy for osteoporosis. Int J Fertil Womens Med 2002; 47:103-115.
12. Tregear GW, Van Rietschoten J, Greene E et al. Bovine parathyroid hormone: minimum chain length of synthetic peptide required for biological activity. Endocrinology 1973; 93:1349-1353.
13. Gardella TJ, Axelrod D, Rubin D et al. Mutational analysis of the receptor-activating region of human parathyroid hormone. J Biol Chem 1999; 266:13141-13146.
14. Jüppner H, Schipani E, Bringhurst FR et al. The extracellular amino-terminal region of the parathyroid hormone (PTH)/PTH-related peptide receptor determines the binding affinity for carboxyl-terminal fragments of PTH-(1-34). Endocrinology 1994; 134:879-884.
15. Moseley JM, Gillespie MT. Parathyroid hormone-related protein. Crit Rev Clin Lab Sci 1995; 32:299-343.
16. Blind E, Raue F, Knappe V et al. Cyclic AMP formation in rat bone and kidney cells is stimulated equally by parathyroid hormone-related protein (PTHrP) 1-34 and PTH 1-34. Exp Clin Endocrinol 1993; 101:150-155.

17. Gronwald W, Schomburg D, Tegge W et al. Assessment by 1H NMR spectroscopy of the structural behaviour of human parathyroid-hormone-related protein(1-34) and its close relationship with the N-terminal fragments of human parathyroid hormone in solution. Biol Chem 1997; 378:1501-1508.
18. Weidler M, Marx UC, Seidel G et al. The structure of human parathyroid hormone-related protein(1-34) in near-physiological solution. FEBS Lett 1999; 444:239-244.
19. Piserchio A., Usdin T, Mierke DF. Structure of tuberoinfundibular peptide of 39 residues. J Biol Chem 2000; 275:27284-27290.
20. Usdin TB, Bonner TI, Hoare SR. The parathyroid hormone 2 (PTH2) receptor. Receptors Channels 2002; 8:211-218.
21. Jonsson KB, John MR, Gensure RC et al. Tuberoinfundibular peptide 39 binds to the parathyroid hormone (PTH)/PTH-related peptide receptor, but functions as an antagonist. Endocrinology 2001; 142:704-709.
22. Fiskin AM, Cohn DV, Peterson GS. A model for the structure of bovine parathormone derived by dark field electron microsocpy. J Biol Chem 1997; 252:8261-8268.
23. Marx UC, Austermann S, Bayer P et al. Structure of human parathyroid hormone 1-37 in solution. J Biol Chem 1995; 270:15194-15202.
24. Barden JA, Kemp BE. NMR solution structure of human parathyroid hormone(1-34). Biochemistry 1993; 32:7126-7132.
25. Klaus W, Dieckmann T, Wray V et al. Investigation of the solution structure of the human parathyroid hormone fragment (1-34) by 1H NMR spectroscopy, distance geometry, and molecular dynamics calculations. Biochemistry 1991; 30:6936-6942.
26. Strickland LA, Bozzato RP, Kronis KA. Structure of human parathyroid hormone(1-34) in the presence of solvents and micelles. Biochemistry 1993; 32:6050-6057.
27. Chorev M, Behar V, Yang Q et al. Conformation of parathyroid hormone antagonists by CD, NMR, and molecular dynamics simulations. Biopolymers 1995; 36:485-495.
28. Pellegrini M, Royo M, Rosenblatt M et al. Addressing the tertiary structure of human parathyroid hormone-(1-34). J Biol Chem 1998; 273:10420-10427.
29. Neugebauer W, Surewicz WK, Gordon HL et al. Structural elements of human parathyroid hormone and their possible relation to biological activities. Biochemistry 1992; 31:2056-2063.
30. Jin L, Briggs SL, Chandrasekhar S et al. Crystal structure of human parathyroid hormone 1-34 at 0.9-A resolution. J Biol Chem 2001; 275:27238-27244.
31. Jones TA, Kjeldgaard M. Electron-density map interpretation. Methods Enzymol 1997; 277:173-208.
32. Brunger AT. X-PLOR version 3.1: A system for crystallography and N MR. New Haven: Yale University Press, 1993.
33. Sheldrick GM, Schneider TR. SHELXL: High-resolution refinement. Methods Enzymol 1997; 277:319-343.
34. Chorev M, Goldman ME, McKee RL et al. Modifications of position 12 in parathyroid hormone and parathyroid hormone related protein: Toward the design of highly potent antagonists. Biochemistry 1993; 29:1580-1586.
35. Gronwald W, Schomburg D, Harder MPF et al. Structure of recombinant human parathyroid hormone in solution using multidimensional NMR spectroscopy. Biol Chem Hoppe-Seyler 1996; 377:175-186.
36. Chen Z, Xu P, Barbier J-R et al. Solution structure of the osteogenic 1-31 fragment of the human parathyroid hormone. Biochemistry 2000; 39:12766-12777.
37. Marx UC, Adermann K, Bayer P et al. Solution structures of human parathyroid hormone fragments hPTH(1-34) and hPTH(1-39) and bovine parathyroid hormone fragment bPTH(1-37). Biochem Biophys Res Comm 2000; 267:213-220.
38. Wray V, Federau T, Gronwald W et al. The structure of human parathyroid hormone from a study of fragments in solution using 1H NMR spectroscopy and its biological implications. Biochemistry 1994; 33:1684-1693.
39. Luecke H, Schobert B, Lanyi JK et al. Crystal structure of sensory rhodopsin II at 2.4 angstroms: Insights into color tuning and transducer interaction. Science; 293:1499-1503.
40. Archer E, Maigret B, Esrieut C et al. Rhodopsin crystal: New template yielding realistic models of G-protein-coupled receptors? Trends Pharmacol Sci 2003; 24:36-40.

41. Belrhali H, Nollert P, Royant A et al. Protein, lipid and water organization in bacteriorhodopsin crystals: A molecular view of the purple membrane at 1.9 A resolution. Structure Fold Des 1999; 7:909-917.
42. Pellegrini M, Bisello A, Rosenblatt M et al. Binding domain of human parathyroid hormone receptor: From conformation to function. Biochemistry 1998; 37:12737-12743.
43. Bisello A, Adams AE, Mierke DF et al. Parathyroid hormone-receptor interactions identified directly by photocross-linking and molecular modeling studies. J Biol Chem 1998; 273:22498-22505.
44. Adams AE, Bisello A, Chorev M et al. Arginine 186 in the extracellular N-terminal region of the human parathyroid hormone 1 receptor is essential for contact with position 13 of the hormone. Mol Endocrinol 1998; 12:1673-1683.
45. Chorev M, Rosenblatt M. Principle of Bone Biology. San Diego: Academic Press, 1996.
46. Gensure RC, Gardella TJ, Juppner H. Multiple sites of contact between the carboxyl-terminal binding domain of PTHrP-(1-36) analogs and the amino-terminal extracellular domain of the PTH/PTHrP receptor identified by photoaffinity cross-linking. J Biol Chem 2001; 276:28650-28658.
47. Gardella TJ, Wilson AK, Keutmann HT et al. Analysis of parathyroid hormone's principal receptor-binding region by site-directed mutagenesis and analog design. Endocrinology 1993; 132:2024-2030.
48. Rolz C, Pellegrini M, Mierke DF. Molecular characterization of the receptor-ligand complex for parathyroid hormone. Biochemistry 1999; 38:6397-6405.
49. Vickery BH, Avnur Z, Cheng Y et al. RS-66271, a C-terminally substituted analog of human parathyroid hormone-related protein (1-34), increases trabecular and cortical bone in ovariectomized, osteopenic rats. J Bone Miner Res 1996; 11:1943-1951.
50. Pellegrini M, Bisello A, Rosenblatt M et al. Conformational studies of RS-66271, an analog of parathyroid hormone-related protein with pronounced bone anabolic activity. J Med Chem 1997; 40:3025-3031.
51. Condon SM, Morize I, Darnbrough S et al. The bioactive conformation of human parathyroid hormone. Structural evidence for the extended helix postulate. J Am Chem Soc 2001; 122:3007-3014.
52. Mierke DF, Maretto S, Schievano E et al. Conformational studies of mono- and bicyclic parathyroid hormone-related protein-derived agonists. Biochemistry 1997; 36:10372-10383.
53. Takasu H, Gardella TJ, Luck MD et al. Amino-terminal modifications of human parathyroid hormone (PTH) selectively alter phospholipase C signaling via the type 1 PTH receptor: Implications for design of signal-specific PTH ligands. Biochemistry 1999; 38:13453-13460.
54. Behar V, Nakamoto C, Greenberg Z et al. Histidine at position 5 is the specificity "switch" between two parathyroid hormone receptor subtypes. Endocrinology 1996; 137:4217-4224.
55. Gardella TJ, Luck MD, Jensen GS et al. Converting parathyroid hormone-related peptide (PTHrP) into a potent PTH-2 receptor agonist. J Biol Chem 1996; 271:19888-19893.
56. Tashjian AH, Jr., Ontjes DA, Munson PL. Alkylation and oxidation of methionine in bovine parathyroid hormone: Effects on hormonal activity and antigenicity. Biochemistry 1964; 3:1175-1182.
57. Frelinger A.L, III, Zull JE. The role of the methionine residues in the structure and function of parathyroid hormone. Arch Biochem Biophys 1986; 244:641-649.
58. Reidhaar-Olson JF, Davis RM, De Souza-Hart JA et al. Active variants of human parathyroid hormone (1-34) with multiple amino acid substitutions. Mol Cell Endocrinol 2000; 160:135-147.
59. Fraher LJ, Avram R, Watson PH et al. Comparison of the biochemical responses to human parathyroid hormone-(1-31)NH2 and hPTH-(1-34) in healthy humans. J Clin Endocrinol Metab 1999; 84:2739-2743.
60. Whitfield JF, Morley P, Willick G et al. Cyclization by a specific lactam increases the ability of human parathyroid hormone (hPTH)-(1-31)NH2 to stimulate bone growth in ovariectomized rats. J Bone Mineral Res 1997; 12:1246-1252.
61. Barbier J-R, Maclean S, Morley P et al. Structure and activities of constrained analogues of human parathyroid hormone and parathyroid hormone-related peptide: implications for receptor-activating conformations of the hormones. Biochemistry 2000; 39:14522-14530.

62. Shimizu M, Shimizu N, Tsang JC et al. Residue 19 of the parathyroid hormone (PTH) modulates ligand interaction with the juxtamembrane region of the PTH-1 receptor. Biochemistry 2002; 41:13224-13233.
63. Piserchio A, Shimizu N, Gardella TJ et al. Residue 19 of the parathyroid hormone: Structural consequences. Biochemistry 2002; 41:13217-13223.
64. Barbier J-R, MacLean S, Whitfield JF et al. Structural requirements for conserved arginine of parathyroid hormone. Biochemistry 2001; 40:8955-8961.
65. Shimizu M, Potts JT, Jr., Gardella T. Minimization of parathyroid hormone. Novel amino-terminal parathyroid hormone fragments with enhanced potency in activating the type-1 parathyroid hormone receptor. J Biol Chem 2000; 275:21836-21843.
66. Shimizu M, Carter PH, Khatri A et al. Enhanced activity in parathyroid hormone-(1-14) and -(1-11): Novel peptides for probing ligand-receptor interactions. Endocrinology 2001; 142:3068-3074.
67. Shimizu N, Guo J, Gardella TJ. Parathyroid hormone (PTH)-(1-14) and -(1-11) analogs conformationally constrained by alpha-aminoisobutyric acid mediate full agonist responses via the juxtamembrane region of the PTH-1 receptor. J Biol Chem 2001; 276:49003-49012.
68. Jung J, Lim S-K, Kim Y et al. NMR structure of a minimum activity domain of human parathyroid peptide hormone: Structural origin of receptor activation. J Peptide Res 2002; 60:239-246.
69. Shimizu N, Petroni BD, Khatri A et al. Functional evidence for an intramolecular side chain interaction between residues 6 and 10 of receptor-bound parathyroid hormone analogues. Biochemistry 2003; 42:2282-2290.

CHAPTER 4

The Calcium Sensing Receptor

Shozo Yano and Edward M. Brown

Abstract

The acute secretory response of parathyroid hormone (PTH) is strictly regulated by the extracellular calcium concentration (Ca^{2+}_o), and the G protein-coupled, calcium-sensing receptor (CaR) located on the chief cells of the parathyroid glands mediates this process. Abnormalities of the Ca^{2+}_o-sensing system lead to diseases that show hypo-/hypersecretion of PTH in addition to relative hyper-/hypocalciuria. Novel signaling pathways, e.g., mitogen-activated protein kinases (MAPK), have been shown to be involved in CaR signaling in addition to "classical" CaR-regulated pathways, e.g., phospholipase C (PLC) and adenylate cyclase. We will discuss the following topics in this chapter: (1) the regulatory mechanisms of Ca^{2+}_o-sensing and PTH secretion, (2) disorders due to mutations in the CaR gene, abnormal CaR expression, or the production of antibodies against the CaR, and (3) the promising utility of drugs acting on the CaR.

Introduction

In mammals, the extracellular calcium (Ca^{2+}_o) concentration is held nearly constant at ~1 mM by the Ca^{2+}_o-regulating hormones, parathyroid hormone (PTH), 1,25-dihydroxyvitamin D_3 (1,25(OH)$_2D_3$) and calcitonin (CT), which, in turn, modulate the functions of their target organs, bone, kidney and intestine. Conversely, Ca^{2+}_o, in turn, regulates the secretion of PTH and CT from the chief cells of the parathyroid gland and the thyroidal C cells, respectively, as well as the production of 1,25(OH)$_2D_3$ by the kidney.[1-3] While fish living in the ocean do not need to raise Ca^{2+}_o in their bodily fluids because the surrounding seawater contains ~8 mM Ca^{2+}_o, mammals are always exposed to the risk of calcium deficiency. Thus, the Ca^{2+}_o-elevating hormones, PTH and 1,25(OH)$_2D_3$, play pivotal roles in calcium homeostasis in mammals.

When mammals are exposed to calcium deficiency, parathyroid cells recognize and respond to the slight decrease in Ca^{2+}_o by secreting PTH. This acute secretory response of pre-formed PTH occurs within seconds and can last for 60-90 minutes. It is strictly regulated by Ca^{2+}_o and is characterized by an inverse sigmoidal curve (Fig. 1).[4] Secreted PTH increases renal tubular reabsorption of Ca^{2+} and augments the release of Ca^{2+} from bone. PTH also enhances the synthesis of 1,25(OH)$_2D_3$ in the renal proximal tubules, which increases intestinal Ca^{2+}_o absorption and renal Ca^{2+} conservation. Raising Ca^{2+}_o also stimulates the secretion of the Ca^{2+}_o-lowering hormone, CT, although CT has little impact on Ca^{2+}_o homeostasis in normal adult humans. Therefore, the coordinated interactions between Ca^{2+}_o and these Ca^{2+}_o-regulating hormones maintain Ca^{2+}_o homeostasis.

Molecular Biology of the Parathyroid, edited by Tally Naveh-Many. ©2005 Eurekah.com and Kluwer Academic / Plenum Publishers.

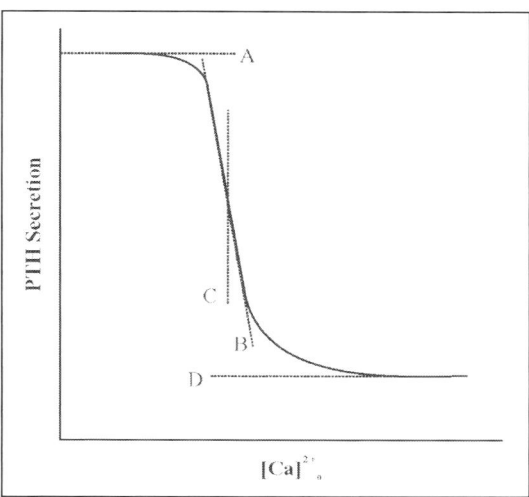

Figure 1. Four parameter models of PTH release. Acute change of PTH secretory rate from parathyroid cells is strictly regulated by Ca^{2+}_o. The Ca^{2+}_o-PTH relationship is characterized by an inverse sigmoidal curve that is represented by four parameters: $Y=\{(A-D)/[1+(X/C)^B]\}+D$. A) the maximal secretory rate, B) the slope of the curve at its midpoint, C) the midpoint or set point, which is defined as the calcium level that produces half of maximal inhibition of PTH release, and D) the minimal secretory rate. [Modified from Brown EM[4]]

Our understanding of how these cells sense Ca^{2+}_o has progressed greatly following the molecular cloning of a Ca^{2+}_o-sensing receptor (CaR).[5] The CaR is expressed mainly on the parathyroid glands, distal tubules of the kidney and the C-cells. Although the receptor is also present in intestinal epithelial cells, bone-forming osteoblasts, several nephron segments other than the distal tubule, and many other tissues and cell lines,[6] we will focus on the role of the CaR in the control of PTH secretion in this chapter.

Biochemical Characteristics of the CaR

Expression cloning in *Xenopus laevis* oocytes enabled isolation of a 5.3 kb cDNA from bovine parathyroid gland encoding a CaR with pharmacological properties similar to those of the Ca^{2+}_o-sensing mechanism in parathyroid cells.[5] Following the isolation of this novel receptor, nucleic acid hybridization-based techniques enabled the cloning of additional full-length CaRs from human[7] and chicken parathyroid;[8] rat,[9] human[10] and rabbit kidney;[11] rat C-cells;[12] and striatum of rat brain.[13] All of these subsequently cloned CaRs were highly homologous to the bovine parathyroid receptor (more than 90% identical in their amino acid sequences), indicating that they all derive from a common ancestral gene. The human CaR gene resides on chromosome 3 (3q13.3-21) as previously predicted by linkage analysis of a disorder of Ca^{2+}_o-sensing, familial hypocalciuric hypercalcemia.[14,15] The CaR belongs to family C of the G protein-coupled receptors (GPCRs), which comprises at least three different subfamilies that share more than 20% amino acid identity over their seven membrane-spanning domains.[16] Recent work has shown that there is some homology with the gamma-aminobutyric acid B, taste and vomeronasal odorant receptors as well as metabotropic glutamate receptors. The CaR exhibits three principal structural domains (Fig. 2): a very large NH_2-terminal extracellular domain, seven transmembrane domains, and an intracellular COOH-terminal tail.[17]

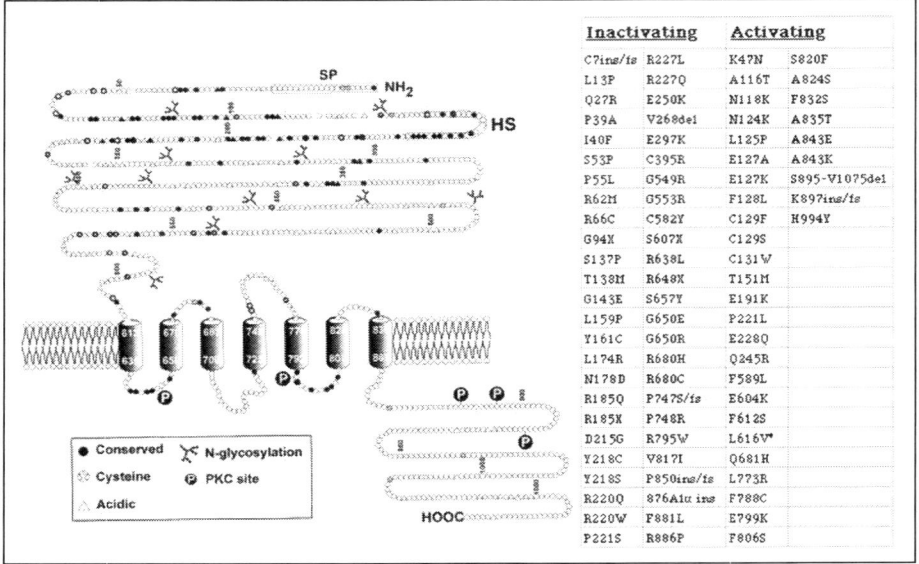

Figure 2. Schema of the human CaR. The extracellular domain contains 612 amino acids, and the transmembrane domain and intracellular domain contain 250 and 216 amino acids, respectively. The PKC sites, N-glycolysation sites and conserved cysteines are shown. SP, signal peptide; HS, hydrophobic segment. Activating and inactivating mutations reported so far are also shown. *L616V CaR was not different from the wild type CaR in a functional assay.

The CaR expressed on the cell surface comprises principally a dimeric form of the receptor when transfected into human embryonic kidney (HEK293) cells. The two monomers within the dimeric CaR interact with one another, as documented by the fact that co-expressing two individually inactive mutant CaRs reconstitutes substantial biological activity.[18] In addition to intermolecular covalent disulfide bonds mediated by Cys-129 and Cys-131, noncovalent, possibly hydrophobic, interactions also contribute to the formation of the dimeric CaR.[19,20] Since the metabotropic glutamate receptor belongs to the same family of GPCRs and the structure of its extracellular domain has been solved,[21] the CaR probably has a similar overall structure. Binding to its ligand presumably induces a conformational change in the dimerized CaR. This current model has been termed the "Venus-flytrap model".[19,22] This structure consists of two lobes each with alpha-helices and beta-sheets connected by a hinge region of three strands.[21]

The cell surface form of the CaR is N-glycosylated with complex carbohydrates.[23] The glycosylation of at least three out of eight sites seems to be essential for the efficient cell surface expression of the CaR.[24] Although treatment with concanavalin A interferes with Ca^{2+}_o-induced CaR signaling, perhaps by cross-linking receptors and interfering with their mobility within the plasma membrane, glycosylation is not critical for ligand binding and signal transduction.[25,26] The CaR has several protein kinase C (PKC) and protein kinase A (PKA) phosphorylation sites within its intracellular domains (the PKA sites are present in all species studied to date except the bovine CaR). Phosphorylation at amino acid 888 in the CaR C-terminal tail reduces its coupling to mobilization of intracellular calcium (Ca^{2+}_i) by interfering with receptor-mediated activation of phospholipase C (PLC).[27] This indicates that the action of PKC on the CaR could contribute, at least in part, to the PKC-mediated inhibition of high Ca^{2+}_o-induced suppression of PTH. The functional significance of the CaR's PKA sites, however, is unknown.

Disorders Presenting with Abnormalities in Calcium Metabolism and in the CaR

PTH-Dependent Hypercalcemia Associated with Hypocalciuria

Familial hypocalciuric hypercalcemia (FHH) is a generally benign, autosomal dominantly inherited condition, which usually presents as asymptomatic hypercalcemia accompanied by an inappropriately low rate of urinary calcium excretion (a calcium to creatinine clearance ratio of <0.01). At least two-thirds of all FHH families harbor heterozygous inactivating mutations within the coding region of the CaR gene on chromosome 3. The chromosomal location of the CaR gene was suggested by earlier linkage analyses of families with FHH.[14,28-31] Two families, however, despite having clinical features similar to those of FHH, have disease genes located at two additional chromosomal locations—one on chromosome 19p13.3 and the other on chromosome 19q13.[28,32] These additional chromosomal locations for an FHH-like condition indicate the existence of factors modulating the expression/function of the CaR or of additional calcium sensing mechanisms. In most FHH cases, serum intact PTH levels are within the normal range or slightly elevated, and are accompanied by mild hypercalcemia. In some FHH cases, however, there can be a greater degree of hypercalcemia (over 3 mM) or an absence of hypocalciuria, which has been accompanied by enlargement of the parathyroid glands in some cases.[33-35] FHH families exhibit a mild increase in set point in in vivo Ca^{2+}_o-PTH dynamic studies, indicating that the CaR plays a central role in setting the level of Ca^{2+}_o.[36,37] Most of the mutations in the CaR gene are missense mutations that result in variable degrees of inactivation of the receptor (Fig. 2).[38,39] Some missense or truncation mutations, however, totally inactivate the receptor. When co-transfected with the wild type receptor, some mutant CaRs can interfere with the normal CaR's function, exerting a so-called dominant negative action that is mediated by the formation of wild type-mutant heterodimers with reduced function.[40] Such dominant negative mutations can cause elevations in serum calcium concentration that are greater than those observed with mutant receptors that do not exert such an effect on the wild type CaR.

Neonatal severe hyperparathyroidism (NSHPT) generally presents within the first few weeks postnatally with symptoms of severe hypercalcemia. The marked hypercalcemia is due to severe primary hyperparathyroidism accompanied by the enlargement of all four parathyroid glands. NSHPT, in some cases, can be fatal unless parathyroidectomy is performed in the neonatal period. Functional examination of parathyroid glands resected from infants with NSHPT have revealed markedly abnormal Ca^{2+}_o-regulated PTH release, with substantial increases in set point and, in some cases, severely impaired inhibition of secretion even at levels of Ca^{2+}_o (e.g., 4 mM) higher than those encountered in vivo. Cases of true NSHPT (e.g., as opposed to milder cases of neonatal hyperparathyroidism) are caused by the presence of homozygous or compound heterozygous mutations in the CaR and, as a result, the lack of any normal CaRs.[29] NSHPT has also been seen in infants heterozygous for CaR mutations in some families with mutations that exert a dominant negative effect of the wild type CaR.

Ablation of the CaR produces clinical and biochemical findings similar to those of FHH and NSHPT in mice heterozygous or homozygous, respectively, for knockout of the CaR gene.[41] In the homozygous CaR knockout mice, severe hypercalcemia is caused by high levels of PTH in association with parathyroid hyperplasia. These mice generally die within the first week of life, similar to infants with NSHPT resulting from homozygous or compound heterozygous inactivating mutations in the CaR. The heterozygous CaR knockout mice also exhibit findings similar to those of patients with FHH. In these heterozygous mice, the expression level of the normal CaR is decreased in parathyroid and kidney, indicating that a normal level of expression of a normal CaR is essential for normal calcium homeostasis.

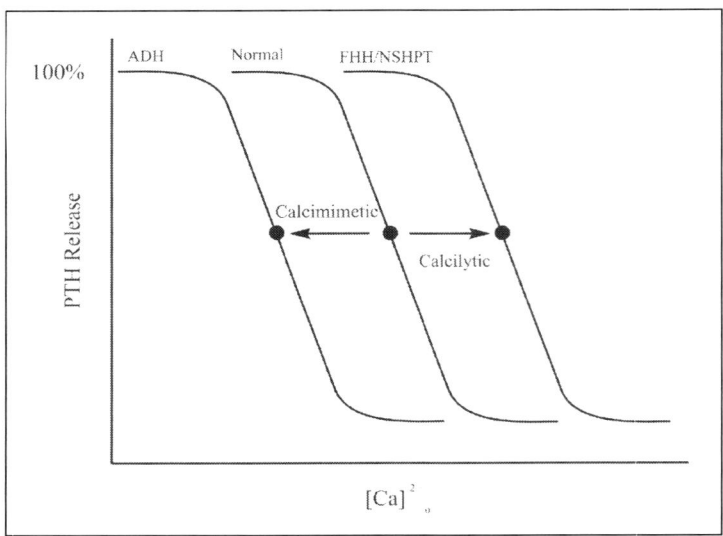

Figure 3. Shift of the Ca^{2+}_o-PTH curve. Inverse sigmoidal Ca^{2+}_o-PTH curves are represented in pathophysiological states. The curve is shifted to the rightward in patients with familial hypocalcuric hypercalcemia (FHH) and neonatal severe hyperparathyroidism (NSHPT). On the other hand, a leftward shift has been observed in autosomal dominant hypoparathyroidism (ADH). Set points are shown by filled circle. The calcimimetic and calcilytic agents cause dose and time-dependent shifts of the set point to the left and to the right, respectively. [Modified from Brown EM and Tfelt-Hansen J[39]]

Hypoparathyroidism Related to CaR Gene Mutations

In contrast to inactivating mutations of the CaR, gain-of-function mutations in the CaR gene produce hypocalcemia.[42] Autosomal dominant hypoparathyroidism (ADH) is characterized by autosomal dominant inheritance of hypocalcemia accompanied by relative hypercalciuria and inappropriately low-normal or low serum PTH levels. Some sporadic hypoparathyroidism cases caused by de novo mutations of the CaR gene have been also reported.[43,44] When vitamin D is administered to these patients, they are prone to the development of marked hypercalciuria, nephrocalcinosis, and renal stone formation, even while still hypocalcemic. Therefore, it is important to monitor them carefully during therapy and to correct their hypocalcemia only to the point when their hypocalcemic symptom(s) is relieved and not beyond. So far, more than 30 activating mutations have been found in the CaR gene, most of which are located in the extracellular domain (Fig. 2).[17,38,45] When appropriate dynamic studies of parathyroid function are carried out and the curve relating PTH to calcium is drawn, inactivating mutations cause a rightward shift in the set point (e.g., the level of calcium at which PTH is halfway between the maximum and minimum of the curve). In contrast, the set point is shifted to the left in patients with activating mutations (Fig. 3).[39] Recent reports suggest that activating mutations of the CaR with high activity can develop Bartter's syndrome through the inhibition of a renal outer medullary potassium channel.[46]

Hyperparathyroidism and Reduced CaR Expression

In primary hyperparathyroidism as well as secondary hyperparathyroidism due to chronic renal failure, an increase in PTH release and increased cell proliferative activity are usually seen. Although the possibility that an abnormality in the CaR gene could produce this pathological progression seemed reasonable, no mutations in the CaR gene were found in human parathyroid tumors.[47] So far, however, CaR expression is known to decrease at the mRNA and protein levels in the parathyroid glands from patients with primary and secondary hyperparathyroidism.[48-50] This alteration in CaR expression could potentially be related to the excessive PTH release and parathyroid cell proliferation seen with these conditions.[51,52] Although the factors regulating CaR expression are not fully understood, vitamin D is thought to be a candidate as suggested by an in vivo study of a renal failure model.[53] In fact, the promoter of the human CaR gene possesses the vitamin D responsive elements through which vitamin D stimulates transcription of the gene.[54] At least in the rat, a single intraperitoneal injection of vitamin D lead to a 2-2.5 fold increase in CaR mRNA in parathyroid, thyroid, and kidney.[54] In addition, since there is a positive correlation between CaR and vitamin D receptor expression in both primary and secondary hyperparathyroidism, decreased expression of the CaR can be explained by a reduction in serum $1,25(OH)_2D_3$ level and/or vitamin D receptor expression.[51,52] However, overexpression of cyclin D1 in the mouse parathyroid gland leads to increases in serum calcium and PTH, as well as enhanced parathyroid cell proliferation, associated with a decrease in CaR expression.[55] In uremic rats after switching from low phosphate diet to high phosphate diet, parathyroid cell proliferation is observed before reduction of the CaR.[56] Taken together, these findings indicate that CaR expression may be related to the cell cycle status of the parathyroid cells.

Autoimmune Hypo/Hyperparathyroidism

Other types of disorders related to the CaR have been described. Li et al found autoantibodies to the extracellular domain of the CaR in patients with autoimmune hypoparathyroidism.[57] The antibodies were present in 14 out of 25 patients with acquired hypoparathyroidism, 17 of whom had type I autoimmune polyglandular syndrome and 8 of whom had autoimmune hypothyroidism. These authors did not find any evidence that these antibodies altered the function of the CaR, however, suggesting that the hypoparathyroidism was not caused by an activating effect of the antibodies on the CaR, analogous to that seen with ADH. Recently, autoimmune hypocalciuric hypercalcemia caused by anti-CaR antibodies has been documented and must be added to the differential diagnosis of hypercalcemia.[58] In this report, four individuals from two kindreds presented with PTH-dependent hypercalcemia. These patients showed findings similar to those in FHH (e.g., hypercalcemia accompanied by relative or absolute hypocalciuria), However, an autoimmune etiology was suspected since one of the patients had sprue and the other three had autoimmune thyroid disease. No mutations were found in their CaR genes by DNA sequencing. Parathyroidectomy was unsuccessful in correcting the hypercalcemia in one case. The patients' sera reacted with the cell surface of bovine parathyroid cells and stimulated PTH release from human parathyroid in vitro. Since autoimmune hypocalciuric hypercalcemia is presumably acquired, the hypercalcemia might cause symptoms, unlike the generally asymptomatic nature of FHH.

Signaling Pathways of the CaR

CaR agonists can activate phospholipase A_2 and D as well as PLC in bovine parathyroid cells.[59] Activation of PLC transiently increases the cytosolic calcium concentration (Ca^{2+}_i) as a result of IP_3-mediated release of intracellular calcium stores, and then produces a sustained increase in Ca^{2+}_i due to calcium influx from the extracellular fluid. High Ca^{2+}_o induces a pertussis toxin-sensitive inhibition of cAMP accumulation in bovine parathyroid cells, suggesting that CaR-mediated inhibition of adenylate cyclase could involve one or more isoforms of the inhibitory G protein, Gi.[60] However, this inhibitory action can result from an indirect mechanism mediated by transient increases in Ca^{2+}_i, which then inhibits a calcium-sensitive adenylate cyclase, in some kidney cells.[61] Recent work has demonstrated that activation of the CaR leads to phosphorylation of the extracellular signal-regulated kinases 1 and 2 (ERK1/2), members of one of the mitogen-activated protein kinase (MAPK) families. This stimulation of ERK1/2 was partially blocked by inhibitors of PKC as well as by pertussis toxin, suggesting the involvement of $G_{q/11}$-mediated activation of PLC and a G_i-dependent pathway, respectively.[62] Activation of this MAPK pathway was also inhibited by tyrosine kinase inhibitors, demonstrating the involvement of tyrosine kinases as well. In normal human parathyroid cells, pretreatment with a specific inhibitor of the ERK1/2 pathway did not modify PTH release at low Ca^{2+}_o but totally abrogated the inhibition of PTH secretion at high Ca^{2+}_o, indicating that the ERK1/2 pathway could play a major role in high Ca^{2+}_o-induced PTH suppression.[63] These workers also showed that CaR-induced ERK activation is mediated by PKC and, to a lesser extent, phosphoinositide 3 kinase (PI3K). CaR-induced MAPK activation is also seen in HEK cells stably transfected with the CaR and in H-500 rat leydig cancer cells, and MAPK activation stimulates the secretion of parathyroid hormone-related protein (PTHrP) production in these cells.[64,65] Furthermore, MAPK activation is involved in high Ca^{2+}_o-induced cell proliferation in rat-1 fibroblasts and other cell types.[66-68] Thus, the CaR can modulate cellular proliferation as well as hormonal secretion through MAPK activation.

Drugs Acting on the CaR

The pharmacology of the CaR is divided into calcimimetics and calcilytics. Calcimimetics mimic or potentiate the action of Ca^{2+}_o. They can be subdivided into type I and type II ligands; type I ligands can activate the CaR in the absence of Ca^{2+}_o and behave as true agonists, while type II ligands act as allosteric activators. Calcilytics are CaR antagonists.

The CaR can be activated by a surprising variety of ligands. These include inorganic di- and trivalent cations (La^{3+}, Gd^{3+}, Ca^{2+}, Ba^{2+}, Sr^{2+}, and Mg^{2+}) as well as organic polycations including spermine, neomycin, polylysine and protamine.[6,69,70] These all behave as type I calcimimetics. In contrast, the phenylalkylamine, NPS R-568, selectively activates the CaR only in the presence of Ca^{2+}_o. Type II calcimimetic compounds like NPS R-568 shift the concentration-response curve for Ca^{2+}_o to the left without affecting the maximal or minimal response (Fig. 3).[71,72] Furthermore, the R-enantiomer is 10- to 100-fold more potent than the corresponding S-enantiomer. The R-enantiomer, NPS R-568, increases Ca^{2+}_i and inhibits PTH secretion from bovine parathyroid cells in vitro at concentrations between 3 and 100 nM, whereas NPS S-568 has no effect on either response at concentrations less than 3 μM.[71] Additional differences between the type I and type II calcimimetic ligands are the regions of the CaR on which they act. The type I calcimimetic ligands are thought to act predominantly in the extracellular domain of the CaR, whereas type II calcimimetics of the phenylalkylamine class bind within the transmembrane domain.[73] Recent studies have shown that certain amino acids, particularly aromatic amino acids, can also potentiate the actions of calcium on the CaR.[74] Since they are markedly less effective at 1.0 mM calcium and below, they can be considered to be type II calcimimetics.[75] In contrast to the phenylalkylamines, however, amino acids have been shown to have a binding site on the extracellular domain of the CaR.[76,77]

Calcimimetic Agents

Oral administration of NPS R-568 to normal rats causes a rapid fall in plasma PTH levels associated with hypocalcemia.[78] Since calcimimetic induced-hypocalcemia results mostly from the ability of the compounds to inhibit PTH secretion, these compounds were recognized to have applicability for the treatment of hypercalcemic disorders caused by PTH excess. In a pilot study, oral administration of NPS R-568 was effective in reducing plasma levels of PTH and Ca^{2+} in patients with primary HPT.[79] The inhibitory effects of NPS R-568 were readily reversible, and plasma PTH returned to pre-dose levels within 2 to 8 hours in a dose dependent manner. The fall in plasma Ca^{2+} level begins within 1 hour and lasts longer at higher doses. In rats with secondary HPT resulting from chronic renal failure, oral administration of NPS R-568 caused a dose dependent decrease in plasma PTH levels.[80] The magnitude and rate of change in plasma PTH levels were similar to what was observed in normal animals. Moreover, activation of the CaR with NPS R-568 markedly lowers circulating PTH levels irrespective of the severity of the secondary HPT or the magnitude of the hyperphosphatemia. Results from studies of calcimimetics in patients with secondary HPT are similar to those in the partially nephrectomized rat model. In seven patients receiving hemodialysis who had secondary HPT, serum PTH level decreased by 40 to 60% 1 to 2 hours after administration of a single dose of NPS R-568 in both the low-dose group (40-80 mg) and the high-dose group (120-200 mg).[81] Although the PTH level increased gradually thereafter, it remained considerably lower than the basal level after 24 hours. In the high-dose group, there was a significant decrease in PTH level even 48 hours after administration. The effects of NPS R-568 were also confirmed by a randomized, double-blind, placebo-controlled study in 21 patients undergoing hemodialysis.[82] Interestingly, NPS R-568 inhibited not only PTH secretion but also parathyroid growth in rats with renal insufficiency.[83-85] In addition, NPS R-568 prevented osteitis fibrosa, which is caused by an excessive action of PTH on bone, in the rat model of renal failure.[86] Despite these encouraging results, NPS R-568 (or R-467) has a couple of problems in terms of their potential commercialization. The major problems with these compounds are their low bioavailability and differences in their metabolism depending on the genotypes within the general population. Since these agents are metabolized primarily by CYP2D6, one of the cytochrome P-450 enzymes, much higher serum concentrations were reached in individuals with a form of this enzyme that has reduced biological activity. Therefore, AMG 073, a newer type II calcimimetic compound, has been developed. This compound is not metabolized by CYP2D6 and has increased bioavailability. In ongoing clinical trials, AMG073 has produced results similar as those observed previously with NPS R-568 in patients with primary and secondary HPT.[87-89]

Calcilytic Agents

PTH is known to have an anabolic action on bone, and PTH derived-peptides have been shown to be effective for the treatment of osteoporosis.[90] PTH has shown this anabolic effect only when administered intermittently (e.g., once daily subcutaneously).[91] Since calcilytics produce a transient increase in PTH secretion by "tricking" the parathyroid into thinking that the serum calcium concentration is low, they could provide another mechanism for producing the transient increase in PTH needed for obtaining the associated anabolic effect. Moreover, calcilytic agents have an advantage over PTH in that they can be taken orally, whereas PTH must be injected. Therefore, an in vivo study was performed to examine the effects of NPS 2143, a calcilytic compound, on bone.[92,93] In aged ovariectomized (OVX) rats, plasma PTH levels increased 5 fold or more within 30 min after administering NPS 2143 orally, which was followed by a remarkable increase in bone turnover.[93] However, there was no net change in bone mineral density (BMD). When NPS 2143 was administered with or without 17β-estradiol, however, OVX rats treated with both agents showed a significant increase in BMD at the distal femur 5 weeks after treatment, compared with the groups treated with either NPS 2143 or 17β-estradiol alone. These findings suggest that calcilytic compounds could increase bone mass in the presence of an antiresorptive agent.

Summary

The CaR has a variety of functions in the numerous tissues in which it is expressed. Its most important functions to date are in maintaining calcium homeostasis, as demonstrated by the human diseases that result from activating or inactivating mutations in the CaR gene or mice with ablation of the gene. The recent development of mice in which the homozygous genotype has been "rescued" by deletion of the PTH gene or the parathyroid glands should provide a useful model for understanding the CaR's role in the numerous tissues expressing it.[94,95] Further studies are also required to understand the biochemical characteristics of the CaR and its signaling pathways, including MAPK. Finally, calcimimetics will undoubtedly provide a very effective way of controlling hyperparathyroidism in patients with primary or uremic hyperparathyroidism, and calcilytics provide an effective means of producing pulses of PTH that could be useful in the treatment of osteoporosis.

Acknowledgments

We thank Mr. Robert R. Butters for his generous help in drawing the figures.

References

1. Brown EM. Extracellular Ca^{2+} sensing, regulation of parathyroid cell function, and role of Ca^{2+} and other ions as extracellular (first) messengers. Physiol Rev 1991; 71:371-411.
2. Fried R, Tashjian AJ. Unusual sensitivity of cytosolic free Ca^{2+} to changes in extracellular Ca^{2+} in rat C-cells. J Biol Chem 1986; 261:7669-7674.
3. Murayama A, Takeyama K, Kitanaka S et al. Positive and negative regulations of the renal 25-hydroxyvitamin D3 1alpha-hydroxylase gene by parathyroid hormone, calcitonin, and 1alpha,25(OH)2D3 in intact animals. Endocrinology 1999; 140:2224-2231.
4. Brown EM. Four-parameter model of the sigmoidal relationship between parathyroid hormone release and extracellular calcium concentration in normal and abnormal parathyroid tissue. J Clin Endocrinol Metab 1983; 56:572-581.
5. Brown EM, Gamba G, Riccardi D et al. Cloning and characterization of an extracellular Ca(2+)-sensing receptor from bovine parathyroid. Nature 1993; 366:575-580.
6. Brown EM, MacLeod RJ. Extracellular calcium sensing and extracellular calcium signaling. Physiol Rev 2001; 81:239-297.
7. Garrett JE, Capuano IV, Hammerland LG et al. Molecular cloning and functional expression of human parathyroid calcium receptor cDNAs. J Biol Chem 1995; 270:12919-12925.
8. Diaz R, Hurwitz S, Chattopadhyay N et al. Cloning, expression, and tissue localization of the calcium-sensing receptor in chicken (Gallus domesticus). Am J Physiol 1997; 273(3 Pt 2):R1008-1016.
9. Riccardi D, Park J, Lee WS et al; Cloning and functional expression of a rat kidney extracellular calcium/polyvalent cation-sensing receptor. Proc Natl Acad Sci USA 1995; 92:131-135.
10. Aida K, Koishi S, Tawata M et al. Molecular cloning of a putative Ca(2+)-sensing receptor cDNA from human kidney. Biochem Biophys Res Commun 1995; 214:524-529.
11. Butters RR Jr, Chattopadhyay N, Nielsen P et al. Cloning and characterization of a calcium-sensing receptor from the hypercalcemic New Zealand white rabbit reveals unaltered responsiveness to extracellular calcium. J Bone Miner Res 1997; 12:568-579.
12. Garrett JE, Tamir H, Kifor O et al. Calcitonin-secreting cells of the thyroid express an extracellular calcium receptor gene. Endocrinology 1995;136:5202-5211.
13. Ruat M, Molliver ME, Snowman AM, et al. Calcium sensing receptor: molecular cloning in rat and localization to nerve terminals. Proc Natl Acad Sci USA 1995; 92:3161-3165.
14. Chou YH, Brown EM, Levi T, et al. The gene responsible for familial hypocalciuric hypercalcemia maps to chromosome 3q in four unrelated families. Nat Genet 1992; 1:295-300.
15. Janicic N, Soliman E, Pausova Z et al. Mapping of the calcium-sensing receptor gene (CASR) to human chromosome 3q13.3-21 by fluorescence in situ hybridization, and localization to rat chromosome 11 and mouse chromosome 16. Mamm Genome 1995; 6:798-801.
16. Kolakowski LF. GCRDb: a G-protein-coupled receptor database. Receptors Channels 1994; 2:1-7.
17. Chattopadhyay N, Mithal A, Brown EM. The calcium-sensing receptor: a window into the physiology and pathophysiology of mineral ion metabolism. Endocr Rev 1996; 17:289-307.
18. Bai M, Trivedi S, Brown EM. Dimerization of the extracellular calcium-sensing receptor (CaR) on the cell surface of CaR-transfected HEK293 cells. J Biol Chem 1998; 273:23605-23610.

19. Ray K, Hauschild BC, Steinbach PJ et al. Identification of the cysteine residues in the amino-terminal extracellular domain of the human Ca^{2+} receptor critical for dimerization. Implications for function of monomeric Ca^{2+} receptor. J Biol Chem 1999; 274:27642-27650.
20. Zhang Z, Sun S, Quinn SJ et al. The extracellular calcium-sensing receptor dimerizes through multiple types of intermolecular interactions. J Biol Chem 2001; 276:5316-5322.
21. O'Hara PJ, Sheppard PO, Thogersen H et al. The ligand-binding domain in metabotropic glutamate receptors is related to bacterial periplasmic binding proteins. Neuron 1993; 11:41-52.
22. Hu J, Mora S, Colussi G et al. Autosomal dominant hypocalcemia caused by a novel mutation in the loop 2 region of the human calcium receptor extracellular domain. J Bone Miner Res 2002; 17:1461-1469.
23. Bai M, Quinn S, Trivedi S et al. Expression and characterization of inactivating and activating mutations in the human Ca2+o-sensing receptor. J Biol Chem 1996; 271:19537-19545.
24. Fan G, Goldsmith PK, Collins R et al. N-linked glycosylation of the human Ca2+ receptor is essential for its expression at the cell surface. Endocrinology 1997; 138:1916-1922.
25. Goldsmith PK, Fan G, Miller JL et al. Monoclonal antibodies against synthetic peptides corresponding to the extracellular domain of the human Ca^{2+} receptor: Characterization and use in studying concanavalin A inhibition. J Bone Miner Res 1997; 12:1780-1788.
26. Ray K, Clapp P, Goldsmith PK et al. Identification of the sites of N-linked glycosylation on the human calcium receptor and assessment of their role in cell surface expression and signal transduction. J Biol Chem 1998; 273:34558-34567.
27. Bai M, Trivedi S, Lane CR et al. Protein kinase C phosphorylation of threonine at position 888 in Ca2+o-sensing receptor (CaR) inhibits coupling to Ca^{2+} store release. J Biol Chem 1998; 273:21267-21275.
28. Heath H 3rd, Jackson CE, Otterud B et al. Genetic linkage analysis in familial benign (hypocalciuric) hypercalcemia: Evidence for locus heterogeneity. Am J Hum Genet 1993; 53:193-200.
29. Pollak MR, Brown EM, Chou YH et al. Mutations in the human Ca(2+)-sensing receptor gene cause familial hypocalciuric hypercalcemia and neonatal severe hyperparathyroidism. Cell 1993; 75:1297-1303.
30. Pollak MR, Chou YH, Marx SJ et al. Familial hypocalciuric hypercalcemia and neonatal severe hyperparathyroidism. Effects of mutant gene dosage on phenotype. J Clin Invest 1994; 93:1108-1112.
31. Trump D, Whyte MP, Wooding C et al. Linkage studies in a kindred from Oklahoma, with familial benign (hypocalciuric) hypercalcaemia (FBH) and developmental elevations in serum parathyroid hormone levels, indicate a third locus for FBH. Hum Genet 1995; 96:183-187.
32. Lloyd SE, Pannett AA, Dixon PH et al. Localization of familial benign hypercalcemia, Oklahoma variant (FBHOk), to chromosome 19q13. Am J Hum Genet 1999; 64:189-195.
33. Bai M, Janicic N, Trivedi S et al. Markedly reduced activity of mutant calcium-sensing receptor with an inserted Alu element from a kindred with familial hypocalciuric hypercalcemia and neonatal severe hyperparathyroidism. J Clin Invest 1997; 15; 99:1917-1925.
34. Carling T, Szabo E, Bai M et al. Familial hypercalcemia and hypercalciuria caused by a novel mutation in the cytoplasmic tail of the calcium receptor. J Clin Endocrinol Metab 2000; 85:2042-2047.
35. Yamauchi M, Sugimoto T, Yamaguchi T et al. Familial hypocalciuric hypercalcemia caused by an R648stop mutation in the calcium-sensing receptor gene. J Bone Miner Res 2002; 17:2174-2182.
36. Auwerx J, Demedts M, Bouillon R. Altered parathyroid set point to calcium in familial hypocalciuric hypercalcaemia. Acta Endocrinol (Copenh) 1984; 106:215-218.
37. Khosla S, Ebeling PR, Firek AF. Calcium infusion suggests a "set-point" abnormality of parathyroid gland function in familial benign hypercalcemia and more complex disturbances in primary hyperparathyroidism. J Clin Endocrinol Metab 1993; 76:715-720.
38. Hendy GN, D'Souza-Li L, Yang B et al. Mutations of the calcium-sensing receptor (CASR) in familial hypocalciuric hypercalcemia, neonatal severe hyperparathyroidism, and autosomal dominant hypocalcemia Hum Mutat 2000; 16:281-296.
39. Tfelt-Hansen J, Yano S, Brown EM et al. The role of the calcium-sensing receptor in human pathophysiology: Current medicinal chemistry—Immunology, endocrine & metabolic agents. 2002; 2:175-193.

40. Bai M, Pearce SH, Kifor O et al. In vivo and in vitro characterization of neonatal hyperparathyroidism resulting from a de novo, heterozygous mutation in the Ca^{2+}-sensing receptor gene: Normal maternal calcium homeostasis as a cause of secondary hyperparathyroidism in familial benign hypocalciuric hypercalcemia. J Clin Invest 1997; 99:88-96.
41. Ho C, Conner DA, Pollak MR et al. A mouse model of human familial hypocalciuric hypercalcemia and neonatal severe hyperparathyroidism. Nat Genet 1995; 11:389-394.
42. Pollak MR, Brown EM, Estep HL et al. Autosomal dominant hypocalcaemia caused by a Ca(2+)-sensing receptor gene mutation. Nat Genet 1994; 8:303-307.
43. Baron J, Winer KK, Yanovski JA et al. Mutations in the Ca^{2+}-sensing receptor gene cause autosomal dominant and sporadic hypoparathyroidism. Hum Mol Genet 1996; 5:601-606.
44. De Luca F, Ray K, Mancilla EE et al. Sporadic hypoparathyroidism caused by de Novo gain-of-function mutations of the Ca^{2+}-sensing receptor. J Clin Endocrinol Metab 1997; 82:2710-2715.
45. Brown EM, Pollak M, Hebert SC. The extracellular calcium-sensing receptor: its role in health and disease. Annu Rev Med 1998; 49:15-29.
46. Watanabe S, Fukumoto S, Chang H et al. Association between activating mutations of calcium-sensing receptor and Bartter's syndrome. Lancet 2002; 360:692-694.
47. Hosokawa Y, Pollak MR, Brown EM et al. Mutational analysis of the extracellular Ca^{2+}-sensing receptor gene in human parathyroid tumors. J Clin Endocrinol Metab 1995; 80:3107-3110.
48. Kifor O, Moore FD Jr, Wang P et al. Reduced immunostaining for the extracellular Ca2+-sensing receptor in primary and uremic secondary hyperparathyroidism. J Clin Endocrinol Metab 1996; 81:1598-1606.
49. Farnebo F, Enberg U, Grimelius L et al. Tumor-specific decreased expression of calcium sensing receptor messenger ribonucleic acid in sporadic primary hyperparathyroidism. J Clin Endocrinol Metab 1997; 82:3481-3486.
50. Gogusev J, Duchambon P, Hory B et al. Depressed expression of calcium receptor in parathyroid gland tissue of patients with hyperparathyroidism. Kidney Int 1997; 51:328-336.
51. Yano S, Sugimoto T, Tsukamoto T et al. Association of decreased calcium-sensing receptor expression with proliferation of parathyroid cells in secondary hyperparathyroidism. Kidney Int 2000; 58:1980-1986.
52. Yano S, Sugimoto T, Tsukamoto T et al. Decrease in vitamin D receptor and calcium-sensing receptor in highly proliferative parathyroid adenomas. Eur J Endocrinol 2003; 148:403-411.
53. Brown AJ, Zhong M, Finch J et al. Rat calcium-sensing receptor is regulated by vitamin D but not by calcium. Am J Physiol 1996; 270(3 Pt 2):F454-460.
54. Canaff L, Hendy GN. Human calcium-sensing receptor gene. Vitamin D response elements in promoters P1 and P2 confer transcriptional responsiveness to 1,25-dihydroxyvitamin D. J Biol Chem 2002; 277:30337-30350.
55. Imanishi Y, Hosokawa Y, Yoshimoto K et al. Primary hyperparathyroidism caused by parathyroid-targeted overexpression of cyclin D1 in transgenic mice. J Clin Invest 2001; 107:1093-1102.
56. Ritter CS, Finch JL, Slatopolsky EA et al. Parathyroid hyperplasia in uremic rats precedes down-regulation of the calcium receptor. Kidney Int 2001; 60:1737-1744.
57. Li Y, Song YH, Rais N et al. Autoantibodies to the extracellular domain of the calcium sensing receptor in patients with acquired hypoparathyroidism. J Clin Invest 1996; 97:910-914.
58. Kifor O, Moore-FD J, Delaney M et al. A syndrome of hypocalciuric hypercalcemia caused by autoantibodies directed at the calcium-sensing receptor. J Clin Endocrinol Metab 2003; 88:60-72.
59. Kifor O, Diaz R, Butters R et al. The Ca^{2+}-sensing receptor (CaR) activates phospholipases C, A2, and D in bovine parathyroid and CaR-transfected, human embryonic kidney (HEK293) cells. J Bone Miner Res 1997; 12:715-725.
60. Chen CJ, Barnett JV, Congo DA et al. Divalent cations suppress 3',5'-adenosine monophosphate accumulation by stimulating a pertussis toxin-sensitive guanine nucleotide-binding protein in cultured bovine parathyroid cells. Endocrinology 1989; 124:233-239.
61. de Jesus Ferreira MC, Helies-Toussaint C, Imbert-Teboul M et al. Co-expression of a Ca^{2+}-inhibitable adenylyl cyclase and of a Ca^{2+}-sensing receptor in the cortical thick ascending limb cell of the rat kidney. Inhibition of hormone-dependent cAMP accumulation by extracellular Ca^{2+}. J Biol Chem 1998; 273:15192-15202.

62. Kifor O, MacLeod RJ, Diaz R et al. Regulation of MAP kinase by calcium-sensing receptor in bovine parathyroid and CaR-transfected HEK293 cells. Am J Physiol Renal Physiol. 2001; 280:F291-302.
63. Corbetta S, Lania A, Filopanti M et al. Mitogen-activated protein kinase cascade in human normal and tumoral parathyroid cells. J Clin Endocrinol Metab 2002; 87:2201-2205.
64. MacLeod RJ, Chattopadhyay N, Brown EM. PTHrP stimulated by the calcium-sensing receptor requires MAP kinase activation. Am J Physiol Endocrinol Metab 2003; 284:E435-442.
65. Tfelt-Hansen J, MacLeod RJ, Chattopadhyay N et al. Calcium-sensing receptor stimulates PTHrP release by PKC-, p38 MAPK-, JNK- and ERK1/2-dependent pathways in H-500 cells. Am J Physiol Endocrinol Metab (in press).
66. McNeil SE, Hobson SA, Nipper V et al. Functional calcium-sensing receptors in rat fibroblasts are required for activation of SRC kinase and mitogen-activated protein kinase in response to extracellular calcium. J Biol Chem 1998; 273:1114-1120.
67. Hobson SA, McNeil SE, Lee F et al. Signal transduction mechanisms linking increased extracellular calcium to proliferation in ovarian surface epithelial cells. Exp Cell Res 2000; 258:1-11.
68. Yamaguchi T, Chattopadhyay N, Kifor O et al. Activation of p42/44 and p38 mitogen-activated protein kinases by extracellular calcium-sensing receptor agonists induces mitogenic responses in the mouse osteoblastic MC3T3-E1 cell line. Biochem Biophys Res Commun 2000; 279:363-368.
69. Quinn SJ, Ye CP, Diaz R et al. The Ca^{2+}-sensing receptor: a target for polyamines. Am J Physiol 1997; 273:C1315-1323.
70. Brown EM, Katz C, Butters R et al. Polyarginine, polylysine, and protamine mimic the effects of high extracellular calcium concentrations on dispersed bovine parathyroid cells. J Bone Miner Res 1991; 6:1217-1225.
71. Nemeth EF and Bennett SA. Tricking the parathyroid gland with novel calcimimetic agents. Nephrol Dial Transplant 1998; 13:1923-1925.
72. Hammerland LG, Garrett JE, Hung BC et al. Allosteric activation of the Ca^{2+} receptor expressed in Xenopus laevis oocytes by NPS 467 or NPS 568. Mol Pharmacol 1998; 53:1083-1088.
73. Hauache OM, Hu J, Ray K et al. Effects of a calcimimetic compound and naturally activating mutations on the human Ca^{2+} receptor and on Ca^{2+} receptor/metabotropic glutamate chimeric receptors. Endocrinology 2000; 141:4156-4163.
74. Conigrave AD, Quinn SJ, Brown EM. L-amino acid sensing by the extracellular Ca^{2+}-sensing receptor. Proc Natl Acad Sci USA 2000; 97:4814-4819.
75. Young SH, Rozengurt E. Amino acids and Ca^{2+} stimulate different patterns of Ca^{2+} oscillations through the Ca^{2+}-sensing receptor. Am J Physiol Cell Physiol 2002; 282(6): C1414-1422.
76. Zhang Z, Jiang Y, Quinn SJ et al. L-phenylalanine and NPS R-467 synergistically potentiate the function of the extracellular calcium-sensing receptor through distinct sites. J Biol Chem 2002; 277:33736-33741.
77. Zhang Z, Qiu W, Quinn SJ et al. Three adjacent serines in the extracellular domains of the CaR are required for L-amino acid-mediated potentiation of receptor function. J Biol Chem 2002; 277:33727-33735.
78. Fox J, Lowe SH, Petty BA et al. NPS R-568: a type II calcimimetic compound that acts on parathyroid cell calcium receptor of rats to reduce plasma levels of parathyroid hormone and calcium. J Pharmacol Exp Ther 1999; 290:473-479.
79. Silverberg SJ, Bone HG, Marriott TB et al. Short-term inhibition of parathyroid hormone secretion by a calcium-receptor agonist in patients with primary hyperparathyroidism. N Engl J Med 1997; 337:1506-1510.
80. Fox J, Lowe SH, Conklin RL et al. The calcimimetic NPS R-568 decreases plasma PTH in rats with mild and severe renal or dietary secondary hyperparathyroidism. Endocrine 1999; 10:97-103.
81. Antonsen JE, Sherrard DJ and Andress DL. A calcimimetic agent acutely suppresses parathyroid hormone levels in patients with chronic renal failure. Kidney Int 1998; 53:223-227.
82. Goodman WG, Frazao JM, Goodkin DA et al. A calcimimetic agent lowers plasma parathyroid hormone levels in patients with secondary hyperparathyroidism. Kidney Int 2000; 58:436-445.
83. Wada M, Furuya Y, Sakiyama J et al. The calcimimetic compound NPS R-568 suppresses parathyroid cell proliferation in rats with renal insufficiency. Control of parathyroid cell growth via a calcium receptor. J Clin Invest 1997; 100:2977-2983.

84. Wada M, Nagano N, Furuya Y et al. Calcimimetic NPS R-568 prevents parathyroid hyperplasia in rats with severe secondary hyperparathyroidism. Kidney Int 2000; 57:50-58.
85. Chin J, Miller SC, Wada M et al. Activation of the calcium receptor by a calcimimetic compound halts the progression of secondary hyperparathyroidism in uremic rats. J Am Soc Nephrol 2000; 11:903-911.
86. Wada M, Ishii H, Furuya Y et al. NPS R-568 halts or reverses osteitis fibrosa in uremic rats. Kidney Int 1998; 53:448-453.
87. Goodman WG, Hladik GA, Turner SA et al. The Calcimimetic agent AMG 073 lowers plasma parathyroid hormone levels in hemodialysis patients with secondary hyperparathyroidism. J Am Soc Nephrol 2002; 13:1017-1024.
88. Lindberg JS, Moe SM, Goodman WG et al. The calcimimetic AMG 073 reduces parathyroid hormone and calcium x phosphorus in secondary hyperparathyroidism. Kidney Int 2003; 63:248-254.
89. Quarles LD, Sherrard DJ, Adler S et al. The calcimimetic AMG 073 as a potential treatment for secondary hyperparathyroidism of end-stage renal disease. J Am Soc Nephrol 2003; 14:575-583.
90. Dempster DW, Cosman F, Parisien M et al. Anabolic actions of parathyroid hormone on bone. Endocr Rev 1993; 14: 690-709.
91. Tam CS, Heersche JNM, Murray TM et al. Parathyroid hormone stimulates the bone apposition rate independently of its resorptive action: differential effects of intermittent and continuous administration. Endocrinology 1982; 110: 516-512.
92. Nemeth EF, Delmar EG, Heaton WL et al. Calcilytic compounds: potent and selective Ca^{2+} receptor antagonists that stimulate secretion of parathyroid hormone. J Pharmacol Exp Ther 2001; 299: 323-331.
93. Gowen M, Stroup GB, Dodds RA et al. Antagonizing the parathyroid calcium receptor stimulates parathyroid hormone secretion and bone formation in osteopenic rats. J Clin Invest 2000; 105:1595-1604.
94. Tu Q, Pi M, Karsenty G, Simpson L et al. Rescue of the skeletal phenotype in CasR-deficient mice by transfer onto the Gcm2 null background. J Clin Invest 2003; 111:1029-1037.
95. Kos CH, Karaplis AC, Peng JB et al. The calcium-sensing receptor is required for normal calcium homeostasis independent of parathyroid hormone. J Clin Invest 2003; 111:1021-1028.

CHAPTER 5

Regulation of Parathyroid Hormone mRNA Stability by Calcium and Phosphate

Rachel Kilav, Justin Silver and Tally Naveh-Many

Abstract

Calcium and phosphate regulate parathyroid hormone (PTH) secretion, gene expression and if prolonged also parathyroid proliferation. The regulation of PTH gene expression by Ca^{2+} and Pi is post-transcriptional, affecting mRNA stability. The regulation of PTH mRNA stability is mediated by the binding of protective *trans* acting factors to a *cis* acting instability element in the 3'UTR. In hypocalcemia there is increased binding that protects the PTH mRNA from degradation by cytosolic ribonucleases resulting in higher levels of PTH mRNA. In hypophosphatemia there is decreased binding, with a subsequent increased degradation of the PTH mRNA resulting in low levels of PTH mRNA. We have identified an AU rich protein binding sequence in the PTH mRNA 3'-UTR and determined its functionality. A 63 nt PTH mRNA 3' UTR element was inserted into a growth hormone (GH) reporter transcript and led to destabilization of the GH transcript. Moreover, this element was sufficient to reproduce the regulation of the reporter transcript's stability by calcium and phosphate in an in vitro degradation assay (IVDA). The PTH mRNA 63 nt element destabilized mRNA for the reporter genes GFP and growth hormone, in transient transfection experiments in HEK293 cells, similar to its effect in the IVDA. Therefore, the PTH RNA-protein binding region is a destabilizing element and is sufficient to confer responsiveness to calcium and phosphate of PTH mRNA. We have identified one of the PTH mRNA 3'-UTR *trans* acting proteins as AU rich binding factor 1 (AUF1). Recombinant AUF1 stabilized the PTH mRNA transcript in the IVDA with parathyroid proteins. In parathyroid extracts from rats fed the different diets there was no difference in AUF1 protein levels by Western blots despite differences in RNA – protein binding. However, 2D gels showed post-translational modification of AUF1 in these extracts, suggesting that calcium and phosphate alter the AUF1 protein and thus its ability to bind and stabilize the PTH mRNA.

Regulation of the Parathyroid Gland by Calcium and Phosphate

PTH has a central role in maintaining normal calcium (Ca^{2+}) and phosphate (Pi) homeostasis as well as bone strength. Dietary induced hypocalcemia markedly increases PTH secretion, mRNA levels and after prolonged stimulation, parathyroid cell proliferation.[1] PTH then acts to correct serum calcium by mobilizing calcium from bone and increasing renal reabsorption of calcium. Dietary induced hypocalcemia leads to a 10X increase in PTH mRNA levels. We have shown that the increase in PTH mRNA levels is post-transcriptional affecting mRNA stability.[3,4]

Molecular Biology of the Parathyroid, edited by Tally Naveh-Many. ©2005 Eurekah.com and Kluwer Academic / Plenum Publishers.

The serum inorganic phosphate is also regulated within a narrow range. 90% of the inorganic phosphate in the serum is ultrafiltrable by the kidneys where there is a sensitive regulation of renal reabsorption by both intrinsic and hormonal (PTH) mechanisms.[5] PTH decreases serum phosphate by increasing renal phosphate excretion. Serum phosphate in turn, has a direct effect to increase PTH secretion, PTH mRNA levels and parathyroid cell proliferation.[4,6] In careful in vivo studies we were able to show that the effect of phosphate on the parathyroid was independent of any changes in serum calcium and 1,25(OH)2D3.[4] This was confirmed by in vitro studies that showed that phosphate had a direct effect on the parathyroid.[7,8] For the in vitro effect of increasing phosphate concentrations on PTH secretion it was imperative to maintain tissue architecture. There was an effect in whole glands or tissue slices but not in isolated cells. This effect was confirmed by Slatopolsky and his colleagues[9] and Olgaard et al,[10] in parathyroid tissue of different sources, showing that phosphate regulates the parathyroid directly.[11]

In vivo studies showed that dietary induced hypophosphatemia leads to a dramatic decrease in PTH gene expression and this effect is post-transcriptional[4] as is the effect of hypocalcemia to increase PTH mRNA levels.[3]

There is a ~60-fold difference in PTH mRNA levels between hypocalcemic and hypophosphatemic rats, and we used these dietary models as tools to define the mechanism of the post-transcriptional regulation of PTH gene expression. We have shown that the post-transcriptional regulation by dietary induced hypocalcemia and hypophosphatemia is mediated by protein-RNA interactions involving protein binding to a specific element in the PTH mRNA 3'-UTR that determine PTH mRNA stability.[12] After a low calcium diet there is increased binding to the PTH mRNA 3'-UTR that protects the RNA from degradation resulting in increased PTH mRNA levels. After a low phosphate diet there is less binding that allows more degradation and leads to the decrease in PTH mRNA levels.

Protein Binding and PTH mRNA Stability

To define the mechanism of the post-transcriptional regulation of PTH gene expression, weanling rats were fed a control diet (0.6% calcium, 0.3% phosphate) or diets deficient in calcium (0.02% calcium, 0.3% phosphate)(low calcium) or phosphate (0.6% calcium, 0.02% phosphate) (low phosphate) for 2 weeks. A low calcium diet resulted in a decrease in serum calcium levels, an increase in serum PTH levels and a 10-fold increase in PTH mRNA levels. A low phosphate diet led to a decrease in serum phosphate levels and PTH levels and a 6-fold decrease in PTH mRNA levels compared to the rats fed a control diet (Fig. 1). Nuclear run on experiments showed that these effects were post-transcriptional.[4,13] Post-transcriptional regulation of gene expression is often associated with protein-RNA interactions that regulate mRNA stability. There was specific binding of parathyroid cytosolic protein extracts to the PTH mRNA transcript using UV cross-linking and RNA mobility shift assays (REMSA). The binding was to the full-length PTH mRNA transcript or to a transcript for the 3' terminal region of the PTH mRNA 3'-UTR but not to transcripts that did not include this region. Protein binding to the PTH mRNA 3'-UTR was increased by hypocalcemia and decreased by hypophosphatemia by REMSA and UV cross-linking gels, in correlation with PTH mRNA levels (Fig. 2). The PTH mRNA binding proteins were also present in other tissues but the binding was regulated by Ca^{2+} and Pi only in the parathyroid and not in other tissues of the same rats.[13] UV cross-linking of parathyroid proteins to transcripts for the full-length and the 3'-UTR showed 3 protein-RNA bands of ~110, 70 and 50 kDa, and the regulation of binding was evident in all 3 bands.

There is no parathyroid cell line. Therefore we have utilized a cell-free mRNA in vitro degradation assay (IVDA) to demonstrate the functionality of the parathyroid cytosolic proteins in determining the stability of the PTH transcript. This assay has been shown to authentically reproduce cellular decay processes.[13,14] Parathyroid proteins were incubated with a ^{32}P labeled full-length PTH mRNA transcript and at timed intervals samples were removed and run on

Calcium Phosphate and PTH mRNA Stability

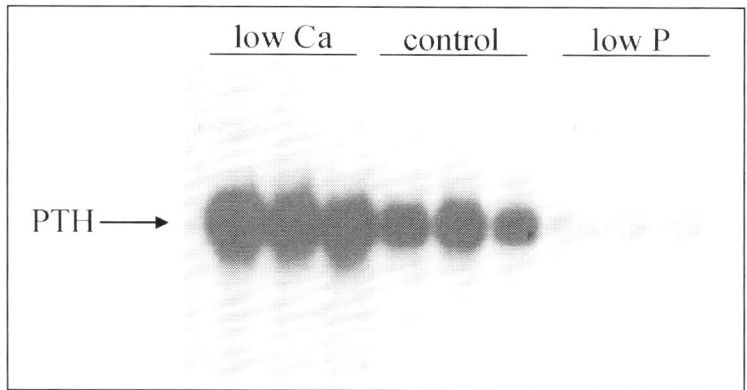

Figure 1. A low calcium diet increases and a low phosphate decreases PTH mRNA levels. Weanling rats were fed control (0.6% calcium, 03% phosphate), low calcium (0.02% calcium, 0.6% phosphate) or low phosphate (0.6% calcium, 0.02% phosphate) diets for 14 days. Total RNA from thyro-parathyroid tissue from each rat was extracted and PTH mRNA levels determined by Northern blots. Each lane represents PTH mRNA from a single rat.

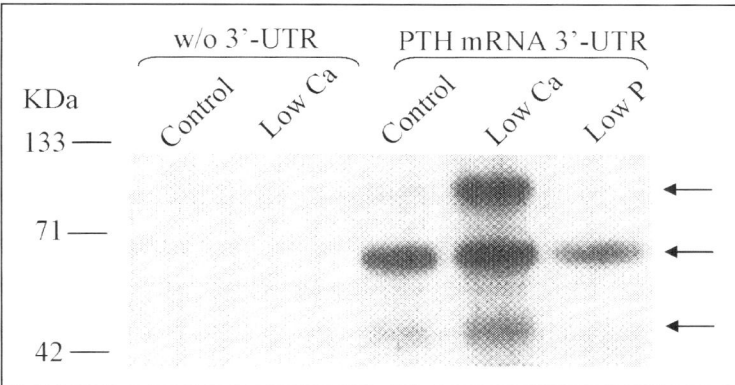

Figure 2. Calcium and phosphate regulate the binding of parathyroid cytosolic proteins to the PTH mRNA 3' UTR. UV cross-linking of S100 parathyroid cytosolic protein extracts (10 µg) from rats fed a normal diet, or a low calcium, or low phosphate diet for 14 days and ^{32}P labeled transcripts for the PTH mRNA 3' UTR or the PTH mRNA without the 3' UTR. The proteins were UV cross linked to the ^{32}P labeled transcripts and then run on a SDS polyacrylamide gel, and visualized by autoradiography. The size of molecular weight markers is shown on the left and protein RNA complexes indicated by the arrows. Reproduced with permission of the Journal of Biological Chemistry.[13]

agarose gels to measure the amount of labeled intact PTH transcript remaining with time. Incubation with parathyroid proteins led to gradual degradation of the PTH transcript, which was a result of degrading and protective factors present in the cytosolic extract. The transcript was intact until 40 min. with parathyroid proteins from control rats, however, with hypocalcemic parathyroid proteins the transcript was not degraded until 180 min. and with hypophosphatemic proteins the transcript was degraded already at 5 min, correlating with mRNA levels in vivo (Fig. 3A).[13] Therefore, the IVDA reproduces the in vivo stabilizing effect of low Ca^{2+} and destabilizing effects of low Pi on PTH mRNA levels. A transcript that did not include the 3'-UTR or just the

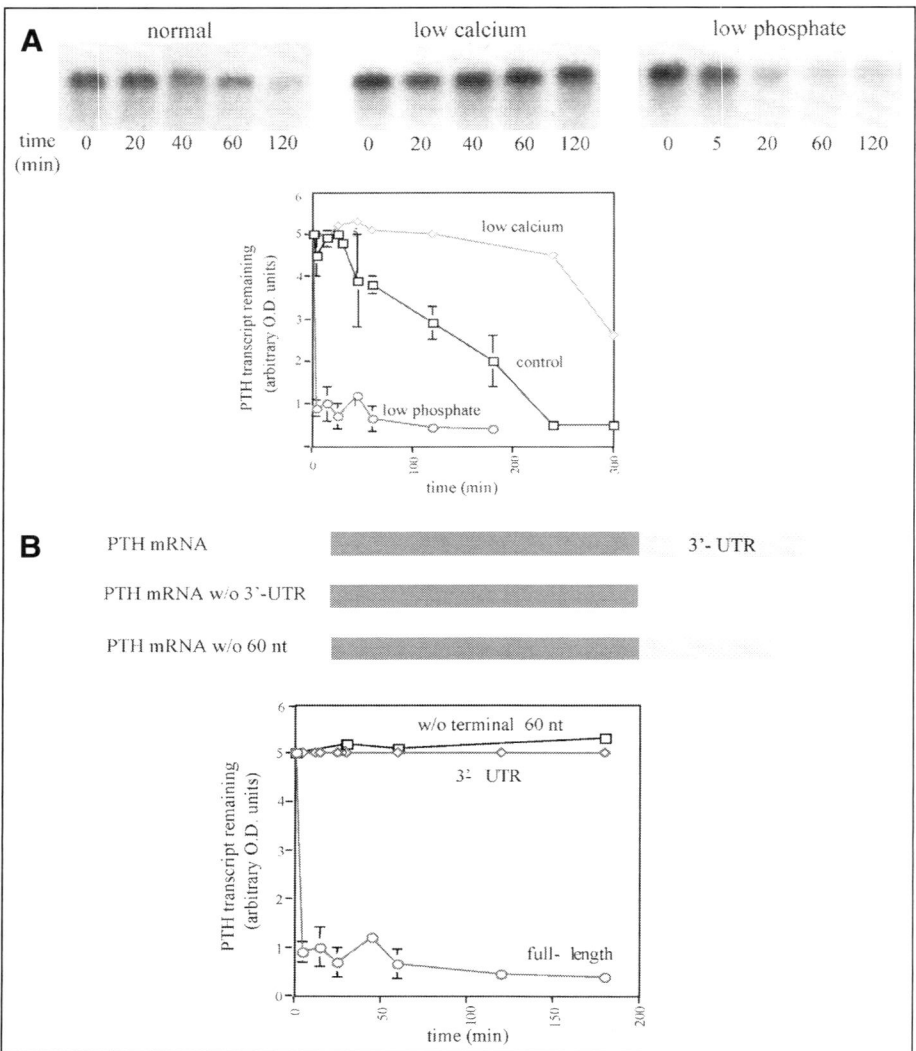

Figure 3. A) Calcium and phosphate regulate PTH RNA degradation in vitro by parathyroid proteins. An in vitro transcribed probe for the full-length PTH mRNA was incubated with cytosolic patathyroid proteins for different time periods and then the RNA was extracted, run on gels, autoradiographed and quantified by densitometry to measure the remaining intact probe at each time point. Representative gels and time response curves of intact full-length PTH mRNA after incubation with parathyroid cytosolic proteins from rats fed control, low calcium, or low phosphate diets. Each point in the curve represents the mean ± SE of 2-4 different experiments. At some points the SE is less than the size of the graphic symbols. The PTH transcript is degraded very rapidly by proteins from -P rats, and is stable with parathyroid proteins from – Ca rats. B) Mapping the region in the PTH 3'-UTR that mediates degradation by proteins from -P rats. The full length PTH mRNA probe or probes without the PTH mRNA 3' UTR or without the terminal 60 nt of the 3' UTR were incubated with parathyroid cytosolic proteins from rats fed a low phosphate diet for the indicated time periods. Transcripts that did not include the terminal region of the PTH mRNA 3' UTR were not degraded even by low phosphate parathyroid proteins that led to rapid degradation of the full-length PTH transcript. Reproduced with permission of the Journal of Biological Chemistry.[13]

60 terminal nt protein-binding region of the PTH mRNA 3'-UTR was more stable than the full-length PTH transcript and was intact for more than 180 min (Fig. 3B) even with the low Pi parathyroid proteins, suggesting that the binding region is an instability element. Moreover, Ca^{2+} (not shown) and Pi (Fig. 3B) did not regulate the stability of the truncated transcript that did not include the protein-binding region.[13] Therefore, the regulation by parathyroid proteins from low Ca^{2+} and Pi rats in the IVDA is dependent upon the protein-binding sequence in the PTH mRNA 3'-UTR.

Identification of the PTH mRNA 3'-UTR Binding Proteins and Their Function

The PTH mRNA binding proteins were purified by PTH RNA 3'-UTR affinity chromatography. One of the purified proteins was sequenced and was identical to AU-rich binding factor (AUF1) (hnRNP D)[15] which is central to the stability of other mRNAs.[16] Recombinant AUF1 bound the PTH mRNA 3'-UTR with a single band at 50 kDa, which corresponds to 1 of 3 protein-RNA bands found with cytosolic parathyroid proteins by UV cross-linking gels.[15] Binding of recombinant AUF1 to the PTH mRNA transcript was also demonstrated by RNA electrophoretic mobility shift assays. Antibodies to AUF1 resulted in super shift of the PTH RNA-parathyroid extract complex, demonstrating that AUF1 is part of the protein-RNA complex.

Brewer et al have cloned this RNA-binding protein which binds with high affinity to a variety of A+U-rich elements (AREs) in the 3'-UTRs of a number of mRNAs.[16] These include mRNAs for cytokines, oncoproteins, and G protein coupled receptors, where AUF1 is associated with instability of these mRNAs.[17] Three classes of AREs have been characterized, two of which contain several scattered or overlapping copies of the pentanucleotide AUUUA.[18,19] The class III AREs lack the AUUUA motif but require a U-rich sequence and possibly other unknown determinants.[20] It has been shown that the AUUUA motif is not required for AUF1 binding. The PTH mRNA 3'-UTRs are rich in A and U ranging from 68-74% of the nucleotides.[21] The PTH mRNA 3'-UTR binding element (see below) is a type III ARE that does not contain any AUUUA sequences.[12]

There are four isoforms for AUF1 that are generated by alternative splicing of the AUF1 transcript.[22] The activity of AUF1 may be determined by the presence of other proteins in the binding complex, such as αCP1 and 2 in stabilizing the α-globin mRNA[23] and in the stabilization of other mRNAs.[24,25]

To study the effect of AUF1 on PTH RNA stability, recombinant AUF1 was added to the IVDA. Addition of recombinant $p40^{AUF1}$ and $p37^{AUF1}$ isoforms of AUF1 stabilized the PTH transcript dose-dependently even by hypophosphatemic proteins, that without AUF1 led to rapid degradation of the full-length PTH transcript. Addition of eluate from the PTH 3' UTR affinity column together with $p40^{AUF1}$ stabilized the PTH transcript at doses that alone had no effect (Fig. 4A).[15] Other control proteins, bovine serum albumin (BSA) and dynein light chain (Mr 8000) (LC8), that also binds to the PTH mRNA 3'-UTR[29] had no effect[15] (Fig. 4B). This result supports the regulatory role of AUF1 in PTH mRNA stability.

The mechanism by which AUF1 stabilizes PTH mRNA is not clear. Low calcium and Pi did not affect the level of AUF1 in the parathyroids by western blots (unpublished results). Preliminary results show that calcium and phosphate lead to differences in post-translational modifications, possibly phosphorylation, rather than in the amount of AUF1 protein (not shown).

Two additional PTH mRNA binding proteins were identified by RNA affinity chromatography (not shown), hnRNP K,[26,27] and Up stream of *nras* (UNR).[28] These proteins bind PTH transcripts by binding assays using recombinant proteins. Specific antibodies to each of the identified proteins led to a super shift of the PTH mRNA 3'-UTR binding complex (not shown), demonstrating that they are part of the protein RNA complex.

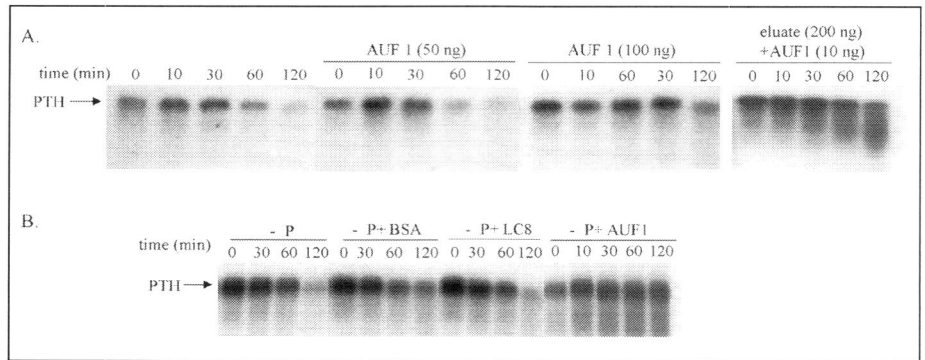

Figure 4. Stabilizing effect of p40[AUF1] on the degradation in vitro of PTH mRNA by hypophosphatemic rat parathyroid proteins. The full-length radiolabeled PTH mRNA was incubated with cytosolic parathyroid protein extracts (10 μg) from hypophosphatemic rats and at timed intervals samples were extracted, run on agarose gels and autoradiographed to measure the intact transcript remaining. A. Degradation of the PTH transcript in the presence of parathyroid extracts and increasing doses of recombinant p40[AUF1]. p40[AUF1] stabilized the PTH transcript dose-dependently. Addition of eluate (200 ng), from the PTH 3' UTR affinity column together with 10 ng of p40[AUF1] stabilized the PTH transcript at doses that alone had no effect. B. Degradation with parathyroid proteins from hypophosphatemic (-P) rats, without and with added BSA (6 μg), LC8 (6 μg) or recombinant p40[AUF1] (200 ng). Recombinant p40[AUF1], but not bovine serum albumin (BSA) or dynein light chain (LC8) stabilized the PTH transcript. Reproduced with permission of the Journal of Biological Chemistry.[15]

An additional protein, dynein light chain (M_r 8000) or LC8 was identified by Northwestern expression cloning and was shown to mediate the binding of the PTH mRNA to microtubules.[29] LC8 may have a role in the intracellular localization of PTH mRNA in the parathyroid cell rather than in the stability of PTH mRNA.

Identification of the Minimal *cis* Acting Protein Binding Element in the PTH mRNA 3'-UTR

The PTH cDNA consists of three exons coding for the 5'-UTR (exon I), the prepro region of PTH (exon II), and the structural hormone together with the 3'-UTR (exon III)[21] (Fig. 5). The rat 3'-UTR is 239 nt long out of the 712 nt of the full-length PTH RNA. The 3'-UTR is 42% conserved between human and rat, while the coding region is 78% conserved.

Binding and competition experiments by REMSA and UV cross-linking gels identified a 26 nt element as the minimal sequence sufficient for parathyroid protein binding.[12] Sequence analysis of the 26 nt element in the PTH mRNA of different species revealed high conservation of the rat element in the PTH mRNA 3'-UTRs of the murine (23 of 26 nt), human (19 of 26 nt) and canine (19 of 26 nt) species, with human and canine being identical (Fig. 5). The human and rat are 73% identical in the 26 nt element, and only 42% identical in their 3'-UTR. The canine and rat are 73% identical in their 26 nt element and 50% identical in their 3'-UTR. The human and canine are 100% identical in the 26 nt of the element and only 70% identical in their 3'-UTR. Comparison of the 26 nt sequence in rat and mouse showed 89% identity, however, their 3'-UTRs show a comparable degree of identity. All in all, this analysis suggests that the binding element may represent a functional unit that has been evolutionarily conserved, but sequencing of the 3'-UTR in many other species is needed to establish this conclusion.

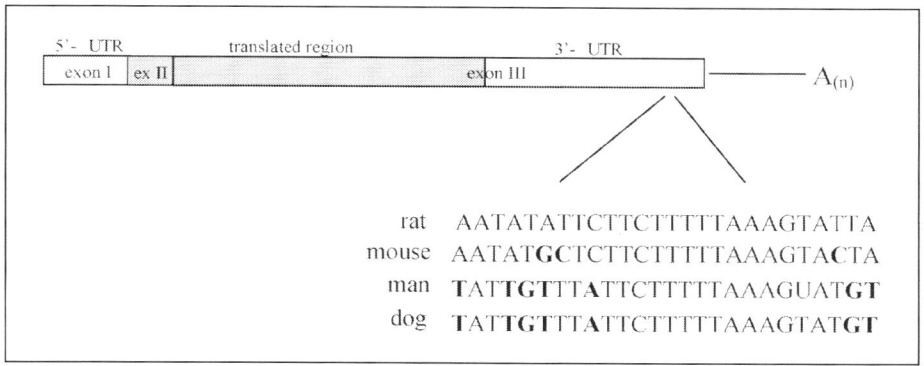

Figure 5. The sequence of the 26 nt element in rat PTH mRNA 3'-UTR and in the 3'-UTR of other species. Diagram of the PTH cDNA and the protein-RNA binding sequence in the 3'-UTR. The 26 nt minimal protein-binding sequence in the PTH mRNA 3' UTR is shown for rat, mouse, man and dog. The nt that differ from rat are shown in bold.

To study the functionality of the protein-binding element in the context of another RNA, a 63 bp fragment coding for the 26 nt of the PTH mRNA 3'- UTR and flanking nt, was fused to growth hormone (GH) reporter gene (Fig. 6). RNAs were transcribed in vitro and transcripts subjected to IVDA with parathyroid proteins. The chimeric GH-PTH 63 nt transcript, as the full-length PTH transcript, was stabilized by parathyroid proteins from rats fed a low calcium diet and destabilized by parathyroid proteins of a low P diet. The native GH transcript was more stable than PTH and the chimeric RNAs and was not affected by parathyroid proteins from the different diet (Fig. 6). Therefore, the PTH RNA protein-binding region destabilized the GH transcript in the presence of parathyroid proteins. Furthermore, this element conferred responsiveness of GH to changes in parathyroid proteins by calcium and phosphate.[12] The results demonstrate that the protein binding region of the PTH mRNA 3'-UTR is both necessary and sufficient for determining RNA stability and for the response to Ca^{2+} and Pi.

We also studied the function of the PTH element in cells, using the heterologous cell line HEK293. cDNAs coding for the protein binding region of the PTH mRNA 3'-UTR were inserted at the 3' end of two reporter cDNAs. 63 bp of the PTH 3'UTR, coding for the *cis* element were inserted into a GH expression construct driven by a S16 ribosomal protein gene promoter and a larger fragment of 100 bp into GFP expression construct driven by a CMV promoter. The plasmids were transiently transfected into HEK293 cells. At 24 h mRNA levels were measured by Northern blot, and protein levels of GFP by immunofluorescence, and secreted GH by radioimmunoassay. There was a marked reduction in the expression of the chimeric genes containing the PTH elements compared to the wt genes. A truncated PTH 40 nt element had no effect on GH reporter gene expression.[30] The 100 and 63 nt transcripts, like the full-length PTH RNA, bound parathyroid proteins by UV and REMSA, but the truncated element did not bind parathyroid proteins.[12] The different constructs for each reporter gene all used the same promoter; therefore these results suggest that insertion of the PTH 3'-UTR element decreased the stability of the reporter transcripts and not their transcription levels. Measurement of mRNA decay after inhibition of transcription by the addition of DRB to the transfected cells, confirmed that the decrease in mRNA levels was post-transcriptional.[30] These results are in agreement with the decreased stability of the GH chimeric transcript in the IVDA with parathyroid proteins (Fig. 6).[12]

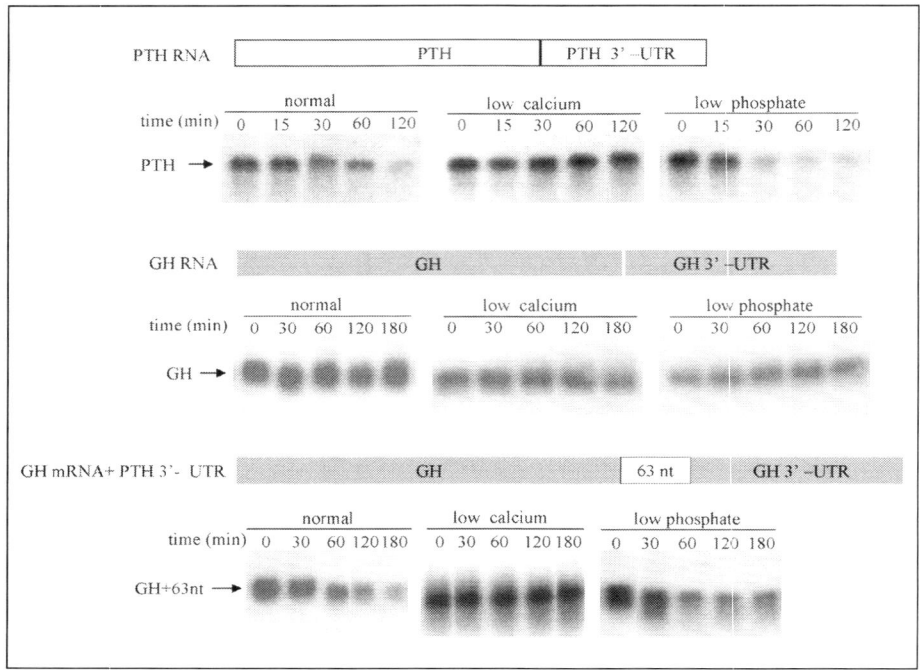

Figure 6. The PTH mRNA 3'-UTR 63 nt protein-binding region confers responsiveness of growth hormone (GH) mRNA to parathyroid proteins from rats fed low calcium or low phosphate diets in an in vitro degradation assay. Schematic representation of the transcripts for PTH mRNA, GH mRNA and the chimeric GH mRNA containing the PTH 3'-UTR 63 nt element inserted at the end of the GH coding region (GH + PTH 3'-UTR 63 nt). For each transcript the corresponding representative gel of IVDA with parathyroid proteins from rats fed normal, low calcium or low phosphate diets is shown. The full-length PTH and the chimeric GH-PTH 63 transcripts were stabilized with low calcium parathyroid proteins and destabilization with low phosphate parathyroid proteins. The GH transcript was more stable than the chimeric transcript and was not affected by parathyroid proteins from the different diets. Reproduced with permission of the Journal of Biological Chemistry.[12]

The Structure of the PTH *cis* Acting Element

The IVDA and the transfection experiments demonstrate the functional importance of the RNA protein binding region in the PTH 3'-UTR. RNA utilizes sequence and structure for its regulatory functions. To understand how the *cis* element functions as an instability element we have analyzed its structure by RNase H, primer extension, mutation analysis and computer modeling.[30] The results indicates that the PTH mRNA 3'-UTR and in particular the region of the *cis* element are dominated by significant open regions with little folded base pairing. Mutation analysis of the 26 nt core binding element demonstrated the importance of defined nts for protein-RNA binding. The same mutations that prevented binding were also ineffective in destabilizing reporter GFP mRNA in HEK293 cells. The PTH mRNA 3'-UTR *cis* acting element is therefore an open region that utilizes the distinct sequence pattern to determine mRNA stability by its interaction with *trans* acting RNA binding factors.

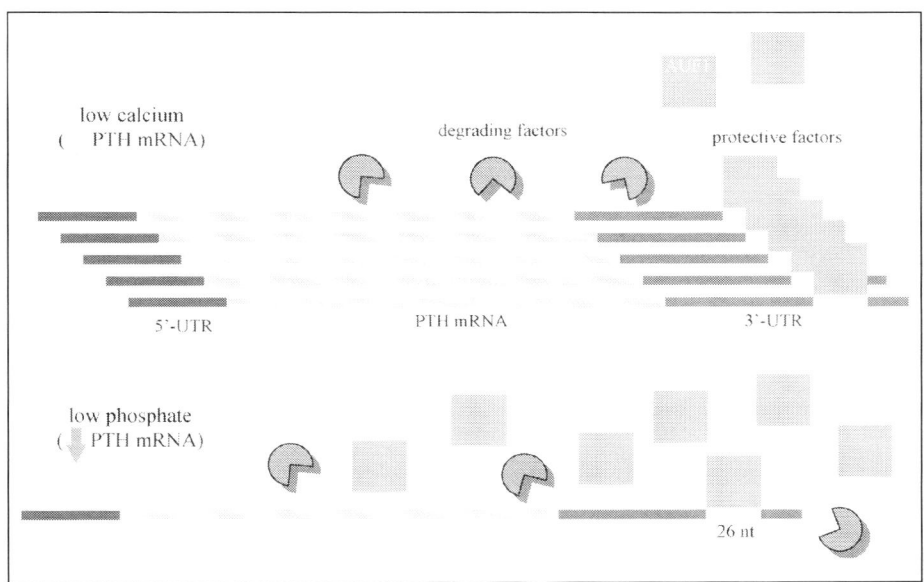

Figure 7. Model for the regulation of PTH mRNA stability by calcium and phosphate. Regulation of PTH mRNA stability involves protective and degrading proteins. Schematic representation of the PTH mRNA with the protective RNA binding proteins (square) and the ribonucleases that degrade the RNA (packman). Protein binding is increased after a low calcium diet resulting in protection of the bound RNA and an increase in PTH mRNA levels. After a low phosphate diet there is less binding, more degradation and a decrease in PTH mRNA levels. The PTH mRNA binding protein, AUF1 and the PTH mRNA 3' UTR conserved 26 nt protein binding element are shown.

Conclusions

Dietary-induced hypocalcemia and hypophosphatemia regulate PTH gene expression post-transcriptionally and this is dependent upon the binding of parathyroid cytosolic proteins to instability regions in the PTH mRNA 3'-UTR. The binding of these cytosolic proteins to the PTH mRNA is increased in hypocalcemia and decreased in hypophosphatemia correlating with PTH mRNA levels in vivo. There is no parathyroid cell line and the stability of PTH transcripts was studied by an IVDA. Parathyroid proteins from hypocalcemic rats lead to an increase in PTH RNA stability in the IVDA and hypophosphatemic proteins to a marked decrease in stability. One of the PTH mRNA binding proteins is AUF1 that stabilizes the PTH mRNA. We have identified a conserved 26 nt element in the PTH mRNA 3'-UTR as the minimal protein binding sequence. The functionality of a 63 nt element that included the 26 nt was studied in reporter RNAs. The stability of the chimeric RNAs was studied in the IVDA with parathyroid proteins of low Ca and P rats and in transfection experiments in HEK 293 cells. This element destabilized the reporter genes and was sufficient to confer responsiveness to calcium and phosphate in the IVDA with parathyroid proteins. The results demonstrate a functional *cis* element in the PTH mRNA 3'-UTR that upon binding to *trans* acting proteins determines PTH mRNA stability. Our model suggests that differences in binding of the *trans* acting factors to this element determine PTH mRNA stability and its regulation by Ca^{2+} and Pi (Fig. 7). The increased protein binding to the PTH mRNA 3' UTR *cis* element after a low calcium diet results in protection of the PTH mRNA and an increase in PTH mRNA levels. After a low phosphate diet there is less binding, more degradation and a decrease in PTH mRNA levels.

Acknowledgments

We thank The United States-Israel Binational Science Foundation, The Israel Science Foundation and The Minerva Foundation.

References

1. Naveh-Many T, Bell O, Silver J et al. Cis and trans acting factors in the regulation of parathyroid hormone (PTH) mRNA stability by calcium and phosphate. FEBS Lett 2002; 529:60-64.
2. Brown EM, Gamba G, Riccardi D et al. Cloning and characterization of an extracellular Ca^{2+}-sensing receptor from bovine parathyroid. Nature 1993; 366:575-580.
3. Moallem E, Silver J, Naveh-Many T. Regulation of parathyroid hormone messenger RNA levels by protein kinase A and C in bovine parathyroid cells. J Bone Miner Res 1995; 10:447-452.
4. Kilav R, Silver J, Naveh-Many T. Parathyroid hormone gene expression in hypophosphatemic rats. J Clin Invest 1995; 96:327-333.
5. Silver J, Kilav R, Naveh-Many T. Mechanisms of secondary hyperparathyroidism. Am J Physiol Renal Physiol 2002; 283:F367-F376.
6. Naveh-Many T, Rahamimov R, Livni N et al. Parathyroid cell proliferation in normal and chronic renal failure rats: the effects of calcium, phosphate and vitamin D. J Clin Invest 1995; 96:1786-1793.
7. Almaden Y, Hernandez A, Torregrosa V et al. High phosphorus directly stimulates PTH secretion by human parathyroid tissue [abstract]. J Am Soc Nephrol 1995; 6:957.
8. Almaden Y, Canalejo A, Hernandez A et al. Direct effect of phosphorus on parathyroid hormone secretion from whole rat parathyroid glands in vitro. J Bone Miner Res 1996; 11:970-976.
9. Slatopolsky E, Finch J, Denda M et al. Phosphate restriction prevents parathyroid cell growth in uremic rats. High phosphate directly stimulates PTH secretion in vitro. J Clin Invest 1996; 97:2534-2540.
10. Nielsen PK, Feldt-Rasmusen U, Olgaard K. A direct effect of phosphate on PTH release from bovine parathyroid tissue slices but not from dispersed parathyroid cells. Nephrol Dial Transplant 1996; 11:1762-1768.
11. Rodriguez M, Almaden Y, Hernandez A et al. Effect of phosphate on the parathyroid gland: direct and indirect? Curr Opin Nephrol Hypertens 1996; 5:321-328.
12. Kilav R, Silver J, Naveh-Many T. A conserved cis-acting element in the parathyroid hormone 3'-untranslated region is sufficient for regulation of RNA stability by calcium and phosphate. J Biol Chem 2001; 276:8727-8733.
13. Moallem E, Silver J, Kilav R et al. RNA protein binding and post-transcriptional regulation of PTH gene expression by calcium and phosphate. J Biol Chem 1998; 273:5253-5259.
14. Levy AP, Levy NS, Goldberg MA. Post-transcriptional regulation of vascular endothelial growth factor by hypoxia. J Biol Chem 1996; 271:2746-2753.
15. Sela-Brown A, Silver J, Brewer G et al. Identification of AUF1 as a parathyroid hormone mRNA 3'-untranslated region binding protein that determines parathyroid hormone mRNA stability. J Biol Chem 2000; 275:7424-7429.
16. Wilson GM, Brewer G. Identification and characterization of proteins binding A + U-rich elements. Methods: A Companion to Methods in Enzymology 1999; 17:74-83.
17. Wilson GM, Brewer G. The search for trans-acting factors controlling messenger RNA decay. Prog Nucleic Acid Res Mol Biol 1999; 62:257-291.
18. Peng SS, Chen CY, Shyu AB. Functional characterization of a non-AUUUA AU-rich element from the c- jun proto-oncogene mRNA: evidence for a novel class of AU-rich elements. Mol.Cell Biol 1996; 16:1490-1499.
19. Nakamaki T, Imamura J, Brewer G et al. Characterization of adenosine-uridine-rich RNA binding factors. J Cell Physiol 1995; 165:484-492.
20. Chen CY, Shyu AB. AU-rich elements: characterization and importance in mRNA degradation. Trends Biochem Sci 1995; 20:465-470.
21. Kemper B. Molecular biology of parathyroid hormone. CRC Crit Rev Biochem 1986, 19:353-379.
22. Wagner BJ, DeMaria CT, Sun Y et al. Structure and genomic organization of the human AUF1 gene: alternative pre-mRNA splicing generates four protein isoforms. Genomics 1998; 48:195-202.

23. Kiledjian M, DeMaria CT, Brewer G et al. Identification of AUF1 (heterogeneous nclear ribonucleoprotein D) as a component of the α-globin mRNA stability complex. Mol Cell Biol 1997; 17:4870-4876.
24. Loflin P, Chen CY, Shyu AB. Unraveling a cytoplasmic role for hnRNP D in the in vivo mRNA destabilization directed by the AU-rich element. Genes Dev. 1999; 13:1884-1897.
25. Xu N, Chen CY, Shyu AB. Versatile role for hnRNP D isoforms in the differential regulation of cytoplasmic mRNA turnover. Mol Cell Biol 2001; 21:6960-6971.
26. Ostareck DH, Ostareck-Lederer A, Wilm M et al. mRNA silencing in erythroid differentiation: hnRNP K and hnRNP E1 regulate 15-lipoxygenase translation from the 3' end. Cell 1997; 89:597-606.
27. Ostareck DH, Ostareck-Lederer A, Shatsky IN et al. Lipoxygenase mRNA silencing in erythroid differentiation: The 3'UTR regulatory complex controls 60S ribosomal subunit joining. Cell 2001; 104:281-290.
28. Triqueneaux G, Velten M, Franzon P et al. RNA binding specificity of Unr, a protein with five cold shock domains. Nucleic Acids Res 1999; 27:1926-1934.
29. Epstein E, Sela-Brown A, Ringel I et al. Dynein light chain (M_r 8000) binds the parathyroid hormone mRNA 3'-untranslated region and mediates its association with microtubules. J Clin Invest 2000; 105:505-512.
30. Kilav R, Bell O, Shun-Yun L et al. The PTH mRNA 3'-UTR AU rich element is an unstructured functional element. J Biol Chem 2004; 279:2109-2116.

CHAPTER 6

In Silico Analysis of Regulatory Sequences in the Human Parathyroid Hormone Gene

Alexander Kel, Maurice Scheer and Hubert Mayer

Introduction

Parathyroid hormone (PTH) is intimately involved in the homeostasis of normal serum concentrations of calcium and phosphate, which, in turn, regulate the synthesis and secretion of PTH. The synthesis and secretion of PTH in the parathyroid gland is regulated at the transcriptional, posttranscriptional and posttranslational levels. A major control mechanism in PTH synthesis occurs at the level of gene transcription by the interaction of transcription factors with DNA. Understanding the molecular mechanisms controlling PTH-specific gene transcription requires the determination of cis-regulatory DNA-elements, and the "trans"-factors—the transcription factors (TF)—proteins that bind to them. Cell cultures have been invaluable for identifying TF-binding sites involved in basal, tissue specific, and regulatory expression of PTH-gene transcription. Animal models largely confirmed the in vitro findings and led to the model of the function of PTH regulated by positive and negative feedback loops through components in the serum and by peripheral organs. These findings of feedback regulatory events suggest that fine-tuned cooperative interactions between different classes of proteins are the basis of normal parathyroid gland function and alteration in their expression level or function leads to an abnormal mineral homeostasis. In an effort to understand the organisation of the PTH promotor we searched the genome of human and mouse by a bioinformatics-based screen in silico for the existence of cis regulatory elements. A detailed understanding of how the PTH synthesis is regulated will help to devise rational therapy for the management of failure conditions, as well as treatment for diseases, such as osteoporosis, in which alterations in PTH expression may play an important role.

In Silico Study of Gene Transcription Regulation

There are several thousands of different transcription factors functioning in human cells. More than 800 of them are well studied and characterized in databases such as SWISSPROT and TRANSFAC.[11] TRANSFAC database collects experimentally identified binding sites for most of the transcription factors in many different genes of human, mouse, fly and other eukaryotic organisms. Even though TRANSFAC contains information about more than 6000 genomic sites (about 1500 in human genes) it is far from being complete. Taking into account 33000 genes in human genome and the fact that every gene might have up to 100 functioning sites in all regulatory regions including promoters, enhancers and far upstream regulatory regions, we could expect millions of sites in human genome. Knowledge about the number, the sequence and the position of all these sites in genome will bring us to a principally new level of

Molecular Biology of the Parathyroid, edited by Tally Naveh-Many. ©2005 Eurekah.com and Kluwer Academic / Plenum Publishers.

In Silico Analysis of Regulatory Sequences in the Human Parathyroid Hormone Gene

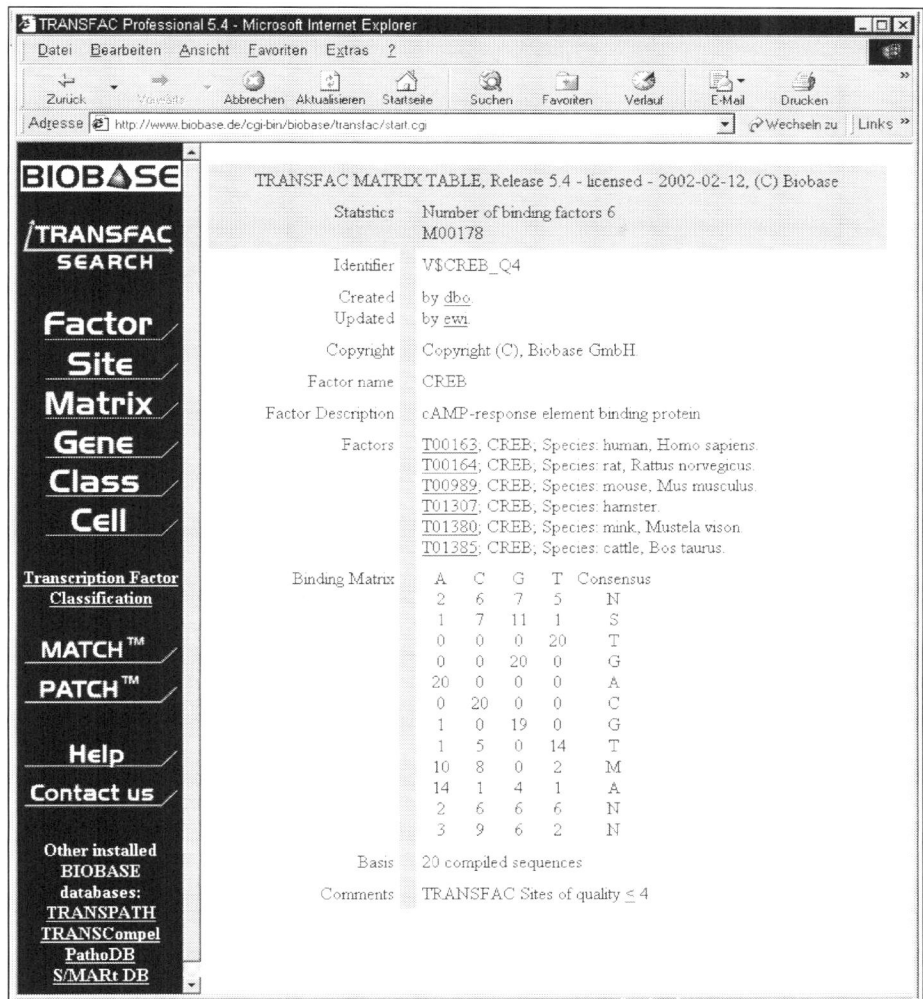

Figure 1. An example of the weight matrix for binding sites for CREB transcription factors that is responsible for activation of gene expression through cAMP signalling pathways. The example is taken from TRANSFAC database. The following information is given: the names of the corresponding transcription factors for different mammalian species, counts for every nucleotide in the positions of the motif and the consensus sequences in the last column.

understanding of how genes are regulated during development of organisms and how genes are dis-regulated is the cases of diseases.

Computer analysis of genome sequences provides means for prediction of binding sites for different transcription factors. Every transcription factor can be characterized by a specific DNA motif that is common for most of their binding sites. Such motifs being quite degenerate typically are described by means of a positional "weight matrix" (see Fig. 1), where counts are given for the observed nucleotides in the corresponding position of known TF binding sites. In Figure 1, an example of the weight matrix for binding sites for CREB

Table 1. Regulatory elements in the human, bovine and chicken PTH gene

Species	Transfac-Acc.	Sequence 5' → 3'	From	To	Binding Factors
HUMAN	R11433	TGAGACaggGTCTCA	-4194	-4180	unidentified complex, Ref-1
HUMAN	R11432	CcatTTGTGTATGCagaa	-3354	-3337	unidentified complex, Ref-1
HUMAN	R10019	tatGTGTCTGctttTGAACCtatagtt	-126	-101	VDR
HUMAN	R11695	TtTGAACCTATagttgagat	-115	-96	unidentified complex
HUMAN	R03738	TGACATCA	-98	-69	CREB
HUMAN	-	AGTCTTTGCATAAG	-223	-216	Oct
BOVINE	-	AGGTTA	-461	-465	?
BOVINE	-	AGTTCC	-449	-444	?
BOVINE	R01196	TGACGTCA	-52	-41	CREB ?
CHICKEN	R10018	GaGGGTCAggaGGGTGTgc	-76	-58	VDR/RXR-alpha

Sequences are sorted by location relative to transcriptional start site and by species. Nucleotide sequence of sense strand is shown in 5´ → 3´ orientation. Accession numbers for TRANSFAC Database which collects information about transcription factors and its DNA binding sites is given too.

transcription factor that is responsible for activation of gene expression through cAMP signalling pathways is shown.

Quite a number of the weight matrices for many different transcription factors have been collected in TRANSFAC database. Many different tools, such as SIGNAL SCAN,[24] MatInspector,[25] Match,[8] and TRANSPLORER (www.biobase.de) use this collection of matrices for searching of potential TF binding sites in any DNA sequence. These tools were applied intensively in the last years for analysis of regulatory regions of many different functional classes of genes. Among them: globin genes,[28] muscle and liver specific genes,[33] genes involved in the regulation of cell cycle[14] and many others. Such "in silico" analysis greatly speed up the process of experimental identification and characterisation of all regulatory sites and clarify mechanisms of their functioning.

PTH Gene: Known Regulatory Elements

For the parathyroid hormone (PTH) gene the tissue-specific analysis of the regulatory elements and especially of the binding transcription factors was complicated by the lack of a parathyroid cell line bearing the normal regulatory function. Nevertheless some transcription factor binding sites have been identified experimentally in this gene (see Table 1).

The most established sites are: CREB site (site for c-AMP responsive element binding factors), Oct site and the site for 1,25-dihydroxyvitamine D response element (VDRE). Presence of these sites in the promoter of PTH gene explains the main features of the regulation of expression of this gene. But it becomes clear that to understand the full regulation we need to know much more about binding sites for different transcription factors acting in the promoter of PTH gene as well as in its far upstream and downstream regulatory regions.

Farrow et al[7,10] has identified 3 DNA sequences upstream of the bovine PTH gene involved in transcriptional regulation. One is located at position −52 to −41 with a palindromic core sequence (TGACGTCA) and the other two are located more upstream from −461 to −465 (AGGTTA) and −449 to −444 (AGTTCC) from transcription start site. For the single proximal element direct binding of CREB is very likely and the palindromic core sequence is also found in other CREs like for example in multiple copies in the rat somatostatin gene promoter.[12] The two distal elements probably bind VDR which was gathered from binding of VDR to two longer promoter fragments containing either hexamer element.[6] They are therefore potential VDREs (vitamin D response elements).[6] Whether vitamin D (via VDR) and cAMP (via CREB), respectively, mediate transcriptional control through these three elements has to be proven experimentally.

In the chicken PTH gene VDRE was identified, and its binding to VDR/RXR-alpha heterodimer as well as its functionality as negative regulatory element after transfection into the opossum kidney OK cell line was demonstrated.[17] It is located in the proximal promoter region (from −76 to −58 to transcription start site) and consists of an imperfect direct repeat of two hexamer half-sites separated by 3 nucleotides (gaGGGTCAggaGGGTGTgc). Interestingly mutation of the last two nucleotides of proximal half-site leads to conversion from a negative to a positive transcriptional element in vitro.[16]

Global Homology of PTH Gene between Human and Mouse

The international initiative of sequencing of human and mouse complete genomes provides a very useful resource for analysing gene structure and regulation. One can retrieve genomic sequences from two major sources: (1) PFP—publicly funded Human Genome Project sequence repository located at NCBI and (2) Celera—complete genome sequence done by the Celera company. Currently, human and mouse genome sequences are available both from PFP and Celera. Figures 2 and 3 are examples of retrieval of human PTH genes sequence from the 11 chromosome using the UCSC genome browser (http://genome.ucsc.edu) and NCBI genome retrieval tool (http://www.ncbi.nlm.nih.gov/entrez/). In the genome browser the location of PTH gene in the chromosome 11 is shown. It is surrounded by two predicted genes with unknown function. Other genes such as: ARNTL (aryl hydrocarbon receptor nuclear translocator-like), TEAD1 (TEA domain family member 1 (SV40 transcriptional enhancer factor)), PARVA (alpha-parvin), TC21 (oncogene TC21), COPB (coatomer protein complex, subunit beta) are located in a proximity of 1MB. There are number of SINE and LINE repeats found in this region (see Fig. 3). In Figure 2, a snapshot of the NCBI genome sequence retrieval tool is representing sequence around PTH gene transcription start. One can see the promoter and first exon of this gene.

The genome resources of human and mouse allows to do a large scale sequence comparison between orthologous genes from human and mouse. Using Celera database we have retrieved human and mouse genomic sequences containing PTH gene together with their 5' and 3' flanking regions (10kb each). The sequence comparison was done by the VISTA tool (http://www-gsd.lbl.gov/VISTA/). The result is shown in Figure 4. The program makes a global alignment and presents the result in a form of plot showing the percent of homology in the 200nt window sliding along the alignment. The positions on the alignment are given according the corresponding positions in the human sequence. So, the position 10000 corresponds to the start of transcription of the human PTH gene. One can see that the regions of the highest homology are located in the coding regions of the second and third exons (see the coloured boxes). Three other regions of high homology (homology > 70%) were revealed in the 5' sequence of the gene. These are regions: 5'C (5431-5639), 5'B (6395-6599), 5'A (9772-9999). It is interesting to observe further regions of relatively high homology in the 5' and 3' regions

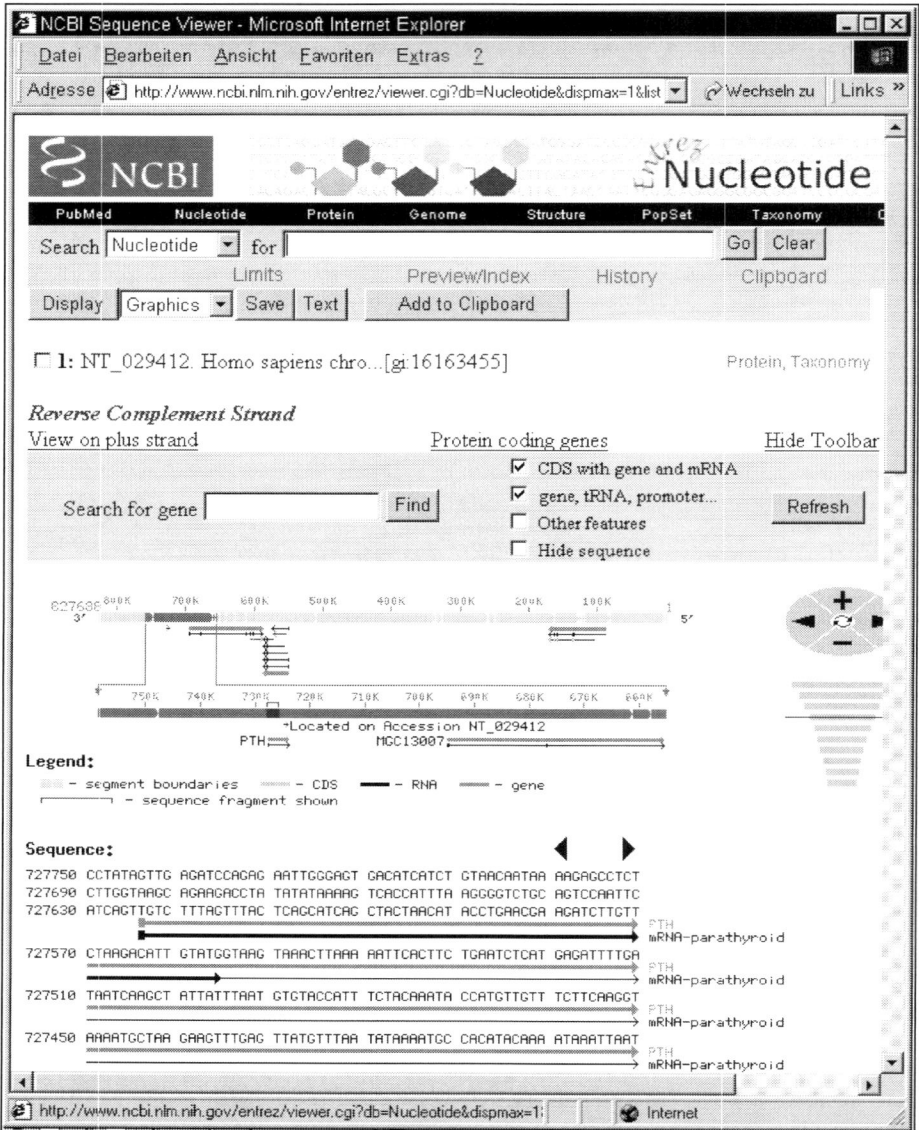

Figure 2. Retrieval of the human PTH gene sequence using the NCBI genome retrieval tool (http://www.ncbi.nlm.nih.gov/entrez/). One can see a snapshot of the NCBI genome sequence retrieval tool representing the sequence around PTH transcription start. This tool provides the possibility to select a region of interest on a human chromosome of your choice and zoom into this region to see the gene. The sequence containing the promoter and first exon of the PTH gene is shown at the bottom of the figure.

In Silico Analysis of Regulatory Sequences in the Human Parathyroid Hormone Gene

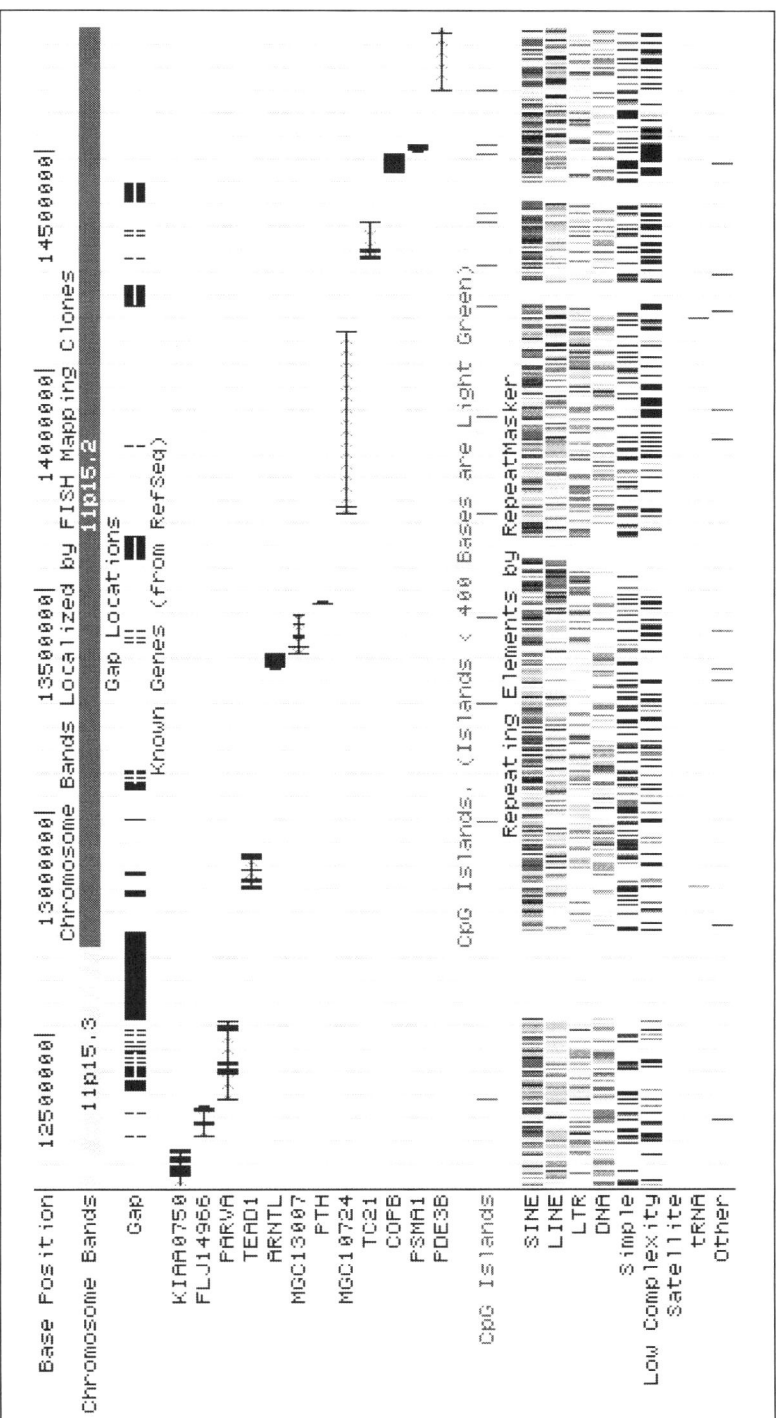

Figure 3. Retrieval of the human PTH gene sequence from the 11 chromosome using UCSC genome browser (http://genome.ucsc.edu). The location of PTH gene in the chromosome is shown by a short bar in the center of the picture. It is surrounded by two predicted genes with unknown function and several other genes. A number of SINE and LINE repeats that were found in this region are shown in the bottom of the figure.

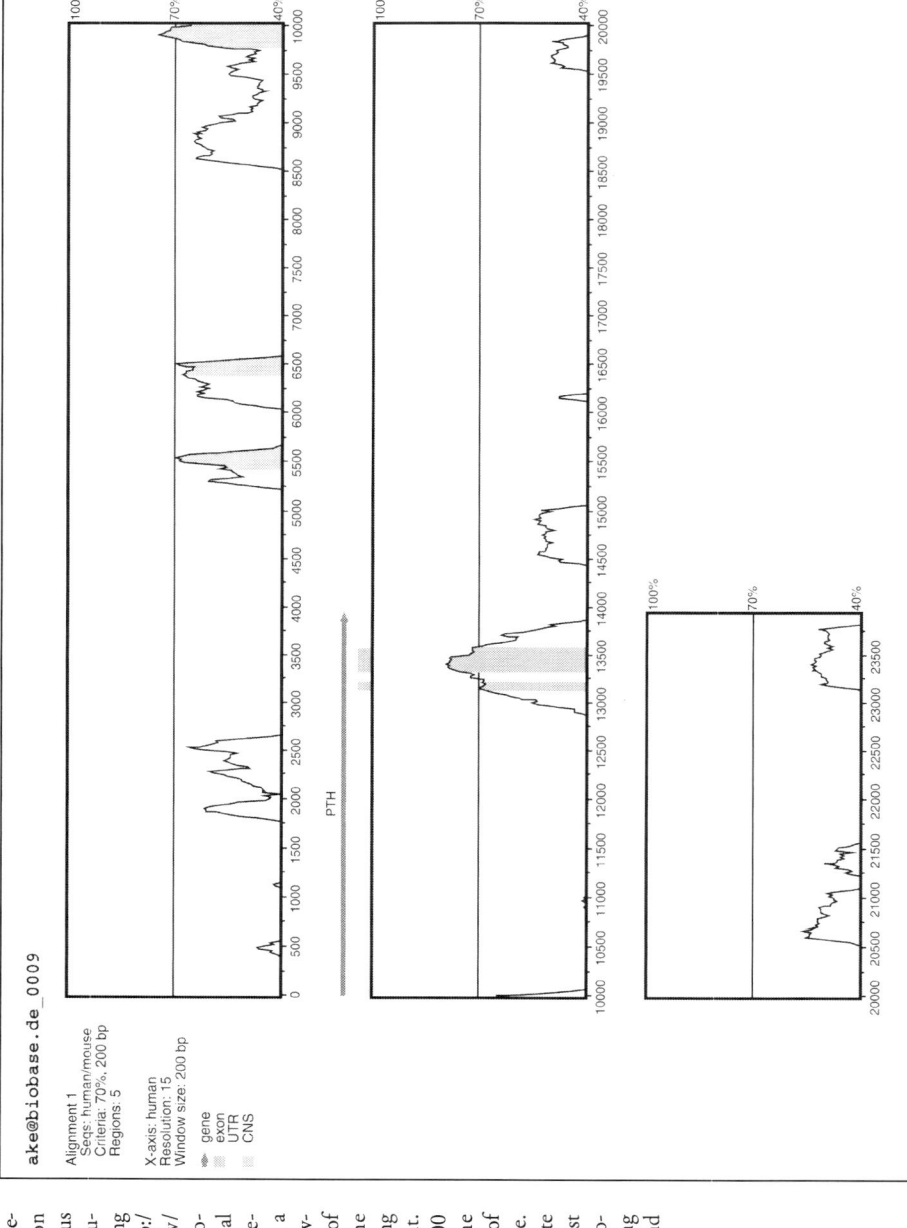

Figure 4. Large scale sequence comparison between orthologous PTH genes from human and mouse using the VISTA tool (http://www-gsd.lbl.gov/VISTA/). The program makes a global alignment and presents the results in a form of a plot showing the percentage of homology in the 200nt window sliding along the alignment. The position of 10000 corresponds to the start of transcription of the human PTH gene. The grey areas indicate regions of the highest homology that are located in the coding regions of the second and third exons.

corresponding to the UTR sequences in mRNA, as well as in the far upstream region, in 3' region of the gene and in introns. It is tempting to speculate that the high homology regions in the non coding regions can correspond to at least basic regulatory sequences that play a similar role in regulation of human and mouse PTH gene expression and therefore are conserved in mammalia during the evolution.

Computer Assisted Search for Potential Cis-Regulatory Elements in PTH Gene

To reveal TF binding sites in the human PTH gene that may contribute to the regulation of this gene we have used TRANSPLORER tool (www.biobase.de). We have searched TF binding sites in the human PTH gene sequences including 10 kb upstream and 10 kb downstream sequences. In Figure 5 the analysed sequence is represented by a ruler showing the nucleotide positions. Position 10000 corresponds to the start of transcription. The green lines above the ruler show location of the PTH gene, its exons, and protein coding regions (CDS). 'Homol_reg' lines show mouse homology regions in the 5' noncoding region that were found by the global sequence comparison in the previous section. By this analysis we were able to confirm the known CREB site in the promoter region of human PTH gene (-96, -69) (see Table 1) and the binding site for Oct factor that was found by Gruendel, H [9] and a number of additional sites for these factors.

In addition, TRANSPLORER reveals a number of new potential binding sites for different transcription factors. These are multiple sites for Sp-1—an ubiquitos activator factor; NF-ATp/c—the family of transcription factors that are activated through Ca^{++} signalling; AP-1—ubiquitous early response factors; c-Ets-1—a mitogenic transcription factor. There were reports about the role of c-Ets transcription factors in development [32] and that activation of these factors might go through Ca-dependent phosphorylation.[23] So, it is tempting to speculate that these factors can play a role in regulation of the expression of the human PTH gene. An experimental verification is necessary to confirm these sites.

We found a number of potential binding sites for transcription factors from nuclear receptor family in many different positions in the 5' region, in introns and in 3' region of the gene. However, generally, it is difficult to distinguish between binding sites for different nuclear receptors, since the DNA sequence of the binding sites is very similar. We found sites for HNF-4 (hepatocyte nuclear factor 4), ER (estrogen receptor) and ROR (RAR-related orphan receptor). Experimental analysis should be done to identify which member of this transcription factor family is involved in the regulation of PTH gene.

The suppressive effect of 1,25(OH)2 D3 on the PTH gene expression is one of the major regulator on the transcriptional level of the PTH gene. Therefore we paid additional attention to search for binding sites of VDR (vitamin-D receptor). To search for potential VDRE elements we have developed a new method based on pattern matching technique. First of all, we have compiled a representative set of known VDRE from human, mouse and rat genes. These sites have a rather complicated structure. Usually, the VDRE consist of two half-sites in different directions—short sequences with a characteristic pattern, that are separated by a 3 nucleotide spacer.[2] But often this spacer can be 4 , 6 or 9 nucleotide long, and one of the half-site can become nearly indistinguishable. That is the reason, why it is practically impossible to use positional weight matrix to recognize VDRE, therefore we applied the pattern matching technique that is used for searching composite elements.[13] We scanned the human PTH gene sequence using a compilation of known VDRE sites from different genes of human, mouse, rat and chicken. In Table 2 a list of 20 potential VDRE found in the genomic sequence of human PTH gene is presented. Surprisingly, after the first search we had not found the known VDRE in position –138 (position: –126 according to a different transcription start described in reference 5. After a careful analysis of the sequence we observed in this region a single nucleotide

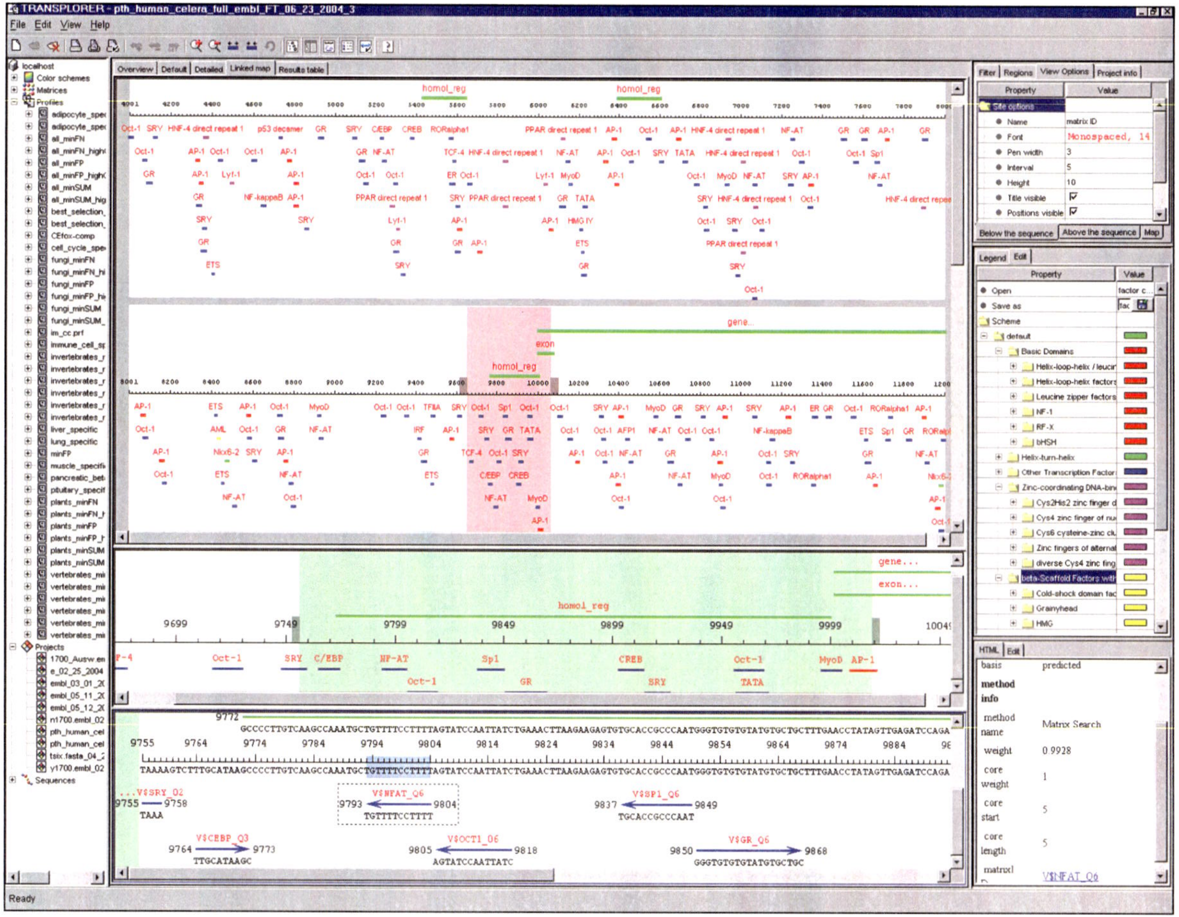

Figure 5. Results of the search of TF binding sites in the human PTH gene sequences including 10 kb upstream and 10 kb downstream sequences and introns. The analysis is done by TRANSPLORER (R) (www.biobase.de). The lines above the ruler show location of the PTH gene, its exons, and protein coding regions (CDS) as well as mouse homology regions in the five noncoding regions that were found by the global sequence comparison.

Table 2. Potential VDRE found in human genomic sequence containing PTH gene (10 kb upstream + 10 kb downstream) by pattern similarity with known VDR sites in other vertebrate genes

Position	Relative to TSS	Strand	Site ID (TRANSFAC)	No. of Mismatches Left	No. of Mismatches Right	Sequence
1747	(-8253)	(+)	HS$CYP24_02	1	1	TCATCTtcaTGCCTC
3778	(-6222)	(-)	MOUSE$CABD28_01	1	1	GAGGGAaagaAGTAGA
5034	(-4966)	(-)	RAT$OC_05	1	0	GATGAgtttGGACA
6132	(-3868)	(-)	RAT$OC_05	1	0	GGTAAgaaaGGACA
6376	(-3624)	(+)	HS$OC_03	1	1	GAGTGAacaaatGAGTGA
8401	(-1599)	(+)	HS$CYP24_01	1	1	AAACCCaatGAACCC
8414	(-1586)	(-)	MOUSE$SPP1_01	0	1	GGTTCAttgGGTTTA
8924	(-1076)	(+)	RAT$OC_05	1	1	GGTGGaccaGGAAA
9862	**(-138)**	**(+)**	**HS$PTH_02**	**0**	**0**	**GTG(T)CTGcttTGAACC**
11965	(1965)	(+)	RAT$OC_01	1	1	GAGTGAatgAGGAAA
12303	(2303)	(-)	RAT$OC_05	1	1	GATGAcagaGTACA
14835	(4835)	(-)	HS$CYP24_02	1	1	TCATCTagtTAACTC
15851	(5851)	(-)	MOUSE$CABD28_01	1	1	GGTGGAaataAGGAAA
15971	(5971)	(+)	CHICK$PTH_01	1	1	TGGTCAtttGAGTGT
17278	(7278)	(+)	CHICK$PTH_01	1	1	GGGGCAtggGGATGT
17284	(7284)	(+)	MOUSE$CABD28_01	1	1	TGGGGAtgttGGGAGA
18547	(8547)	(+)	RAT$OC_05	0	1	GGTGAgaggTGACA
19445	(9445)	(-)	HS$CYP24_02	1	1	TCATCTctcTGCCTC
19628	(9628)	(-)	MOUSE$SPP1_01	1	1	GGTTCTgcaGGCTCA
21786	(11786)	(+)	HS$P21WAF1_05	1	1	AGGGATccaGTTTCA

Known functional VDRE in this gene is shown in bold. The deletion of nucleotide T in the VDRE is shown in brackets.

difference (insertion) between the genomic sequence of the used databases and the sequence of the PTH promoter published herein.[5] This mismatch is located within the left flanking region of the VDRE and distorts this half-site. A further experimental analysis is necessary to verify whether this mutation corresponds to a SNP (single nucleotide polymorphism) in the human population. SNP of the PTH gene was observed earlier by generation of restriction fragments (restriction fragment length polymorphism).[30,18]

The region around the VDRE near the position −138 attracts our attention since in the work of Darwish and DeLuca[4] a new regulatory site was found in this region. This site is overlapping with the VDRE element and it is required for high transcriptional activity of the human PTH gene. Transcription factor that binds to this element was not identified in the paper. We have analysed this region using TRANSPLORER and found two adjacent potential binding sites as target sites for a transcription factor of NF-AT family and another for AP-1 factors (Fig. 6). It is known that these two transcription factor binding sites can be organized into so called "composite elements"—that provide the possibility for synergistic interaction of two factors binding to the adjacent sites.[15] The NF-AT/AP-1 composite element with a high score in the PTH gene can be additionally confirmed by a special program previously devel-

```
                        ---------->V$NFAT_Q6(0.66)
                                 --------->V$AP1FJ_Q2(0.75)
    CCGCCCAATGGGTGTGTGTATGTGCTGCTTTGAACCTATAGTTGAGATCCAGAGAATTGG    60
                                VDRE

Site1 (+/-) Location Weight|  Site2 (+/-) Location Weight| dist     | Sequence
---------------------------|------------------------------|----------|-----------------------
NFATp  (+) 26 .. 39  1.07  |  AP-1  (+) 36 .. 49   0.93   | dist=  8 | tgctttTGAACctaTAGTTGAgatc
```

Figure 6. Results of analysis of the region around the VDR site at position –126. With the help of TRANSPLORER we have found two adjacent potential binding sites: for a transcription factor of NF-AT family and for AP-1 factors. These two transcription factor binding sites can be organized into a composite elements that provide possibility for synergistic interaction of two factors binding. The result of the search for NF-AT/AP-1 composite elements by a special software tool (Kel et al, 1999) is sown in the bottom of the figure.

oped (Fig. 6) . NF-AT/AP-1 composite elements can be found in many genes.[13] So, it can be assumed that transcription factors of NF-AT family in cooperation with AP-1 factors can play a significant role in the regulation of PTH gene transcription.

Phylogenetic Footprint: Identification of TF Binding Sites by Comparison of Regulatory Regions of PTH Gene of Different Organisms

By comparison of human PTH gene 5' sequence with the corresponding mouse sequence we have revealed three major regions of homology (5'C (5431-5639), 5'B (6395-6599), 5'A (9772-9999)) (see above). We have analysed these three regions in more details using a new computer tool: "Phylogenetic Footprint". This tool works as follows. First, it makes an pair-vice or multiple alignment of the corresponding regions from the sequences of different organisms. Then, it provides a search for potential TF binding sites in the aligned sequences and selects only colocalized matches for the same transcription factors (or factors of the same family). The program filters out all potential sites that are found only in one sequence, and shows conservative binding sites found in both sequences. The conservation of binding sites during evolution is used as an argument for functionality.

The result of applying of the "Phylogenetic Footprint" tool to the proximal part of the human/mouse homology region 5'A is shown in Figure 7. The beginning of the fist exon in both sequences are shadowed. Three known sites (CREB, VDRE and OCT) that have been described previously for this region in human PTH gene are marked. We found that the CREB site is very conservative among human and mouse. We have found a number of other conservative sites in both sequences, among them these are the OCT site near start of transcription, a couple of NF-AT sites, SP-1 sites, GATA sites and some others.

It is interesting to mention that in the broader Phylogenetic Footprint the OCT, CREB, SP-1 and GATA sites are conserved among the promoter sequences of four different mammals: human, mouse, rat and bovine (data not shown). It seems that these sites provide modules for the basic regulation of PTH gene for all mammals, whereas the other sites are involved more in species-specific regulation.

In the more distal regions of homology: 5'C (5431-5639: -4569 to -4361), 5'B (6395-6599: -3605 to -3401) we have revealed conservative sites for the following transcription factors: NF-AT, IK-2, CREB, USF, GATA, AP-1, RAR-related orphan receptor, c-MYB, OCT, HNF3 and ER. An experimental verification should be done to confirm the role of these factors in the regulation of PTH gene transcription.

```
                                            <==========                 ...V$OCT1_06(0.88)
pth_human_T-ATAAA------------------TTCAGATTCATTAATCC------ACATA------     -279
                                            <==========V$OCT1_02(0.80)
pth_mouse_TCATGAAGCAGAGGCAGGTGGATCTCTTGAGTTTTAGCCATCCTGGTCTACATATTGAGT    -311
           *  **  **               **  ** ** *        ****        *****
================================================================================
                                ===========V$OCT1_06(0.88)
pth_human_------------------------GAATTTTTCTCGATGGTAT-------------A        -259
pth_mouse_TCCAGACCAGCCAGAGTTACATAGTGAGACTTTGTCTCAAAGATTTTTTTAAAATATAA      -251
                                  **  ***  ****  *  *                *
================================================================================
                                              OCT                       ==V$CEBPB_01(0.94)
pth_human_ATTCTGTATT--------TGTTAAAA--------------------GTCTTTGCATAAGCC    -226
                                                                       V$CEBPB_01(0.92)
pth_mouse_ACTCAGTATTATCGATTATGTCAAAAATTTCTAGGTGGACACTTAGTATTTGCTACATCT    -191
           *  **  *****           ***  ****              **  *****  *   *  *
================================================================================
                                <==========V$NFAT_Q6(0.98)
                                                         V$GATA_C(0.91)
                                                  -------V$STAT_01(0.82)
                                                  ==========...V$NFAT_Q6(0.82)
pth_human_CCTTGTCAA--------GCCAAA----TGCTGTTTTCCTTTTAGTATCCAATTATCTGAA     -178
                                <==========V$NFAT_Q6(0.87)
                                                         V$GATA_C(0.92)
                                                  -------V$STAT_01(0.82)
                                                  ==========...V$NFAT_Q6(0.82)
pth_mouse_TCTTTTGCACTTTCTTTGTCAGAACCCTGCTCATTTTCATCTACTATCTAGTTATCTGAA     -131
           *** *  *         *  **          ****  ***  *  ** **** *  *********
================================================================================
                          <==========V$SP1_Q6(0.89)
           V$NFAT_Q6(0.82)                         VDRE
pth_human_ACTTAAGAAGAGTGTGCACCGCCCAATGGGTGTGTGTATGTGCTGCTTTGAACCTATAGT     -118
                          <==========V$SP1_Q6(0.86)
           V$NFAT_Q6(0.82)
pth_mouse_ACTTTAGAGGAGTGGGCACCACCCCATGAGGGTATGT---GGCTGTTCTGATCCTGTGAT     -74
           ****  ***  *****  *****  ***  ***  ***           ****  *  ***  ***  *   *
================================================================================
                      <==============V$T3R_01(0.83)           ==========...V$NF1_Q6(0.81)
                                      V$AP1_Q2(0.91)
                                    ==========>V$CREB_Q4(0.86)
                                               =========>V$SRY_02(0.96)
                      CREB                      =========>V$TATA_C(0.93)
pth_human_TGAGATCCAGAGAATTGGGAGTGACATCAtCTGTAACAATAAAAGAGCCTCTCTTGGTAA     -58
                      <==============V$T3R_01(0.84)           ==========...V$NF1_Q6(0.81)
                                      V$AP1_Q2(0.91)
                                    ==========>V$CREB_Q4(0.86)
                                               =========>V$SRY_02(0.97)
                                                =========>V$TATA_C(0.92)
pth_mouse_TGAGAGCCAGAGAACCAGGAGTGACATCAtCCTTAACAATAAAATA-CTCCTCTTGGTGA     -15
           ***** ********   ***************   ***********   * *   ******** *
================================================================================
           ========>V$NF1_Q6(0.81)
                              <==========V$OCT1_06(0.86)
                                                      <==========...V$MYOD_Q6(0.93)
pth_human_GCAGAAGACCTATATATAAAAGTCACCATTTAAGGGGTCTGCAGTCCAATTCATCAGTTG     +2
           ========>V$NF1_Q6(0.81)
                              <==========V$OCT1_06(0.92)
                                                      <==========...V$MYOD_Q6(0.93)
pth_mouse_GCAAAAAGCCTGCATATGAAACTCAGACTTGAAGAA--CTGCAGTCCAGTTCATCAGCTG     +43
           ***  **    ***    ****  ***  ***    **  ***        *********  ********  **
================================================================================
           ==V$MYOD_Q6(0.93)
                             ==   =========>V$AP4_Q5(0.84)
pth_human_TCT---TTAGT-----TTACT-CAGCATCAGCTACTAACATACCTGAACGAAGATCTTGT    +53
                                      =========>V$AP4_Q5(0.80)
           ==V$MYOD_Q6(0.93)
pth_mouse_TCTGGTTTACTCCAGCTTACTACAGCATCAGTTTGTG-CATCCCCGAAGGATCCCCT--T    +106
           ***   *** *        *****  **********  *   *   ***  **  *** **    *
================================================================================
                   <===========V$SRY_02(0.87)
pth_human_TCTAAGACATTGTATGGTAAG
                   <===========V$SRY_02(0.88)
pth_mouse_TGAGAGTCATTGTATGGTAAG
                *   **  **************
```

Figure 7. The result of applying the Phylogenetic Footprint tool to the proximal part of the human/mouse homology region 5 A. Three sites, CREB, VDRE and OCT have been described previously for this region in human PTH gene (marked in the sequence). Conservative sites found by this analysis are shown by the arrows above each sequence.

Discussion

We performed a computer analysis of the PTH genomic sequence and have revealed a number of potential TF binding sites that may contribute to the regulation of this gene. Some of the found sites are new and some correspond to already known binding sites in the promoter of this gene in different mammalian genomes.

An early investigation of the promoter region of the human PTH gene identified several putative recognition elements in the region up to −394 upstream to the start of transcription by using a data collection of transcription factor binding sites.[34] Among these sequences one has high similarity to the consensus (5′-TGACGTCA) of the cAMP-response element (CRE) with only a single nucleotide deviation. Fusion of promoter regions that contain this element and even the element alone to the CAT reporter gene and transfection of these constructs into different cell lines showed clear stimulation of reporter gene activity in response to cAMP-level enhancing agents. Thus cAMP-responsiveness of this element was clearly shown. Furthermore DNase I protection analysis showed protein binding in a region between −98 and −69 from transcription start site which comprises this putative CRE and an adjacent CCAAT element. A good candidate for this binding protein is the CREB (cAMP-response element binding protein)—a transcription factor containing a basic leucine zipper. This conclusion was made because gel retardation assays showed competing binding of this protein to the CRE in PTH gene and to the well-characterized CRE site in the rat somatostatin promoter. In addition, a CREB site with exactly the same sequence was found in plasminogen activator gene.[26]

It was interesting to observe that this CREB site was clearly recognized by the computer analysis and a strong conserved among all studied mammalian genomes. It seems that not only the presence of this site but also its relative position in the promoter is important for the function.

Beside this cAMP-response element the human PTH gene contains a 1,25-dihydroxyvitamin D response element (VDRE) in its promoter.[5] Gel shift competition analysis clearly showed binding of the1,25-dihydroxyvitamin D receptor (VDR)—a nuclear hormone receptor—to this element. The element is located between −126 to −101 to transcription start site. It contains a reverse complementary copy of a hexamer motif GGTTCA which is similar to one of the two half-sites in the direct hexamer-repeat of the VDRE of human calbindin D gene promoter[31] and both half-sites of the VDRE of the mouse osteopontin gene promoter.[19] It should be mentioned that Demay with co-authors[5] specify a sequence for this element with a single-nucleotide deletion (upstream of the above mentioned hexamer motif) compared to the publically available sequence stored in EMBL database and GENBANK database, respectively. Functional analysis of this promoter 25 bp-element suggests a role as down-regulatory element because when it was fused upstream to a viral promoter it caused transcriptional repression in response to 1,25-dihydroxyvitamin D in GH4C1 cells. Since this repression was not observed in another cell line tested (ROS 17/2.8 cells) the requirement for additional cellular factors other than VDR was supposed for the transcriptional repression.[5] To clarify the role of this negative regulatory element in vivo, especially wether it mediates 1,25-dihydroxyvitamin D induced transcriptional repression in the context of the promoter of the PTH gene, further work is necessary.

Computer search for VDR sites confirmed the site located in the positions −126 to −101 of human PTH promoter. In the phylogenetic footprint analysis this site does not show any conservation between human and mouse sequence. The corresponding region in the mouse promoter shows many changes. There were two reports about 3 new potential VDR sites in the human promoter of PTH gene. These are elements in positions: -714 to −692,[29] -347 to −321 and −293 to −266.[9] We cannot confirm these three sites by the search of similarity to a collection of known VDR elements in other genes. The second site (-347 to −321) revealed a good similarity to the AP-1 consensus, and the third one (-293 to -266) – to the Oct consensus sequence. We have identified several other potential VDR sites in other regions of the gene (see Table 2), but no one revealed any conservation between human and mouse genomic sequence. It is clear that more analysis will be needed to reveal all sites responsible for VDR regulation of PTH gene.

Interestingly overlapping with the 3′ end of this negative regulatory element and sharing the above mentioned single copy motif (GGTTCA) there is a sequence (-115 to −96) which binds a yet unidentified factor complex.[4] The latter does not seem to contain VDR because monoclonal antibodies directed against VDR do not influence mobility of the complex in gel mobility shift assays. But some physical interactions between components of the complex and VDR may exist because removal of VDR from cellular extract used in gel retardation assay abolished formation of this DNA-binding complex.[4] We have analysed this particular region in more details and found potential NF-AT/ AP-1 composite element overlapped with the known VDR site. Such a complex "composite element" that include three adjacent binding sites for NF-AT, AP-1 and VDR factors is known in the human interleukin-2 gene.[1] It regulates the alternative switching of the gene activity by different signals going through VDR and NF-AT. It is known that NF-AT factors are activated by increasing of Ca^{2+} level. Therefore, we can propose that the similar molecular mechanism through the composite element VDRE/NF-AT/AP-1 can be implicated in the regulation of expression of PTH gene that is intimately involved in the homeostasis of calcium concentration.

The human PTH promoter also contains at least two negative calcium regulatory elements (nCARE) which are located far upstream of the transcription start site at positions −4194 to 4180 and −3354 to −3337, respectively.[21] The distal one is an inverted palindrome with two hexamer half-site seperated by 3 nucleotides (TGAGACaggGTCTCA) in contrast to the proximal one where such a symmetry is not observed (ccatTTGTGTATGCagaa). Reporter gene assay with promoter fragments show that both elements lead to decrease in transcriptional activity which is abolished when they are mutated. This transcriptional repressive activity was dependent on the extracellular Ca^{2+}-level.[21] Gel retardation analysis indicated that both elements are bound by the redox factor protein Ref-1. This is a remarkable finding in regard to the fact that Ref-1 was originally cloned as a apurinic/apyrimidinic endonuclease (AP endonuclease) with DNA repair activity.[27] A slightly later discovered other catalytic activity of Ref-1, which also was eponymous, was the redox regulation activity, more precisely the reducing activity on other transcription factors like AP-1 or NF-kappa B.[35] This later function yet suggested a role as a trancriptional auxiliary protein for Ref-1. A so far unidentified protein complex, which is present in various mammalian cell lines and different from Ref-1, probably bound to both PTH nCAREs.[20,22] The same protein complex seems to bind to the nCARE of the rat atrial natriuretic peptide gene (ANP) which interestingly shows high sequence similarity (GGAGACaggGTCTCA) to the distal and palindromic nCARE of the PTH gene.[20]

Conclusion

In silico analysis of the regulatory sequences provides a very efficient means for investigating the regulation of eukaryotic genes. We have applied a number of computer tools for searching potential TF binding sites and their organisation in the PTH gene. We found many potential binding sites for different transcription factors that may contribute to the regulation of the PTH gene. With this approach the full organisation of regulatory elements of a gene can be described and the results bring us to a principally new level of understanding of how genes are regulated during development and functioning in the organism and how genes are dis-regulated is the cases of diseases.

References

1. Alroy I, Towers TL, Freedman LP. Transcriptional repression of the interleukin-2 gene by vitamin D3: Direct inhibition of NFATp/AP-1 complex formation by a nuclear hormone receptor. Mol Cell Biol 1995; 15:5789-5799.
2. Carlberg C. "Critical analysis of 1a,25-dixydroxyvitamin D3 response element" in Vitamin D. In: Norman AW, Bouillon R, Thomasset M, eds. Chemistry, Biology and Clinical Applications of the Steroid Hormone 1997:268-275.
3. Celera database www.celera.com
4. Darwish HM, DeLuca HF. Identification of a transcription factor that binds to the promoter region of the human parathyroid hormone gene. Arch Biochem Biophys 1999; 365:123-130.
5. Demay MB, Kiernan MS, DeLuca HF et al. Sequences in the human parathyroid hormone gene that bind the 1,25-dihydroxyvitamin D3 receptor and mediate transcriptional repression in response to 1,25-dihydroxyvitamin D3. Proc Natl Acad Sci USA 1992; 89:8097-8101.
6. Englander EW, Wilson SH. Protein binding elements in the human beta-polymerase promoter. Nucleic Acids Res 1990; 18:919-928.
7. Farrow SM, Hawa NS, Karmali R et al. Binding of the receptor for 1,25-dihydroxyvitamin D3 to the 5'-flanking region of the bovine parathyroid hormone gene. J Endocrinol 1990; 126:355-359.
8. Goessling E, Kel-Margoulis OV, Kel AE et al. MATCH^{TM} - a tool for searching transcription factor binding sites in DNA sequences. Application for the analysis of human chromosomes. Proceedings of the German Conference on Bioinformatics (GCB2001), October 7-10, Braunschweig: 2001:158-160.
9. Gruendel H. Untersuchungen zur Transcriptionsregulation des humanen Parathormongens, Dissertation TU Braunschweig, Techn. Univ., Diss., 1995, 127p.
10. Hawa NS, O'Riordan JL, Farrow SM. Binding of 1,25-dihydroxyvitamin D3 receptors to the 5'-flanking region of the bovine parathyroid hormone gene. J Endocrinol 1994; 142:53-60.
11. Heinemeyer T, Chen X, Karas H et al. Expanding the TRANSFAC database towards an expert system of regulatory molecular mechanisms. Nucleic Acids Res 1999; 27(1):318-322.
12. Kanei-Ishii C, Ishii S. Dual enhancer activities of the cyclic-AMP responsive element with cell type and promoter specificity. Nucleic Acids Res 1989; 17:1521-1536.
13. Kel A, Kel-Margoulis O, Babenko V et al. Recognition of NFATp/AP-1 composite elements within genes induced upon the activation of immune cells. J Mol Biol 1999; 288:353-376.
14. Kel AE, Kel-Margoulis OV, Farnham PJ et al. Computer-assisted identification of cell cycle-related genes - new targets for E2F transcription factors. J Mol Biol 2001; 309:99-120.
15. Kel-Margoulis OV, Romashchenko AG, Kolchanov NA et al. COMPEL: A database on composite regulatory elements providing combinatorial transcription regulation. Nucleic Acids Res 2000; 28(1):311-315.
16. Koszewski NJ, Ashok S, Russell J. Turning a negative into a positive: Vitamin D receptor interactions with the avian parathyroid hormone response element. Mol Endocrinol 13:455-465.
17. Liu SM, Koszewski N, Lupez M et al. Characterization of a response element in the 5''-flanking region of the avian (chicken) PTH gene that mediates negative regulation of gene transcription by 1,25-dihydroxyvitamin D3 and binds the vitamin D3 receptor. Mol Endocrinol 1999; 10:206-215.
18. Miric A, Levine MA. Analysis of the preproPTH gene by denaturing gradient gel electrophoresis in familial isolated hypoparathyroidism. J Clin Endocrinol Metab 1992; 74:509-516.
19. Noda M, Vogel RL, Craig AM et al. Identification of a DNA sequence responsible for binding of the 1,25-dihydroxyvitamin D3 receptor and 1,25-dihydroxyvitamin D3 enhancement of mouse secreted phosphoprotein 1 (SPP-1 or osteopontin) gene expression. Proc Natl Acad Sci USA 1990; 87:9995-9999.
20. Okazaki T, Ando K, Igarashi T et al. Conserved mechanism of negative gene regulation by extracellular calcium. Parathyroid hormone gene versus atrial natriuretic polypeptide gene. J Clin Invest 1992; 89:1268-1273.
21. Okazaki T, Chung U, Nishishita T et al. A redox factor protein, ref1, is involved in negative gene regulation by extracellular calcium. J Biol Chem 1994; 269:27855-27862.
22. Okazaki T, Zajac JD, Igarashi T et al. Negative regulatory elements in the human parathyroid hormone gene. J Biol Chem 1991; 266:21903-21910.

23. Pognonec P, Boulukos KE, Gesquiere JC et al. Mitogenic stimulation of thymocytes results in the calcium-dependent phosphorylation of c-ets-1 proteins. EMBO J 1988; 7:977-983.
24. Prestridge DS. SIGNAL SCAN: A computer program that scans DNA sequences for eukaryotic transcriptional elements. CABIOS 1991; 7:203-206.
25. Quandt K, Frech K, Karas H et al. MatInd and MatInspector - New fast and versatile tools for detection of consensus matches in nucleotide sequence data. Nucleic Acids Researh 1995; 23:4878-4884.
26. Rickles RJ, Darrow AL, Strickland S. Differentiation-responsive elements in the 5' region of the mouse tissue plasminogen activator gene confer two-stage regulation by retinoic acid and cyclic AMP in teratocarcinoma cells. Mol Cell Biol 1989; 9:1691-1704.
27. Robson CN, Hickson ID. Isolation of cDNA clones encoding a human apurinic/apyrimidinic endonuclease that corrects DNA repair and mutagenesis defects in E. coli xth (exonuclease III) mutants. Nucleic Acids Res 1991; 19:5519-5523.
28. Hardison RC, Oeltjen J, Miller W. Long Human-Mouse Sequence Alignments Reveal Novel Regulatory Elements: A Reason to Sequence the Mouse Genome. Genome Research 1997; 7:959-966.
29. Rupp E, Mayer H, Wingender E. The promoter of the human parathyroid hormone gene contains a functional cyclic AMP-response element. Nucleic Acids Res 1990; 18:5677-5683.
30. Schmidtke J, Pape B, Krengel U et al. Restriction fragment length polymorphisms at the human parathyroid hormone gene locus. Hum Gent 1984; 67:428-431.
31. Schrader M, Nayeri S, Kahlen JP et al. Natural vitamin D3 response elements formed by inverted palindromes: Polarity-directed ligand sensitivity of vitamin D3 receptor-retinoid X receptor heterodimer-mediated transactivation. Mol Cell Biol 1995; 15:1154-1161.
32. Sumarsono SH, Wilson TJ, Tymms MJ et al. Down's syndrome-like skeletal abnormalities in Ets2 transgenic mice. Nature1996; 379:534-537.
33. Wasserman WW, Fickett JW. Identification of regulatory regions which confer muscle-specific gene expression. J Mol Biol 1998; 278:167-181.
34. Wingender E. Compilation of transcription regulating proteins. Nucleic Acids Res 1988; 16:1879-1902.
35. Xanthoudakis S, Miao G, Wang F et al. Redox activation of Fos-Jun DNA binding activity is mediated by a DNA repair enzyme. EMBO J 1992; 11:3323-3335.

CHAPTER 7

Regulation of Parathyroid Hormone Gene Expression by 1,25-Dihydroxyvitamin D

Tally Naveh-Many and Justin Silver

Abstract

Vitamin D's active metabolite $1,25(OH)_2D_3$ acts on the parathyroid to markedly decrease PTH gene transcription. It does this by binding to its specific receptor in the parathyroid which then binds to a defined sequence, the vitamin D response element (VDRE) in the parathyroid hormone (PTH) gene promoter. Retinoic acid amplifies the effect of $1,25(OH)_2D_3$ to decrease PTH mRNA levels suggesting that a VDR-RXR heterodimer binds to the VDRE. $1,25(OH)_2D_3$ may amplify its effect on the PTH gene by increasing the concentration of the VDR and the calcium receptor (CaR) in the parathyroid. Calreticulin prevents the binding of the VDR to the VDRE. The effect of $1,25(OH)_2D_3$ to decrease the synthesis and secretion of PTH is used therapeutically to prevent the secondary hyperparathyroidism of patients with chronic renal failure.

Transcriptional Regulation of the PTH Gene by $1,25(OH)_2D_3$

The action of $1,25(OH)_2D_3$ or its analogues to decrease PTH secretion is now a well-established axiom in clinical medicine for the suppression of the secondary hyperparathyroidism of patients in chronic renal failure. So much so, that it is worthwhile to reflect upon its scientific basis. That is the purpose of the present review. There have been more recent developments spurred by pharmaceutical companies to discover drugs that have more selective actions on the parathyroid.

PTH regulates serum concentrations of calcium and phosphate, which, in turn, regulate the synthesis and secretion of PTH. 1,25-dihydroxyvitamin D ($1,25(OH)_2D_3$) or calcitriol has independent effects on calcium and phosphate levels, and also participates in a well defined feedback loop between calcitriol and PTH. PTH increases the renal synthesis of calcitriol. Calcitriol then increases blood calcium largely by increasing the efficiency of intestinal calcium absorption. Calcitriol also potently decreases the transcription of the PTH gene. This action was first demonstrated in vitro in bovine parathyroid cells in primary culture, where calcitriol led to a marked decrease in PTH mRNA levels[1,2] and a consequent decrease in PTH secretion[3-5] The physiological relevance of these findings was established by in vivo studies in rats.[6] The localization of the VDR mRNA to the parathyroids was demonstrated by in situ hybridization studies of the thyro-parathyroid and duodenum. VDR mRNA was localized to the parathyroids in the same concentration as in the duodenum, calcitriol's classic target organ (Fig. 1).[7] Rats injected with amounts of calcitriol which did not increase serum calcium had marked decreases in PTH mRNA levels, reaching <4% of control at 48 hours (Fig. 2). There

---

Molecular Biology of the Parathyroid, edited by Tally Naveh-Many. ©2005 Eurekah.com and Kluwer Academic / Plenum Publishers.

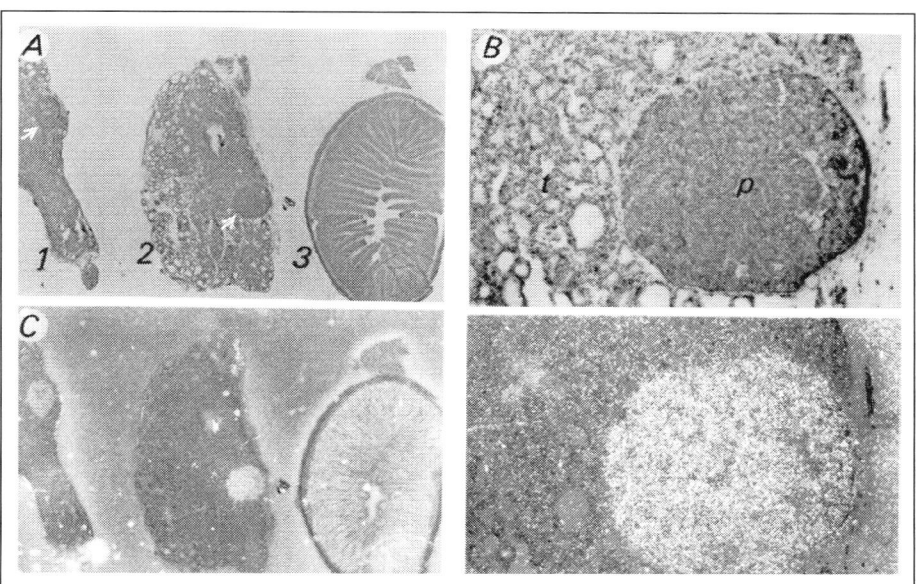

Figure 1. The 1,25(OH)$_2$vitamin D receptor (VDR) is localized to the parathyroid in a similar concentration to that found in the duodenum indicating that the parathyroid is a physiological target organ for 1,25(OH)$_2$D. In situ hybridization with the VDR probe in rat parathyroid-thyroid and duodenum sections. A1) Parathyroid-thyroid tissue from a control rat. A2) Parathyroid-thyroid from a 1,25(OH)$_2$D$_3$ treated rat (100 pmol at 24 h). A3) Duodenum from the 1,25(OH)$_2$D$_3$ treated rat. The white arrows point at the parathyroid glands. B) A higher power view of A2 that shows the parathyroid gland (p) and thyroid follicles (t). Top figures were photographed under bright field illumination, whereas bottom figures show dark-field illumination of the same sections. (Reproduced with permission of the American Society of Clinical Investigation.[7])

was also a decrease in calcitonin mRNA levels in these rats.[8] The effect of 1,25(OH)$_2$D$_3$ on the PTH and calcitonin genes was shown to be transcriptional both in in vivo studies in rats[6] and in in vitro studies with primary cultures of bovine parathyroid cells.[9] Interestingly, in rats given large doses of vitamin D with a resultant hypercalcemia there was still a decrease in calcitonin mRNA levels despite the elevated serum calcium which is a secretagogue for calcitonin.[10] When 684 base pairs of the 5'-flanking region of the human PTH gene was linked to a reporter gene and transfected into a rat pituitary cell line (GH4C1), gene expression was lowered by 1,25(OH)$_2$D$_3$.[11] These studies suggest that 1,25(OH)$_2$D$_3$ decreases PTH transcription by acting on the 5'-flanking region of the PTH gene. The effect of 1,25(OH)$_2$D$_3$ may involve heterodimerization with the retinoid acid recptor. This is because, 9 cis-retinoic acid, which binds to the retinoic acid receptor, when added to bovine parathyroid cells in primary culture, led to a decrease in PTH mRNA levels.[12] Moreover, combined treatment with 1×10^{-6} M retinoic acid and 1×10^{-8} M 1,25(OH)$_2$D$_3$ more effectively decreased PTH secretion and preproPTH mRNA than did either compound alone.[12] Alternatively, retinoic acid receptors might synergize with VDRs through actions on distinct sequences.

1,25(OH)$_2$D$_3$ negatively regulates expression of the avian PTH (aPTH) gene transcript, and Liu et al identified a vitamin D response element (VDRE) near the promoter of the aPTH gene.[13] Koszewski et al converted the negative activity imparted by the aPTH VDRE to a positive transcriptional response through selective mutations introduced into the element.[14]

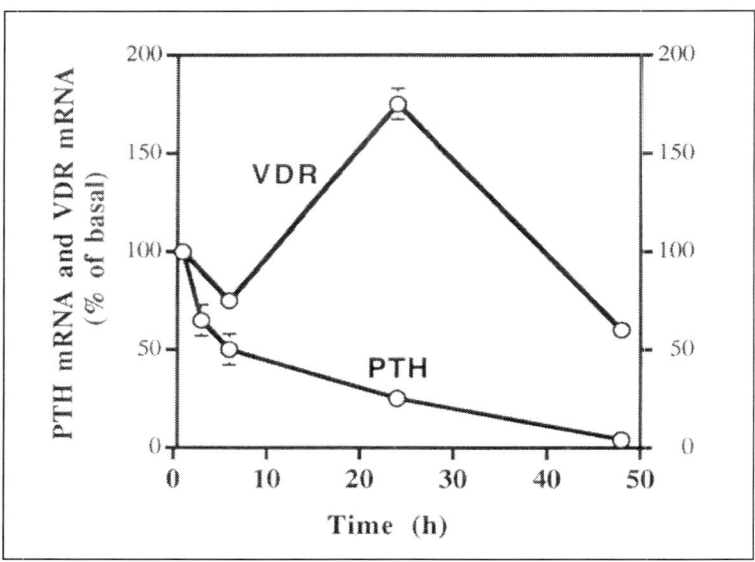

Figure 2. Time course for the effect of 1,25(OH)$_2$D$_3$ on mRNA levels for PTH and the 1,25(OH)$_2$D$_3$ receptor (VDR) in rat thyroparathyroidglands. Rats were injected with either a single dose of 100pmol 1,25(OH)$_2$D$_3$, or 50pmol 1,25(OH)$_2$D$_3$ at 0 and 24h. The arrow represents the second injection of 1,25(OH)$_2$D$_3$. The data represents the mean±SE for four rats. (From Naveh-Many et al.[7] By copyright permission of the American Society for Clinical Investigation.)

The tested sequences were derived from individual and combined mutations to 2 bp in the 3'-half of the direct repeat element, GGGTCAggaGGGTGT. Cold competition experiments using mutant and wild-type oligonucleotides in the mobility shift assay revealed minor differences in the ability of any of these sequences to compete for binding to a heterodimer complex comprised of recombinant proteins. Ethylation interference footprint analysis for each of the mutants produced unique patterns over the 3'-half-sites that were distinct from the weak, wild-type footprint. Transcriptional outcomes evaluated from a chloramphenicol acetyltransferase reporter construct utilizing the aPTH promoter found that the individual T→A mutant produced an attenuated negative transcriptional response while the G→C mutant resulted in a reproducibly weak positive transcriptional outcome. The double mutant, however, yielded a 4-fold increase in transcription, similar to the 7-fold increase observed from an analogous construct using the human osteocalcin VDRE. UV light crosslinking to gapped oligonucleotides assessed the polarity of heterodimer binding to the wild-type and double mutant sequences and was consistent with the vitamin D receptor preferentially binding to the 5'-half of both elements. Finally, DNA affinity chromatography was used to immobilize heterodimer complexes bound to the wild-type and double mutant sequences as bait to identify proteins that may preferentially interact with these DNA-bound heterodimers. This analysis revealed the presence of a p160 protein that specifically interacted with the heterodimer bound to the wild-type VDRE, but was absent from complexes bound to response elements associated with positive transcriptional activity. Thus, the sequence of the individual VDRE appears to play an active role in dictating transcriptional responses that may be mediated by altering the ability of a vitamin D receptor heterodimer to interact with accessory factor proteins. Darwish et al identified a transcription factor that binds to the promoter region of the human PTH gene adjacent to the negative vitamin D responsive element (VDRE).[15] Deletion and mutation analysis revealed that the binding site for this factor overlapped with the proximal repeat element of the VDRE. It includes additional nucleotides at the 3' end of the VDRE. This site has the sequence TTTGAACCTATAGTTGAGAT and a core sequence TGAACCTAT needed for binding of the factor. Experiments with specific anti-vitamin D receptor (VDR) antibodies demonstrated that VDR is not found in the factor/DNA complex. However, removing the VDR from the nuclear extract by immunoprecipitation eliminated the binding complex, and the addition of recombinant VDR to the depleted extract did not restore the factor's ability to bind to the DNA, suggesting that the factor and VDR are closely associated. Transfection experiments with various reporter constructs indicated that the factor is required for the high transcriptional activity of the human PTH gene. This high activity is significantly suppressed by $1,25(OH)_2D_3$. This factor is expressed in several cell types including rat osteoblasts and pituitary.

Kimmel-Jehan et al have shown that the vitamin D receptor-retinoid X receptor (VDR-RXR) heterodimers induces a DNA bend upon binding to various vitamin D response elements (VDRE) by circular permutation and phasing analysis.[16] The VDREs used included the hPTH gene. As shown by circular permutation analysis, VDR-RXR induced a distortion in DNA fragments containing various VDREs. The distortions took place with or without a $1,25-(OH)_2D_3$ ligand. The centers of the apparent bend were found in the vicinity of the midpoint of the VDRE. Phasing analysis revealed that VDR-RXR heterodimers induced a directed bend of 26 degrees, not influenced by the presence of hormone. Therefore, similar to other members of the steroid and thyroid nuclear receptor superfamily, VDR-RXR heterodimers induce DNA bending.

A further level at which $1,25(OH)_2D_3$ might regulate the PTH gene would be at the level of the $1,25(OH)_2D_3$ receptor. $1,25(OH)_2D_3$ acts on its target tissues by binding to the $1,25(OH)_2D_3$ receptor, which regulates the transcription of genes with the appropriate recognition sequences. The concentration of the $1,25(OH)_2D_3$ receptor in the $1,25(OH)_2D_3$ target sites could allow a modulation of the $1,25(OH)_2D_3$ effect, with an increase in receptor concentration leading to an amplification of its effect and a decrease in receptor concentration dampening the $1,25(OH)_2D_3$ effect.

Naveh-Many et al[7] injected $1,25(OH)_2D_3$ to rats and measured the levels of the $1,25(OH)_2D_3$ receptor mRNA (VDRmRNA) and PTHmRNA in the parathyro-thyroid tissue. They showed that $1,25(OH)_2D_3$ in physiologically relevant doses led to an increase in VDRmRNA levels in the parathyroid glands in contrast to the decrease in PTH mRNA levels (Fig. 2). This increase in VDR mRNA occurred after a time lag of 6 h, and a dose response showed a peak at 25 pmol. Weanling rats fed a diet deficient in calcium were markedly hypocalcemic at 3 weeks and had very high serum $1,25(OH)_2D_3$ levels. Despite the chronically high serum $1,25(OH)_2D_3$ levels there was no increase in VDR mRNA levels, and furthermore PTH mRNA levels did not fall and were markedly increased. The low calcium may have prevented the increase in parathyroid VDR levels and this may partially explain the PTH mRNA suppression. Whatever the mechanism, the lack of suppression of PTH synthesis in the setting of hypocalcemia and increased serum $1,25(OH)_2D_3$ is crucial physiologically, because it allows an increase in both PTH and $1,25(OH)_2D_3$ at a time of chronic hypocalcemic stress. Russell et al[17] studied the parathyroids of chicks with vitamin D deficiency and confirmed that $1,25(OH)_2D_3$ regulates PTH and VDR gene expression in the avian parathyroid gland. The chicks in this study were fed a vitamin D deficient diet from birth for 21 days and had established secondary hyperparathyroidism. These hypocalcemic chicks were then fed a diet with different calcium contents (0.5, 1.0 and 1.6%) for 6 days. The serum calciums were all still low (5, 6 and 7 mg/dl) with the expected inverse relationship between PTH mRNA and serum calcium. There was also a direct relationship between serum calcium and VDR mRNA levels.

This result suggests either that VDR mRNA was not upregulated in the setting of secondary hyperparathroidism or that calcium directly regulates the VDR gene. Garfia et al[18] injected a small dose of $1,25(OH)_2D_3$ to hypercalcemic rats to match the serum $1,25(OH)_2D_3$ levels of hypocalcemic rats. Parathyroid gland VDR mRNA and VDR protein were increased in hypercalcemic rats as compared with hypocalcemic rats. Increasing doses of $1,25(OH)_2D_3$ upregulated VDR mRNA and VDR only in hypercalcemic rats. Additional experiments showed that the decrease in VDR in hypocalcemic rats prevented the inhibitory effect of $1,25(OH)_2D_3$ on PTH mRNA. They concluded that extracellular Ca regulates VDR expression by parathyroid cells independently of $1,25(OH)_2D_3$ and that by this mechanism hypocalcemia may help prevent the feedback of $1,25(OH)_2D_3$ on the parathyroids. Brown et al[19] studied vitamin D deficient rats and confirmed that $1,25(OH)_2D_3$ upregulated the parathyroid VDR mRNA and that in secondary hyperparathyroidism with hypocalcemia the PTH mRNA was upregulated without change in the VDR mRNA.[7] All these studies show that $1,25(OH)_2D_3$ increases the expression of its receptor's gene in the parathyroid gland, which would result in increased VDR protein synthesis and increased binding of $1,25(OH)_2D_3$ (Fig. 2). This ligand-dependent receptor upregulation would lead to an amplified effect of $1,25(OH)_2D_3$ on the PTH gene, and might help explain the dramatic effect of $1,25(OH)_2D_3$ on the PTH gene.

Koszewski et al studied by interference footprinting protocols the interactions of the vitamin D receptor (VDR) with either a positive or a negative VDRE.[20] A sequence from the human osteocalcin (hOC) gene was chosen for the prototypical positive VDRE, while an analogous sequence linked to the avian parathyroid hormone gene (aPTH) was used as the negative VDRE. Both types of response elements were examined for phosphate backbone contacts, as well as base-specific interactions with guanine and thymine residues. Sources of VDR included partially purified canine intestinal preparations, as well as extracts of recombinant human VDR and retinoid X receptor alpha prepared from baculovirus-infected Sf9 insect cells. Cold competition experiments using variable amounts of these oligonucleotides in the mobility shift assay revealed that the hOC element was a five-fold better competitor for heterodimer complex binding than the negative VDRE. Interference footprints revealed extensive strong contacts to the phosphate backbone and individual guanine and thymine nucleotides of the hOC element. The composite hOC footprint was asymmetric for the number and strength of interactions observed over each of the respective direct repeat half-sites. In contrast, the aPTH VDRE footprints revealed fewer points of DNA contact that were limited to the hexanucleotide repeat regions and were strikingly weaker in nature. The alignment of DNA contact points for both elements produced a 5' stagger that was indicative of successive major groove interactions, and consistent with dimer binding. DNA helical representations indicate that the heterodimer contacts to these response elements are substantially different and provide insight into functional aspects of each complex.

To determine what phenotypic abnormalities observed in vitamin D receptor (VDR)-ablated mice are secondary to impaired intestinal calcium absorption rather than receptor deficiency, Li et al normalized mineral ion levels by dietary means.[21] VDR-ablated mice and control littermates were fed a diet rich in calcium lactate that has been shown to prevent secondary hyperparathyroidism in vitamin D-deficient rats. This diet normalized growth and random serum ionized calcium levels in the VDR- ablated mice. The correction of ionized calcium levels prevented the development of parathyroid hyperplasia and the increases in PTH mRNA synthesis and in serum PTH levels. VDR-ablated animals fed this diet did not develop rickets or osteomalacia. However, alopecia was still observed in the VDR-ablated mice with normal mineral ions, suggesting that the VDR is required for normal hair growth. This study demonstrates that normalization of mineral ion homeostasis can prevent the development of hyperparathyroidism, osteomalacia, and rickets in the absence of the genomic actions of 1,25-dihydroxyvitamin D_3. Van Cromphaut et al[22,23] have also generated mice with deletions of the VDR and showed that the secondary hyperparathyroidism of these VDR-KO mice could be corrected by a high calcium diet.

Vitamin D may also amplify its effect on the parathyroid by increasing the activity of the calcium receptor (CaR). Canaff et al showed that in fact there are VDREs in the human CaR's promoter.[24] The calcium-sensing receptor (CaR), expressed in parathyroid chief cells, thyroid C-cells, and cells of the kidney tubule, is essential for maintenance of calcium homeostasis. They showed that parathyroid, thyroid, and kidney CaR mRNA levels increased 2-fold at 15 h after intraperitoneal injection of $1,25(OH)_2D_3$ in rats. Human thyroid C-cell (TT) and kidney proximal tubule cell (HKC) CaR gene transcription increased approximately 2-fold at 8 and 12 h after $1,25(OH)_2D_3$ treatment. The human CaR gene has two promoters yielding alternative transcripts containing either exon 1A or exon 1B 5'-untranslated region sequences that splice to exon 2 some 242 bp before the ATG translation start site. Transcriptional start sites were identified in parathyroid gland and TT cells; that for promoter P1 lies 27 bp downstream of a TATA box, whereas that for promoter P2, which lacks a TATA box, lies in a GC-rich region. In HKC cells, transcriptional activity of a P1 reporter gene construct was 11-fold and of P2 was 33-fold above basal levels. 10^{-8} M $1,25(OH)_2D_3$ stimulated P1 activity 2-fold and P2 activity 2.5-fold. Vitamin D response elements (VDREs), in which half-sites (6 bp) are separated by three nucleotides, were identified in both promoters and shown to confer $1,25(OH)_2D_3$ responsiveness to a heterologous promoter. This responsiveness was lost when the VDREs were mutated. In electrophoretic mobility shift assays with either in vitro transcribed/translated vitamin D receptor and retinoid X receptor-alpha, or HKC nuclear extract, specific protein-DNA complexes were formed in the presence of $1,25(OH)_2D_3$ on oligonucleotides representing the P1 and P2 VDREs. In summary, functional VDREs have been identified in the CaR gene and provide the mechanism whereby $1,25(OH)_2D_3$ up-regulates parathyroid, thyroid C-cell, and kidney CASR expression.

The use of calcitriol is limited by its hypercalcemic effect, and therefore a number of calcitriol analogs have been synthesized which are biologically active but are less hypercalcemic than calcitriol. The ability of calcitriol to decrease PTH gene transcription is used therapeutically in the management of patients with chronic renal failure. They are treated with calcitriol in order to prevent the secondary hyperparathyroidism of chronic renal failure. The poor response in some patients who do not respond, may well result from poor control of serum phosphate, decreased vitamin D receptor concentration,[25] an inhibitory effect of a uremic toxin(s) on VDR-VDRE binding.[26] or tertiary hyperparathyroidism with monoclonal parathyroid tumors.[27]

Patel et al[26] have studied the mechanism of the resistance to the action of calcitriol in chronic renal failure. They used the electrophoretic mobility shift assay to compare the ability of VDRs from normal and renal failure rats to bind to the osteocalcin gene VDRE. VDRs from renal failure rats had only half the DNA binding capacity as VDRs from control rats, despite identical calcitriol binding. Furthermore, incubation of normal VDRs with a uremic plasma ultrafiltrate resulted in a loss of > 50% of the binding sites for the osteocalcin VDRE. The inhibitory effect of the uremic ultrafiltrate was due to a specific interaction with the VDR, not retinoid X receptors. They concluded that an inhibitory effect of a uremic toxin(s) on VDR-VDRE binding could underlie the calcitriol resistance of renal failure.

Calreticulin and the Action of $1,25(OH)_2D_3$ on the PTH Gene

Another possible level at which $1,25(OH)_2D_3$ might regulate PTH gene expression involves calreticulin. Calreticulin is a calcium binding protein which is present in the endoplasmic reticulum of the cell, and also may have a nuclear function. It regulates gene transcription via its ability to bind a protein motif in the DNA-binding domain of nuclear hormone receptors of sterol hormones. It has been shown to prevent vitamin D's binding and action on the osteocalcin gene in vitro.[28] Sela-Brown et al showed that calreticulin might inhibit vitamin D's action on the PTH gene.[29] Both rat and chicken VDRE sequences of the PTH gene were incubated with recombinant VDR and retinoic acid receptor (RXR) proteins in a gel retardation assay and showed a clear retarded band. Purified calreticulin inhibited binding of the VDR-RXR complex to the VDREs in gel retardation assays. This inhibition was due to direct protein- protein interactions between the VDR and calreticulin. Opossum kidney (OK) cells were transiently cotransfected with calreticulin expression vectors (sense and antisense) and either rat or chicken PTH gene promoter-CAT constructs. The cells were then assayed for $1,25(OH)_2D_3$-induced CAT gene expression. $1,25(OH)_2D_3$ decreased PTH promoter-CAT transcription. Cotransfection with sense calreticulin, which increases calreticulin protein levels, completely inhibited the effect of $1,25(OH)_2D_3$ on the PTH promoters of both rat and chicken. Cotransfection with the antisense calreticulin construct did not interfere with vitamin D's effect on PTH gene transcription. Sense calreticulin expression had no effect on basal CAT mRNA levels. In order to determine a physiological role for calreticulin in the regulation of the PTH gene, the levels of calreticulin protein were determined in the nuclear fraction of rat parathyroids. The rats were fed either a control diet or a low calcium diet, which leads to increased PTH mRNA levels despite high serum $1,25(OH)_2D_3$ levels that would be expected to inhibit PTH gene transcription.[29] It was postulated that high calreticulin levels in the nuclear fraction would prevent the effect of $1,25(OH)_2D_3$ on the PTH gene. In fact, the hypocalcemic rats had increased levels of calreticulin protein, as measured by Western blots, in their parathyroid nuclear faction. This may help explain why hypocalcemia leads to increased PTH gene expression despite high serum $1,25(OH)_2D_3$ levels, and may also be relevant to the refractoriness of the secondary hyperparathyroidism of many chronic renal failure patients to $1,25(OH)_2D_3$ treatment. These studies, therefore, indicate a role for calreticulin in regulating the effect of vitamin D on the PTH gene, and suggest a physiological relevance to these studies.[29]

PTH Degradation

A further level of control of serum PTH is at the level of PTH degradation. Preproparathyroid hormone (prepro-PTH) is abundantly synthesized by parathyroid chief cells; yet under normal growth conditions, little or no prepro-PTH can be detected in these cells. The addition of proteasome inhibitors to primary cultures of bovine PT cells caused the accumulation of prepro-PTH and pro-PTH.[30] Proteasome-mediated degradation of PTH precursors therefore may be important in the regulation of the levels of these precursors and hence PTH secretion. However, it is not known whether calcium or vitamin D regulate this process. PTH may be degraded in the parathyroid to carboxy and amino terminal fragments in both the parathyroid as well as in other organs such as the liver and kidney.[31] In the situation of hypercalcemia as much as 90% of the synthesized PTH may be degraded. Enzymes that are involved include furin[32] and protein convertase 1, 2 and 7 which are all expressed in the PT.[33] However, both calcium and $1,25(OH)_2D_3$ did not regulate furin or protein convertase 7 mRNA levels.[33]

Secondary Hyperparathyroidism and Parathyroid Cell Proliferation

Chronic changes in the physiological milieu often lead to both changes in parathyroid cell proliferation and PTH gene regulation, as discussed by Silver et al.[34] In such complicated settings, the regulation of PTH gene expression may well be controlled by mechanisms that differ from those in non-proliferating cells. Further, the effects of change in cell number and activity of individual cells can be complicated and difficult to dissect. Nevertheless, such chronic changes represent commonly observed clinical circumstances that require examination. The expression and regulation of the PTH gene has been studied in two models of secondary hyperparathyroidism: (1) rats with experimental uremia due to 5/6 nephrectomy and (2) rats with nutritional secondary hyperparathyroidism due to diets deficient in vitamin D and/or calcium.

Rats with 5/6 nephrectomy had higher serum creatinine levels and also appreciably higher levels of parathyroid gland PTH mRNA.[35] Their PTH mRNA levels decreased after single injections of $1,25(OH)_2D_3$, a response similar to that of normal rats.[35] Interestingly, the secondary hyperparathyroidism is characterized by an increase in parathyroid gland PTH mRNA but not in VDR mRNA. This suggests that in 5/6 nephrectomy rats there is relatively less VDR mRNA per parathyroid cell, or a relative down-regulation of the VDR, as has been reported in VDR binding studies.

The second model of experimental secondary hyperparathyroidism studied was that due to dietary deficiency of vitamin D (-D) and/or calcium (-Ca), as compared to normal vitamin D (ND) and normal calcium (NCa).[36] These dietary regimes were selected to mimic the secondary hyperparathyroidism in which the stimuli for the production of hyperparathyroidism are the low serum levels of $1,25(OH)_2D_3$ and ionized calcium. Weanling rats were maintained on the diets for 3 weeks and then studied. Rats on diets deficient in both vitamin D and calcium (-D, -Ca) exhibited a 10-fold increase in PTH mRNA as compared to controls (ND, NCa) together with much lower serum calcium levels and also lower serum $1,25(OH)_2D_3$ levels. Calcium deficiency alone (-Ca, ND) led to a 5-fold increase in PTH mRNA levels, whereas a diet deficient in vitamin D alone (-D, NCa) led to a 2-fold increase in PTH mRNA levels.

Because renal failure and prolonged changes in blood calcium and $1,25(OH)_2D_3$ can affect both parathyroid cell number and the activity of each parathyroid cell, the change in both these parameters must be assessed in each model in order to understand the various mechanisms of secondary hyperparathyroidism. Parathyroid cell number was determined in thyroparathyroid tissue of normal rats and -D, -Ca rats. To do this, the tissue was enzymatically digested into an isolated cell population, which was then passed through a flow cytometer [fluorescence-activated cell sorter (FACS)] and separated by size into two peaks. The first peak of smaller cells contained parathyroid cells as determined by the presence of PTH mRNA, and the second peak contained thyroid follicular cells and calcitonin-producing cells that hybridized positively for thyroglobulin mRNA and calcitonin mRNA but not PTH mRNA. There was a 1.6-fold increase in cells from the -D, -Ca rats than from the normal rats, and a 10-fold increase in PTH mRNA. Therefore, this model of secondary hyperparathyroidism is characterized by increased gene expression per cell, together with a smaller increase in cell number.

Further studies by Naveh-Many et al[37] have clearly demonstrated that hypocalcemia is a stimulus for parathyroid cell proliferation. They studied parathyroid cell proliferation by staining for proliferating cell nuclear antigen (PCNA) and found that a low calcium diet led to increased levels of PTH mRNA and a 10-fold increase in parathyroid cell proliferation. The secondary hyperparathyroidism of 5/6 nephrectomized rats was characterized by an increase in both PTH mRNA levels and PCNA-positive parathyroid cells. Therefore, both hypocalcemia and uremia induce parathyroid cell proliferation in vivo. The effect of $1,25(OH)_2D_3$ on parathyroid cell proliferation was also studied in this dietary model of secondary hyperparathyroidism. $1,25(OH)_2D_3$ at a dose (25 pmol for 3 days) that lowered PTH mRNA levels had no effect on the number of PCNA-positive cells. Higher doses (100 pmol for 7 days) dramatically decreased the the number of proliferating cells (unpublished). These findings emphasize the importance of a normal calcium in the prevention of parathyroid cell hyperplasia. The importance of the CaR to PT cell proliferation is also evident in that the calcimimetic NPS R-568 largely prevents the PT cell proliferation in rats with experimental uremia.[38,39] However, the role of the CaR is not that clear in view of the interesting findings of Lewin et al.[40] Experimental severe secondary hyperparathyroidism (HPT) is reversed within 1 wk after reversal of uremia by an isogenic kidney transplantation in uremic rats. In view of the reports that abnormal

PTH secretion in uremia is related to down-regulation of CaR[41] and vitamin D receptor (VDR) in the parathyroid glands,[42] they studied the expression of CaR and VDR genes after reversal of uremia and hyperparathyroidism in rats given isogenic kidney transplantation. After kidney transplantation previously uremic rats, the secondary hyperparathyroidism was reversed with normal serum PTH levels. However, both CaR mRNA and VDR mRNA remained severely reduced (CaR, 39 +/- 7%; VDR, 9 +/- 3%; $P < 0.01$) compared with normal rats. In conclusion, circulating plasma PTH levels normalized rapidly after kidney transplantation, despite persisting down-regulation of CaR and VDR gene expression. This indicates that up-regulation of CaR mRNA and VDR mRNA is not necessary to induce the rapid normalization of PTH secretion from hyperplastic parathyroid glands. In addition, Imanishi et al created transgenic mice with the cyclin D1 gene specifically targeted to the parathyroid.[43] In the parathyroids of these rats with hyperparathyroidism there was a down-regulation of the CaR. These results indicate that the changes in the CaR may be secondary to the proliferative state and not causative. In patients with both primary and nodular secondary hyperparathyroidism due to chronic renal failure there is a decrease in VDR mRNA and protein levels.[25,44,45] In hyperparathyroidism there is a decrease in the cyclin kinase inhibitors p21 and p27 with an increase in TGFα in the parathyroids.[45-47] Treatment with vitamin D metabolites increase p21 levels and prevent the decrease in TGFα levels and prevent the parathyroid cell proliferation.

Conclusion

PTH gene expression is powerfully regulated by $1,25(OH)_2D_3$. This is a transcriptional effect and results in a marked decrease in PTH secretion and serum PTH. The effect of $1,25(OH)_2D_3$ on the parathyroid is used in the treatment of many patients in chronic renal failure to prevent secondary hyperparathyroidism.

References

1. Silver J, Russell J, Sherwood LM. Regulation by vitamin D metabolites of messenger ribonucleic acid for preproparathyroid hormone in isolated bovine parathyroid cells. Proc Natl Acad Sci USA 1985; 82:4270-4273.
2. Russell J, Silver J, Sherwood LM. The effects of calcium and vitamin D metabolites on cytoplasmic mRNA coding for pre-proparathyroid hormone in isolated parathyroid cells. Trans Assoc Am Physicians 1984; 97:296-303.
3. Cantley LK, Russell J, Lettieri D et al. 1,25-Dihydroxyvitamin D3 suppresses parathyroid hormone secretion from bovine parathyroid cells in tissue culture. Endocrinol 1985; 117:2114-2119.
4. Karmali R, Farrow S, Hewison M, Barker S, O'Riordan JL. Effects of 1,25-dihydroxyvitamin D3 and cortisol on bovine and human parathyroid cells. J Endocrinol 1989; 123:137-142.
5. Chan YL, McKay C, Dye E et al. The effect of 1,25 dihydroxycholecalciferol on parathyroid hormone secretion by monolayer cultures of bovine parathyroid cells. Calcif Tissue Int 1986; 38:27-32.
6. Silver J, Naveh-Many T, Mayer H et al. Regulation by vitamin D metabolites of parathyroid hormone gene transcription in vivo in the rat. J Clin Invest 1986; 78:1296-1301.
7. Naveh-Many T, Marx R, Keshet E et al. Regulation of 1,25-dihydroxyvitamin D3 receptor gene expression by 1,25-dihydroxyvitamin D3 in the parathyroid in vivo. J Clin Invest 1990; 86:1968-1975.
8. Naveh-Many T, Silver J. Regulation of calcitonin gene transcription by vitamin D metabolites in vivo in the rat. J Clin Invest 1988; 81:270-273.
9. Russell J, Lettieri D, Sherwood LM. Suppression by 1,25(OH)2D3 of transcription of the pre-proparathyroid hormone gene. Endocrinol 1986; 119:2864-2866.
10. Fernandez-Santos JM, Utrilla JC, Conde E et al. Decrease in calcitonin and parathyroid hormone mRNA levels and hormone secretion under long-term hypervitaminosis D3 in rats. Histol Histopathol 2001; 16:407-414.

11. Okazaki T, Igarashi T, Kronenberg HM. 5'-flanking region of the parathyroid hormone gene mediates negative regulation by 1,25-(OH)2 vitamin D3. J Biol Chem 1988; 263:2203-2208.
12. MacDonald PN, Ritter C, Brown AJ et al. Retinoic acid suppresses parathyroid hormone (PTH) secretion and PreproPTH mRNA levels in bovine parathyroid cell culture. J Clin Invest 1994; 93:725-730.
13. Liu SM, Koszewski N, Lupez M et al. Characterization of a response element in the 5'-flanking region of the avian (chicken) parathyroid hormone gene that mediates negative regulation of gene transcription by 1,25-dihydroxyvitamin D_3 and binds the vitamin D_3 receptor. Mol Endocrinol 1996; 10:206-215.
14. Koszewski NJ, Ashok S, Russell J. Turning a negative into a positive: vitamin D receptor interactions with the avian parathyroid hormone response element. Mol Endocrinol 1999; 13:455-465..
15. Darwish HM, DeLuca HF. Identification of a transcription factor that binds to the promoter region of the human parathyroid hormone gene. Arch Biochem Biophys 1999; 365:123-130.
16. Kimmel-Jehan C, Darwish HM, Strugnell SA et al. DNA bending is induced by binding of vitamin D receptor-retinoid X receptor heterodimers to vitamin D response elements. J Cell Biochem 1999; 74:220-228.
17. Russell J, Bar A, Sherwood LM et al. Interaction between calcium and 1,25-dihydroxyvitamin D3 in the regulation of preproparathyroid hormone and vitamin D receptor messenger ribonucleic acid in avian parathyroids. Endocrinol 1993; 132:2639-2644.
18. Garfia B, Canadillas S, Canalejo A et al. Regulation of parathyroid vitamin d receptor expression by extracellular calcium. J Am Soc Nephrol 2002; 13:2945-2952.
19. Brown AJ, Zhong M, Finch J et al. The roles of calcium and 1,25-dihydroxyvitamin D3 in the regulation of vitamin D receptor expression by rat parathyroid glands. Endocrinol 1995; 136:1419-1425.
20. Koszewski NJ, Malluche HH, Russell J. Vitamin D receptor interactions with positive and negative DNA response elements: an interference footprint comparison. J Steroid Biochem Mol Biol 2000; 72:125-132.
21. Li YC, Amling M, Pirro AE et al. Normalization of mineral ion homeostasis by dietary means prevents hyperparathyroidism, rickets, and osteomalacia, but not alopecia in vitamin D receptor-ablated mice. Endocrinol 1998; 139:4391-4396.
22. Van Cromphaut SJ, Dewerchin M, Hoenderop JG et al. Duodenal calcium absorption in vitamin D receptor-knockout mice: functional and molecular aspects. Proc Natl Acad Sci USA 2001; 98:13324-13329.
23. Bouillon R, Van Cromphaut S, Carmeliet G. Intestinal calcium absorption: Molecular vitamin D mediated mechanisms. J Cell Biochem 2003; 88:332-339.
24. Canaff L, Hendy GN. Human calcium-sensing receptor gene. Vitamin D response elements in promoters P1 and P2 confer transcriptional responsiveness to 1,25-dihydroxyvitamin D. J Biol Chem 2002; 277:30337-30350.
25. Fukuda N, Tanaka H, Tominaga Y et al. Decreased 1,25-dihydroxyvitamin D3 receptor density is associated with a more severe form of parathyroid hyperplasia in chronic uremic patients. J Clin Invest 1993; 92:1436-1443.
26. Patel SR, Ke HQ, Vanholder R et al. Inhibition of calcitriol receptor binding to vitamin D response elements by uremic toxin. J Clin Invest 1995; 96:50-59.
27. Arnold A, Brown MF, Urena P et al. Monoclonality of parathyroid tumors in chronic renal failure and in primary parathyroid hyperplasia. J Clin Invest 1995; 95:2047-2053.
28. Wheeler DG, Horsford J, Michalak M et al. Calreticulin inhibits vitamin D3 signal transduction. Nucleic Acids Res 1995; 23:3268-3274.
29. Sela-Brown A, Russell J, Koszewski NJ et al. Calreticulin inhibits vitamin D's action on the PTH gene in vitro and may prevent vitamin D's effect in vivo in hypocalcemic rats. Mol Endocrinol 1998; 12:1193-1200.
30. Sakwe AM, Engstrom A, Larsson M et al. Biosynthesis and secretion of parathyroid hormone are sensitive to proteasome inhibitors in dispersed bovine parathyroid cells. J Biol Chem 2002; 277:17687-17695.

31. Silver J, Kilav R, Naveh-Many T. Mechanisms of secondary hyperparathyroidism. Am J Physiol Renal Physiol 2002; 283:F367-F376.
32. Hendy GN, Bennett HP, Gibbs BF et al. Proparathyroid hormone is preferentially cleaved to parathyroid hormone by the prohormone convertase furin. A mass spectrometric study. J Biol Chem 1995; 270:9517-9525.
33. Canaff L, Bennett HP, Hou Y et al. Proparathyroid hormone processing by the proprotein convertase-7: comparison with furin and assessment of modulation of parathyroid convertase messenger ribonucleic acid levels by calcium and 1,25-dihydroxyvitamin D3. Endocrinol 1999; 140:3633-3642.
34. Silver J, Naveh-Many T, Kronenberg HM. Parathyroid hormone: molecular biology. In: Bilezikian JB, Raisz LG, Rodan GA, eds. Principles of bone biology. San Diego: Academic Press, 2002:407-422.
35. Shvil Y, Naveh-Many T, Barach P et al. Regulation of parathyroid cell gene expression in experimental uremia. J Am Soc Nephrol 1990; 1:99-104.
36. Naveh-Many T, Silver J. Regulation of parathyroid hormone gene expression by hypocalcemia, hypercalcemia, and vitamin D in the rat. J Clin Invest 1990; 86:1313-1319.
37. Naveh-Many T, Rahamimov R, Livni N et al. Parathyroid cell proliferation in normal and chronic renal failure rats: the effects of calcium, phosphate and vitamin D. J Clin Invest 1995; 96:1786-1793.
38. Wada M, Furuya Y, Sakiyama J et al. The calcimimetic compound NPS R-568 suppresses parathyroid cell proliferation in rats with renal insufficiency. Control of parathyroid cell growth via a calcium receptor. J Clin Invest 1997; 100:2977-2983.
39. Chin J, Miller SC, Wada M et al. Activation of the calcium receptor by a calcimimetic compound halts the progression of secondary hyperparathyroidism in uremic rats. J Am Soc Nephrol 2000; 11:903-911.
40. Lewin E, Garfia B, Recio FL et al. Persistent downregulation of calcium-sensing receptor mRNA in rat parathyroids when severe secondary hyperparathyroidism is reversed by an isogenic kidney transplantation. J Am Soc Nephrol 2002; 13:2110-2116.
41. Gogusev J, Duchambon P, Hory B et al. Depressed expression of calcium receptor in parathyroid gland tissue of patiens with hyperparathyroidism. Kidney Int 1997; 51:328-336.
42. Drueke TB. The pathogenesis of parathyroid gland hyperplasia in chronic renal failure. Kidney Int 1995; 48:259-272.
43. Imanishi Y, Hosokawa Y, Yoshimoto K et al. Dual abnormalities in cell proliferation and hormone regulation caused by cyclin D1 in a murine model of hyperparathyroidism. J Clin Invest 2001; 107:1093-1102.
44. Carling T, Rastad J, Szabo E et al. Reduced parathyroid vitamin D receptor messenger ribonucleic acid levels in primary and secondary hyperparathyroidism. J Clin Endocrinol Metab 2000; 85:2000-2003.
45. Tokumoto M, Tsuruya K, Fukuda K et al. Reduced p21, p27 and vitamin D receptor in the nodular hyperplasia in patients with advanced secondary hyperparathyroidism. Kidney Int 2002; 62:1196-1207.
46. Gogusev J, Duchambon P, Stoermann-Chopard C et al. De novo expression of transforming growth factor-alpha in parathyroid gland tissue of patients with primary or secondary uraemic hyperparathyroidism. Nephrol Dial Transplant 1996; 11:2155-2162.
47. Cozzolino M, Lu Y, Finch J et al. p21WAF1 and TGF-alpha mediate parathyroid growth arrest by vitamin D and high calcium. Kidney Int 2001; 60:2109-2117.

CHAPTER 8

Vitamin D Analogs for the Treatment of Secondary Hyperparathyroidism in Chronic Renal Failure

Alex J. Brown

Abstract

Secondary hyperparathyroidism (2°HPT) is a common complication in patients with chronic renal failure. The pathogenesis of 2°HPT is attributed primarily to phosphate retention and low serum $1,25(OH)_2D_3$. Replacement therapy with calcitriol [1,25-dihydroxyvitamin D_3 or $1,25(OH)_2D_3$] or its precursor alfacalcidol [$1\alpha(OH)D_3$] often produces hypercalcemia, especially when combined with calcium-based phosphate binders. In addition, these natural vitamin D compounds can aggravate the hyperphosphatemia in these patients. Furthermore a high Ca x P product has been correlated with the severity of vascular calcification leading to coronary artery disease, the most common cause of mortality in renal patients. Several vitamin D analogs have now been developed that retain the direct suppressive action of $1,25(OH)_2D_3$ on the parathyroid glands but have less calcemic activity, thereby offering a safer and more effective means of controlling 2°HPT. 22-Oxa-$1,25(OH)_2D_3$ (22-oxacalcitriol or OCT), 19-nor-$1,25(OH)_2D_2$(19-norD_2) and $1\alpha(OH)D_2$ are currently available in Japan (OCT) and the United States (19-norD_2 and $1\alpha(OH)D_2$). The mechanisms by which these analogs exert their selective actions on the parathyroid glands are under investigation. The low calcemic activity of OCT has been attributed to its rapid clearance which prevents sustained effects on intestinal calcium absorption and bone resorption, but still allows a prolonged suppression of PTH gene expression. The selectivity of 19-norD_2 and $1\alpha(OH)D_2$ is achieved by other mechanisms. A clear understanding of how these compounds exert their selective actions on the parathyroid glands may allow the design of more effective analogs in the future.

Pathogenesis of Secondary Hyperparathyroidism in Chronic Renal Failure

Secondary hyperparathyroidism (2°HPT) is a common disorder in patients with chronic renal failure.[1,2] The increased PTH secretion and parathyroid gland hyperplasia are attributed primarily to the retention of phosphate and the decreased capacity to produce 1,25-dihydroxyvitamin D_3 [$1,25(OH)_2D_3$][3,4] (Fig. 1). Low serum $1,25(OH)_2D_3$ reduces intestinal calcium transport, and high serum phosphate further drives down the levels of serum calcium. The resulting hypocalcemia is a strong stimulus for PTH synthesis and secretion as

---

Molecular Biology of the Parathyroid, edited by Tally Naveh-Many. ©2005 Eurekah.com and Kluwer Academic / Plenum Publishers.

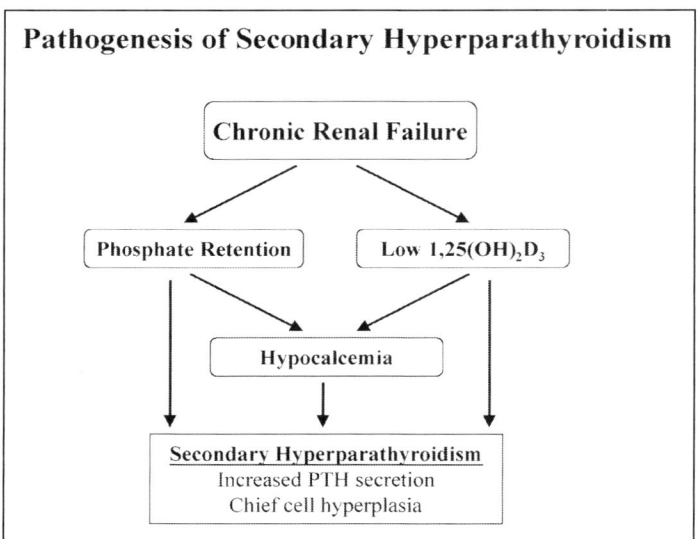

Figure 1. Pathogenesis of secondary hyperparathyroidism in chronic renal failure.

well as parathyroid gland hyperplasia. The parathyroid glands respond directly to phosphate and 1,25(OH)$_2$D$_3$ as well. Phosphate can increase PTH synthesis directly,[5-7] enhance PTH mRNA stability[8-10] and stimulate parathyroid gland hyperplasia.[11-13] 1,25(OH)$_2$D$_3$ inhibits PTH gene transcription[14-16] and parathyroid hyperplasia.[12,17] Initially, the parathyroid responds to the low calcium by increasing PTH secretion and synthesis, but prolonged hypocalcemia, hyperphosphatemia and decreased 1,25(OH)$_2$D$_3$ lead to parathyroid gland hyperplasia. Thus, the low levels of 1,25(OH)$_2$D$_3$ in these patients enhance PTH gene expression.

The major impact of chronically elevated PTH is on the bone.[18-20] PTH stimulates bone turnover, increasing both formation and resorption. The resulting bone, referred to as woven bone, is of poor quality and easily fractured. PTH-mediated high turnover bone disease is a major cause of morbidity in renal patients.

Treatment of Secondary Hyperparathyroidism

Prevention and treatment of secondary hyperparathyroidism requires both the control of serum phosphate and the restoration of 1,25(OH)$_2$D$_3$ levels. Phosphate is commonly controlled by retarding dietary phosphate absorption with the use of phosphate binders. The most commonly used binders at present are calcium salts. Replacement therapy with 1,25(OH)$_2$D$_3$ has been shown to aid in the suppression of PTH both by its direct action on the parathyroid glands and by increasing calcium absorption. However, the therapeutic window for 1,25(OH)$_2$D$_3$ is relatively narrow, with a small separation between the doses that are effective in suppressing PTH and those that elevate serum calcium. Combining 1,25(OH)$_2$D$_3$ therapy with oral calcium binders further increases the risk of hypercalcemia.

Hypercalcemia is especially problematic in renal patients. It is now clear that the combined increase in calcium and phosphate (Ca x P product) is a key factor in vascular calcification and coronary artery disease, the major cause of mortality in renal failure patients. In addition, hypercalcemia can lead to oversuppression of PTH. The bone of renal failure patients is resistant to PTH, and reducing serum PTH back to near normal leads to another form of renal osteodystrophy referred to as low turnover or adynamic bone disease, in which the bones fail to remodel sufficiently and gradually lose their strength. For these reasons, analogs of vitamin D that retain the direct suppressive effect of $1,25(OH)_2D_3$ on the parathyroid glands but that have lower calcemic and phosphatemic activity could provide a safer and more effective means of controlling secondary hyperparathyroidism.

Development of Vitamin D Analogs for Secondary Hyperparathyroidism

Several vitamin D analogs with lower calcemic activity have been examined for their ability to suppress PTH. The general approach has involved identifying analogs with low calcemic activity that still retained high VDR affinity and the ability to suppress PTH secretion in cultured parathyroid cells. Many vitamin D analogs with these properties were found to be ineffective in suppressing PTH in vivo for reasons that will be discussed in the subsequent section on mechanisms. The vitamin D analogs discussed in this section have been proven effective in both experimental animal models and in clinical trials, and three are now approved for use in patients with secondary hyperparathyroidism in renal failure.

22-Oxacalcitriol (OCT)

One of the first analogs shown to exert a selective action on the parathyroid glands was $22\text{-oxa-}1,25(OH)_2D_3$ (22-oxacalcitriol or OCT). Developed by Chugai Pharmaceuticals in Japan, OCT differs from $1,25(OH)_2D_3$ only by substitution of an oxygen in place of carbon 22 in the side chain (Fig. 2). OCT was reported to have very low calcemic activity in experimental animals, yet could potently differentiate leukemia (HL-60) cells in vitro.[21] OCT also mimicked $1,25(OH)_2D_3$ in the suppression of PTH secretion in vitro in cultures of bovine parathyroid.[22] The mechanism for PTH suppression by OCT was similar to that of $1,25(OH)_2D_3$ in that the analog decreased PTH mRNA to the same extent (60 to 80%) as $1,25(OH)_2D_3$ in normal rats,[22] suggesting that it was also acting at the level of gene transcription. Furthermore, the mRNA levels were measured 48 hours after injection of a relatively small dose (40 ng); the significance of this prolonged suppression with respect to the mechanism of the selectivity of OCT is discussed below.

Studies in animal models of renal failure demonstrated the greater therapeutic index of OCT in suppressing PTH. As shown in Figure 3, OCT was able to suppress PTH over a wide dose range with no change in serum calcium. In contrast, doses of $1,25(OH)_2D_3$ just above those that suppress PTH produced a significant increase in serum calcium. Naveh-Many and Silver found that OCT was less active than $1,25(OH)_2D_3$ in lowering PTH mRNA in uremic rats, but that it was much less calcemic indicating a superior therapeutic index.[23] Similar findings in uremic dogs were reported by Brown and coworkers[24] who showed that a single injection of a noncalcemic dose of OCT could suppress serum PTH for up to 69 hours.

The ability of OCT to effectively reduce PTH suggested that it could also ameliorate the high-turnover bone disease. Hirata et al[25] examined this directly in rats with glycopeptide-induced progressive nephritis that developed osteitis fibrosa over an 8-month period. The rats were then treated with two fixed doses OCT (0.03 or 0.15 µg/kg body weight) or vehicle three times per week for 15 weeks. As expected, OCT lowered PTH substantially. Bone histomorphometry revealed that OCT dose-dependently decreased the rates of both bone formation and resorption and reduced the fibrosis volume of the bone. At the doses used, there was no hypercalcemia or adynamic bone disease.

The effect of OCT on renal osteodystrophy was also examined by Monier-Faugere et al[26] in dogs made uremic by subtotal nephrectomy. After 14 weeks of renal insufficiency, the dogs

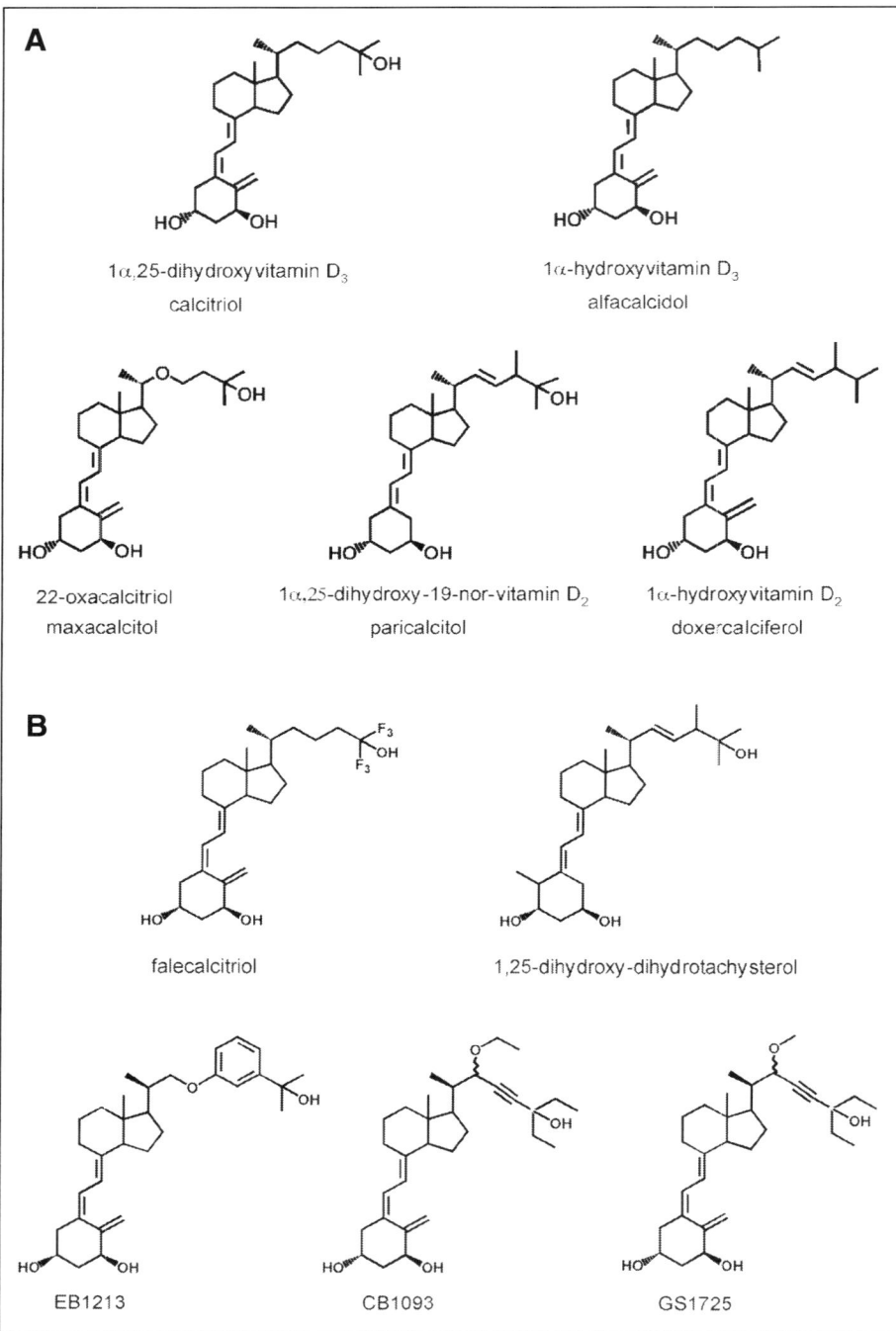

Figure 2. Structures of vitamin D analogs currently approved for treatment of secondary hyperparathyroidism.

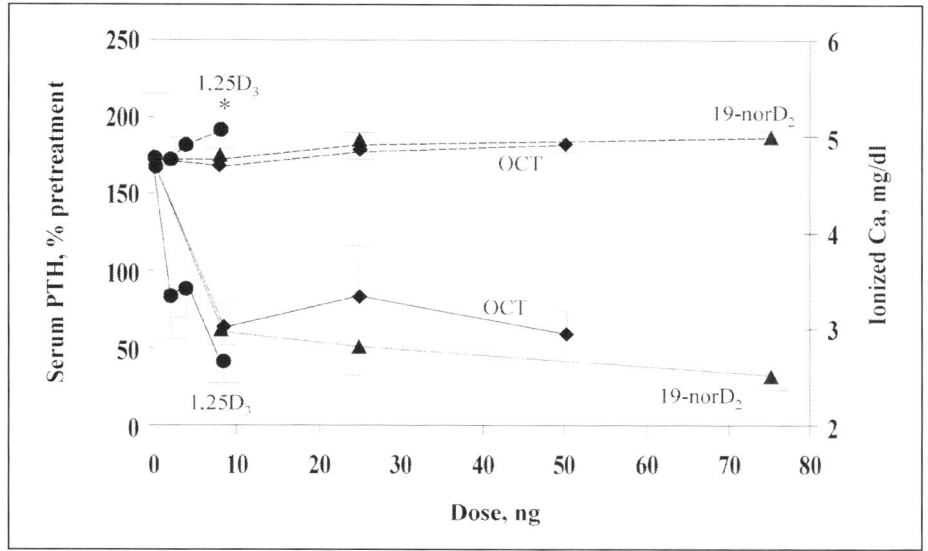

Figure 3. Suppression of PTH by OCT and 19-norD2 in uremic rats. Partially nephrectomized rats were maintained for one month on a high (0.9% P) phosphate diet to promote development of secondary hyperparathyroidism. The rats were then injected IP every other day for 8 days with the specified dose of $1,25(OH)_2D_3$, OCT or 19-norD$_2$. Serum was analyzed for ionized calcium and PTH 24 hours after the last injection. PTH values are expressed as % of pretreatment levels; serum calcium levels are post-treatment. Data are expressed as mean ± S.E.M. (n = 12-15).

were treated 3 times per week with OCT or vehicle for 60 weeks. In this study, the dose of OCT was adjusted to maximize PTH suppression but to avoid hypercalcemia. OCT signicantly decreased PTH levels. The analog reversed abnormalities in bone formation, including woven osteoid and fibrosis. However, no change in the rate of bone turnover was observed. While hypercalcemic episodes occurred, OCT did not induce low turnover bone disease.

Most recently, Tsukamoto et al[27] studied the effect of OCT on bone histology in a small group of patients with severe secondary hyperparathyroidism. The analog was administered 3 times per week at the end of the hemodialysis session. After 24 weeks of treatment, OCT reduced PTH modestly (35%) overall, but considerable variation in response was observed; 5 of the 10 patients did not respond to doses that did not produce hypercalcemia, while the other patients showed a PTH suppression of over 50%. Six patients, 5 of which were responders, agreed to a second, post-treatment bone biopsy. OCT significantly reduced bone marrow fibrosis, and decreased markers of bone turnover although signficant decreases were seen only in osteoid volume and bone formation rate.

The results of the preclinical and clinical trials indicated that OCT can effectively suppress PTH and is less calcemic than $1,25(OH)_2D_3$. The relative therapeutic windows of OCT and $1,25(OH)_2D_3$ in patients remains to be established. OCT was approved in the year 2000 for the treatment of secondary hyperparathyroidism in patients in Japan.

19-nor-1,25(OH)$_2$D$_2$

The promising findings with OCT prompted the development of other analogs for treatment of secondary hyperparathyroidism. 19-nor-1,25(OH)$_2$D$_2$ (19-norD$_2$), developed by Abbott Laboratories in the United States, has also proven to be an effective and safe therapy, and was the first of the new less calcemic vitamin D analog to be approve for use in patients. This compound lacks the exocyclic carbon 19 and has a vitamin D_2 side chain instead of the vitamin D_3 side chain of calcitriol (see Fig. 2). Initial studies in normal rats showed 19-norD_2 to be much less calcemic than $1,25(OH)_2D_3$.[28] Nine daily injections of various doses of 19-norD_2 and $1,25(OH)_2D_3$ indicated that 19-norD_2 was approximately 10 times less calcemic. Initial experiments in primary cultures of bovine parathyroid cells, on the other hand, showed that 19-norD_2 was equipotent to $1,25(OH)_2D_3$ in suppressing steady-state PTH secretion in vitro.[28]

Preclinical studies performed in uremic rats demonstrated that 19-norD_2 effectively suppressed PTH over a wide dose range without inducing hypercalcemia.[28] Just as important, the analog was found to be less phosphatemic as well. Analysis of the parathyroid glands of uremic rats treated with 19-norD_2 showed that it decreased PTH mRNA, suggesting that the analog suppressed PTH via the same transcriptional repression mechanism as $1,25(OH)_2D_3$.

Treatment with 19-norD_2 can both prevent and arrest the development of high turnover bone disease in uremic rats. Slatopolsky et al reported that prophylactic treatment of uremic rats with 19-norD_2 for the first two months following partial nephrectomy largely prevented increases in bone formation rate, unmineralized osteoid, bone resorption and fibrosis.[29] Of greater clinical relevance, these investigators found that 19-norD_2 could reverse the abnormal bone metabolism in rats with chronic renal failure and established parathyroid bone disease.[29] Thus, correction of PTH by 19-norD_2 can prevent and correct the PTH-induced high turnover bone disease.

Calcitriol treatment has been shown to block parathyroid hyperplasia in uremic rats,[17] but this effect could have been attributed to the mild hypercalcemia produced at the doses used. Takahashi et al[30] found that 19-norD_2, when administered from the time of induction of renal failure, reduced the rate of parathyroid gland growth in uremic rats, suggesting that it may have utility in preventing or arrresting parathyroid hyperplasia in patients. The mechanism for the antiproliferative effect of 19-norD_2 and calcitriol in the parathyroids is under investigation. Cozzolino et al reported that both vitamin D compounds, as well as high calcium and low phosphate, suppressed the expression of transforming growth factor alpha (TGFα) and induced the cell cycle inhibitor p21 in the parathyroid glands of uremic rats.[31]

The success of the preclinical studies led to trials in renal failure patients. Martin et al[32] reported the results of a placebo-controlled, multicenter trial in which 19-norD_2 was compared to placebo (Fig. 4). After a 4-week washout period, 19-norD_2 (or placebo) was administered three times per week after dialysis for 12 weeks. The dose of 19-norD_2 was started at 0.04 µg/kg and increased by 0.04 µg/kg every two weeks until a target reduction of 30% was achieved. Of the patients receiving 19-norD_2, 27 out of 40 had at least a 30% decrease in intact PTH, compared to 3 of 38 patients receiving placebo. There was no increase in serum calcium or phosphate before reaching this designated goal. Out of 414 determinations of serum calcium in the 19-norD_2-treated group, only 8 exceeded 11 mg/dl compared to 4 in the placebo group. These findings demonstrated the efficacy and safety of 19-norD_2 in the treatment of secondary hyperparathyroidism.

Similar results were reported in other studies. Llach et al[33] examined the effects of fixed doses of 19-norD_2 in a small number of patients. After a washout period, renal patients received 19-norD_2 at 0.04, 0.08, 0.16 or 0.24 µg/kg three times per week after dialysis for 8 weeks, and the 30% target reduction was achieved in 4/6, 1/4, 5/6 and 5/6 of the patients in the four groups, respectively. Calcium and phosphate changes were not presented for each group, but the combined data showed no signficant effects compared to placebo. Martin et al[34] determined the dose equivalency of $1,25(OH)_2D_3$ and 19-norD_2 in 29 patients that were on a stable dose of $1,25(OH)_2D_3$. Administration of a 4-fold higher dose of 19-norD_2 maintained PTH levels at their suppressed levels with no significant difference between the pre- and post-crossover levels. Serum calcium and phosphate levels also did not change. These clinical

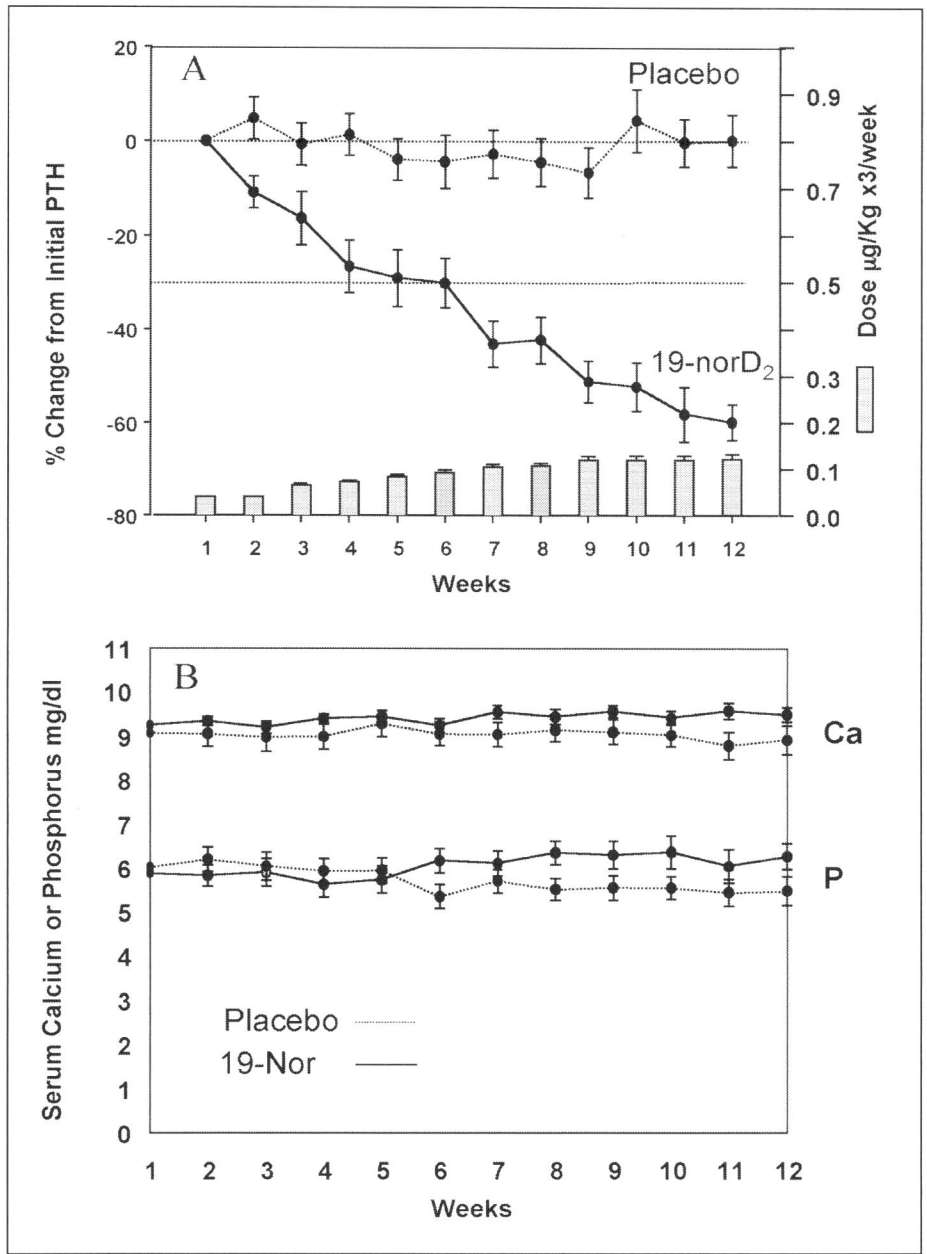

Figure 4. Efficacy of 19-norD$_2$ in the suppression of PTH in renal failure patients. Patients with moderate secondary hyperparathyroidism were treated 3 times per week with placebo or 19-norD$_2$ at doses starting a dose of 0. 04 ug/kg and increasing, as shown in solid bars, until the target 30% reduction in PTH levels was achieved. Panel A) PTH values in patients treated with placebo (dotted line) or 19-norD$_2$ (solid line). Panel B) Calcium and phosphate levels in patients treated with placebo (dotted line) or 19-norD$_2$ (solid line). From Martin et al[32] by permission.

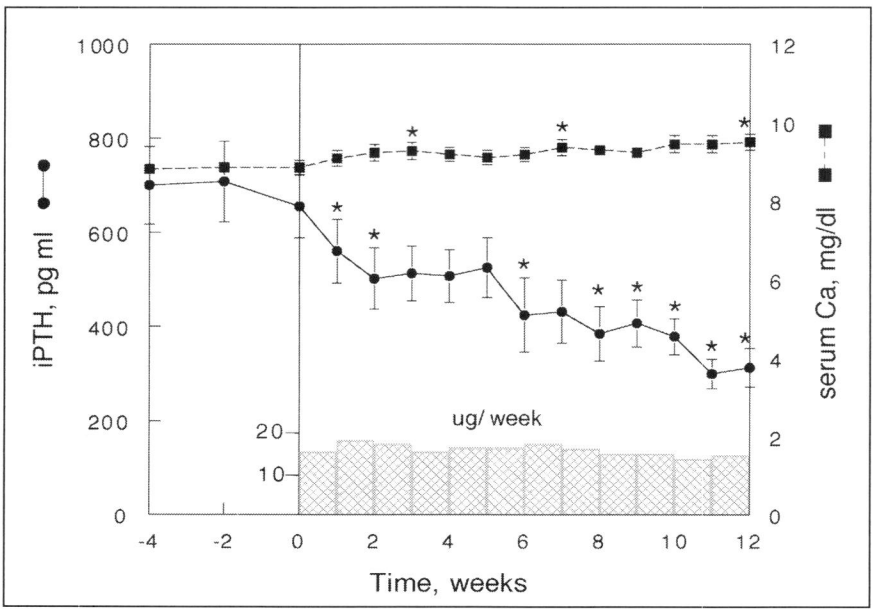

Figure 5. Efficacy of $1\alpha(OH)D_2$ in the suppression of PTH in renal failure patients. PTH levels in 24 hemodialysis patients during the last 4 weeks of a washout a period and during a 12-week treatment with $1\alpha(OH)D_2$, with the mean weekly dosage shown in the solid bars. Data are given as mean ±S.E.M. * $p < 0.05$ vs. baseline. From Tan et al[35] by permission.

trials demontrated that 19-norD_2 is safe and effective in reducing PTH levels. This analog, under the name Zemplar, was approved for treatment of secondary hyperparathyroidism in renal failure patients in the United States in 1998. As with the other analogs, a direct comparison of therapeutic windows for 19-norD_2 and $1,25(OH)_2D_3$ in patients has not been published.

$1\alpha(OH)D_2$

The most recent vitamin D analog to become available for renal failure patients is $1\alpha(OH)D_2$ (Fig. 1). This compound is a prodrug and, like its vitamin D_3 counterpart alfacalcidol, must be activated in vivo. $1\alpha(OH)D_2$, developed by Bone Care International in the United States, was tested directly in dialysis patients; studies in experimental animal models of renal failure have not been reported. The results of a trial in renal failure patients with moderate to severe secondary hyperparathyroidism (intact PTH > 400 pg/ml) was reported by Tan et al (see Fig. 5).[35,36] After an 8-week washout period, oral $1\alpha(OH)D_2$ was administered initially at a dose of 4 μg/day or 4 μg thrice weekly, and the dose was adjusted to maintain serum intact PTH levels between 130 and 250 pg/ml. The target goal was reached by 21 of the 24 patients during the course of the study, although 12 patients reached levels below the target range. At the end of the study, PTH levels were within the target range in 8 patients, below the 130 pg/ml in 5 patients, and above 250 pg/ml in 11 patients (7 of these had stopped treatment temporarily for dose adjustment). The rate of response varied, with two patients showing a 50% reduction in PTH within 2 weeks, while the others required an average of 6.1 weeks.

The average serum calcium rose slightly from 8.8 to 9.5 mg/dl during the 12 week study. There were 13 episodes of hypercalcemia (Ca > 10.5 mg/dl) in 10 patients. For the group, there were 4.7 episodes of hypercalcemia per 100 weeks of treatment compared to 0.53 episodes per week during the washout period. The average serum phosphate levels were unchanged by $1\alpha(OH)D_2$ therapy. There were 30 episodes of hyperphosphatemia (P > 6.9 mg/dl) in 18 patients during treatment compared to 13 episodes in 6 patients during the washout; the rates of hyperphosphatemic episodes, 6.9/100 weeks during the washout versus 10.1/100 weeks during treatment, were not significantly different. There was no alteration of phosphate binders during the treatment phase.

A comparison of daily oral versus intermittent oral $1\alpha(OH)D_2$ by Frazao et al[37] revealed comparable efficacies of daily and thrice-weekly. The incidences of hypercalcemia and hyperphosphatemia were very low in this small group of patients. A larger study was performed with 99 patients that consisted of an 8-week washout period, 16 weeks of open-label treatment with oral $1\alpha(OH)D_2$ and 8 weeks of double-blind crossover to either placebo or continued $1\alpha(OH)D_2$.[38] Dosage was intiated at 10 µg $1\alpha(OH)D_2$ 3 times per week after each dialysis and adjusted to maintain PTH between 150 and 300 pg/ml while avoiding hypercalcemia or hyperphosphatemia. PTH levels fell rapidly (35%) within 2 weeks and more slowly throughout the 16-week open-label treatment to a nadir of 44.7% of the baseline level. The target PTH range was achieved in 82 of the 99 patients. Crossover to placebo resulted in a rapid rebound of PTH, while PTH levels remained suppressed in patients continued on $1\alpha(OH)D_2$. There were modest increases in serum calcium (+ 0.7 mg/dl) and phosphate (0.8 mg/dl) in the open-label period. Hypercalcemia (Ca > 10.5 mg/dl) was detected in 15.3% of the measurements during the treatment period compared to 3.6% during the washout. The prevalence of hyperphosphatemia (P > 6.9 mg/dl) was 18.9% and 6.6% in the treatment and washout periods, respectively. Oral $1\alpha(OH)D_2$ was judged to be safe and effective in treating secondary hyperparathyroidism and was approved for use in patients in the United States in 1999.

Intravenous $1\alpha(OH)D_2$ has also been introduced. Maung et al[39] entered patients from the intermittent oral study described above[38] into a trial to compare the efficacy of intravenous $1\alpha(OH)D_2$. Following an 8-week washout,[38] patients were given placebo and 32 patients were treated with intravenous $1\alpha(OH)D_2$ for 12 weeks. The initial intravenous dose was 4 µg three times per week after dialysis, compared to the 10 µg dose of oral $1\alpha(OH)D_2$ in the previous study, due to the 2.5 times higher bioavailability of the intravenous $1\alpha(OH)D_2$. PTH levels decreased progressively during 12 weeks of intravenous $1\alpha(OH)D_2$ as it had with oral $1\alpha(OH)D_2$. Comparison of the PTH levels during the intravenous trial and the first 12 weeks of the oral trial revealed similar rates of decrease. Serum calcium and phosphate levels rose slightly with intravenous $1\alpha(OH)D_2$ administration, but both the absolute and percent increases were lower than with oral $1\alpha(OH)D_2$. Incidents of hypercalcemia and hyperphsophatemia were also lower with intravenous $1\alpha(OH)D_2$. Thus, it appears that intravenous administration of $1\alpha(OH)D_2$ may be safer than the oral formulation. Intravenous $1\alpha(OH)D_2$ is now available for use in patients.

Falecalcitriol

Falecalcitriol is an analog in which the hydrogens on carbons 26 and 27 have been substituted with fluorine atoms. This compound displays higher activity than $1,25(OH)_2D_3$ in vivo due to its slower metabolism (see below). In an initial trial by Nishizawa et al,[40] 43 renal failure patients were given daily oral falecalcitriol starting at a dose of 0.05 µg/d. This dose was increased by 0.05 µg/d every two weeks, unless hypercalcemia (Ca > 10.5 mg/dl) was produced. At the end of the 12-week protocol, there was a mean reduction in intact PTH of 25%, with only a very small change in serum calcium (8.79 ± 0.12 to 9.09 ± 0.13 mg/dl). More recently, Akiba et al[41] compared the efficacy of falecalcitriol to alfacalcidol ($1\alpha(OH)D_3$) in a small group of renal failure patients with moderate to severe secondary hyperparathyroidism. The doses for falecalcitriol and alfacalcidol were 0.15-0.30 µg/d and 0.25-0.50 µg/d, respectively. Both compounds were administered daily in oral form for 24 weeks. Under these conditions, falecalcitriol was more effective than alfacalcidol in reducing PTH. While both increased serum calcium slightly at the doses used, the change was small and not significantly different for the two vitamin D analogs. Further testing in larger groups of patients will be necessary to adequately evaluate the efficacy of falecalcitriol.

Other Analogs

The search for more effective vitamin D analogs for treating secondary hyperparathyroidism continues. Several have been tested in animal models. Fan et al[42] compared 1,25-dihydroxy-dihydrotachysterol [$1,25(OH)_2DHT$], $1,25(OH)_2D_3$, and OCT for their abilities to suppress PTH in the uremic rat model. Five daily doses of 1 ng of $1,25(OH)_2D_3$ caused significant hypercalcemia and a 50% decrease in PTH. In contrast, $1,25(OH)_2DHT$ and OCT at doses of 25 and 50 ng were equipotent in suppressing PTH, and did not increase serum calcium. Thus, $1,25(OH)_2DHT$ may be another useful analog for the treatment of secondary hyperparathyroidism (see Fig. 2B).

Analogs with inverted stereochemistry at carbon 20 have been demonstrated to be more potent than the natural diastereomers in a number of in vitro biological assays. Three of these were tested in the uremic rat model by Hruby et al.[43] Following five daily injections of 250 ng/kg, CB 1093 (see Fig. 1) suppressed PTH by 73% with no increase in serum calcium and with a slight decrease in serum phosphate (3.72 ± 0.65 to 2.95 ± 0.98 mM). In contrast, the same dose of $1,25(OH)_2D_3$ reduced PTH by 82%, but part of this decrease could be attributed to the large increase in ionized calcium (1.17 ± 0.04 to 1.55 ± 0.14 mM) produced by this excessively high dose of $1,25(OH)_2D_3$. EB 1213 at a dose of 1250 ng/kg-d lowered PTH by 61% with no change in ionized calcium, and GS 1725 at a much lower dose of 25 ng/kg-d suppressed PTH by 80% with only a slight, but not significant, increase in ionized calcium. Clearly, there are many promising new analogs in development for the treatment of secondary hyperparathyroidism.

Mechanisms for the Selectivity of Vitamin D Analogs

The analogs developed for secondary hyperparathyroidism are slightly less active than $1,25(OH)_2D_3$ in suppressing PTH, but are considerably less calcemic. The molecular basis for this selectivity is currently under investigation. Conceptually, it is important to note that vitamin D compounds interact with only a few proteins in vivo (see Fig. 6). In general, the biological profile of vitamin D compounds will be determined by their interaction with four proteins or classes of proteins:

1. the vitamin D receptor (VDR) that mediates the genomic actions;
2. the serum vitamin D binding protein (DBP) and perhaps lipoproteins that transport vitamin D compounds and control their access to the target cell;
3. the vitamin D-24-hydroxylase (24-OHase) and perhaps other minor enzymes that metabolize and/or deactivate the analogs; and
4. cell surface receptors that mediate the rapid, nongenomic actions of vitamin D compounds.

A plausible explanation for the selectivity of OCT, based on these protein interactions, has been proposed. The mechanisms responsible for the actions of 19-norD_2 and $1\alpha(OH)D_2$ are not as clear. This section discusses the current understanding of how these analogs exert their selectivity on the parathyroid glands.

22-Oxacalcitriol (OCT)

The basis for the selectivity of OCT is probably the best understood of the new vitamin D analogs. Its affinity for the VDR is about 8 times lower than that of $1,25(OH)_2D_3$, consistent with its lower activity in suppressing PTH. However, its very low calcemic activity can

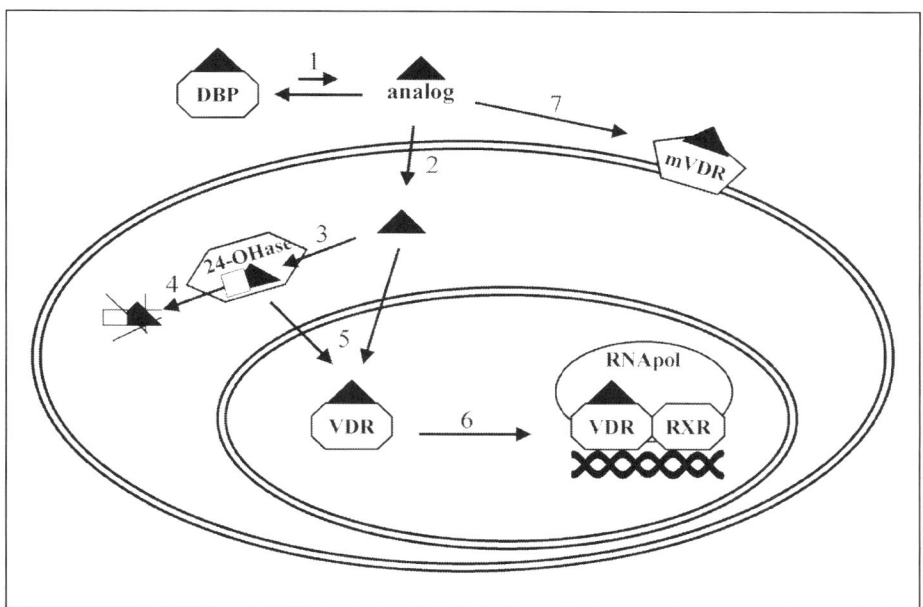

Figure 6. Potential points for differential handling of vitamin D analogs that can produce cell- or gene-specific actions. The steps in the vitamin D activation pathways at which differences in vitamin D analog action could lead to selective activities in vivo are shown. The steps diagrammed include 1) binding to the serum vitamin D binding protein (DBP), 2) cellular uptake, 3) conversion to active metabolites, 4) catabolic inactivation, 5) interaction with the nuclear vitamin D receptor, 6) recruitment of components of the transcriptional complex, and 7) activation of nongenomic pathways by interaction with the membrane receptor. See text for details.

be attributed primarily to its diminished affinity for DBP which is approximately 500 times less than that of $1,25(OH)_2D_3$.[44,45] DBP affinity is a major effector of the pharmacokinetics of vitamin D compounds. Studies in rats and dogs demonstrated that OCT is rapidly cleared from the circulation and achieves lower maximal levels (Cmax) in the blood when administered intraperitoneally.[44,46,47] Despite the lower Cmax, the amount of OCT associated with the intestinal VDR[46] and parathyroid glands[47] at early times post-injection were higher than those of $1,25(OH)_2D_3$, but fell rapidly after the analog was cleared from the circulation. This "pulse" of OCT in its target tissues elicited only transient increases in intestinal calcium transport and bone mobilization while the effects of $1,25(OH)_2D_3$ were much more prolonged. However, in spite of the short time of residence of OCT in the parathyroid glands, the analog elicited a sustained suppression of PTH gene expression. The molecular basis for the differences in the durations of the effects in the parathyroids versus those in the intestine and bone are unclear, but the findings indicate that stimulation of intestinal calcium absorption and bone resorption are short-lived responses that require continuous exposure to vitamin D compounds. On the other hand, even a short exposure of the parathyroid glands to vitamin D compounds leads to a prolonged suppression of PTH. Thus, OCT appears to exert its selectivity via its rapid clearance, which exploits the differences in the biological half-lives of the desirable (PTH suppression) and undesirable (increased calcium) responses (see Fig. 7).

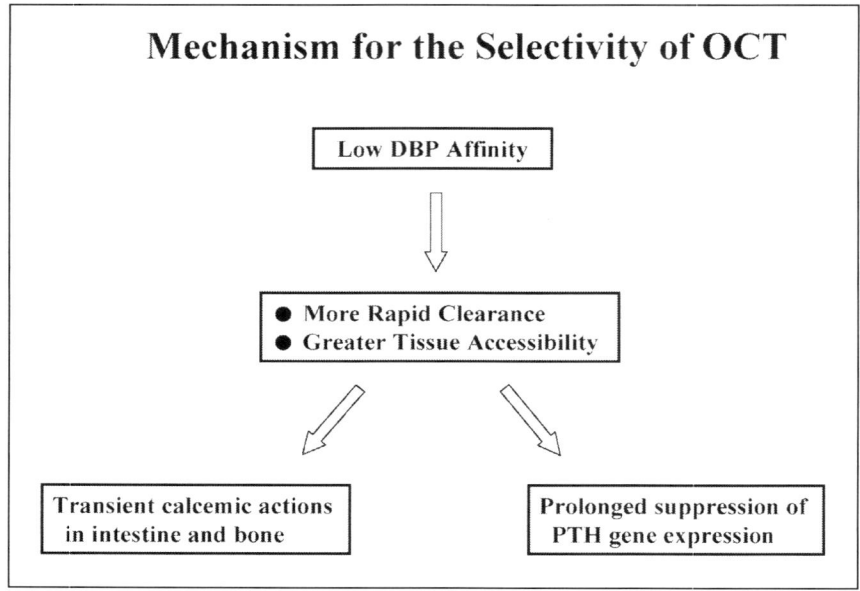

Figure 7. Proposed mechanism for the selective action of OCT on PTH.

19-nor-1,25(OH)$_2$D$_2$

19-norD$_2$ exerts its selectivity in a manner distinct from that of OCT and does not involve pharmacokinetics. The VDR affinity of 19-norD$_2$ is about 3 times less than that of 1,25(OH)$_2$D$_3$, consistent with its slightly lower potency in suppressing PTH. Unlike OCT, 19-norD$_2$ has relatively high DBP affinity: only three time less than that of 1,25(OH)$_2$D$_3$.[48] Not surprisingly, then, the clearance rates and tissue localization of 19-norD$_2$ and 1,25(OH)$_2$D$_3$ are similar.[48] Thus, it would appear paradoxical that 19-norD$_2$ has lower calcemic activity than 1,25(OH)$_2$D$_3$ when its VDR affinity and ability to localize to the bone and intestine are very similar to 1,25(OH)$_2$D$_3$. In fact, direct measurement of the intestinal calcium transport and bone mobilization at 24 hours after a single injection of 19-norD$_2$ or 1,25(OH)$_2$D$_3$ revealed equivalent, dose-dependent effects of the two compounds.[48] Only after chronic treatment were the differences in the effects of 19-norD$_2$ and 1,25(OH)$_2$D$_3$ on intestine and bone apparent. The analog was shown to be significantly less active in stimulating intestinal calcium absorption and bone calcium mobilization after 7 daily injections (Fig. 8).[48] Again, this could not be attributed to altered pharmacokinetics with chronic treatment;there was no difference in the tissue localization of 19-norD$_2$ and 1,25(OH)$_2$D$_3$ following the 7 daily injections. These findings suggested an induced resistance to 19-norD$_2$ with repeated treatments. This could not be attributed to differences in intestinal VDR content, since VDR binding activity was not different following the 7-day treatment with 19-norD$_2$ or 1,25(OH)$_2$D$_3$.

Direct effects of 19-norD$_2$ on osteoclast maturation and osteoclastic bone mobilization were assessed in vitro using the mouse bone marrow culture system. It is well established that inthis model, active vitamin D compounds can promote the differentiation of osteoclast precursors present in the bone marrow to multinucleated cells expressing tartrate-resistant acid phosphatase (TRAP) and other markers of mature osteoclasts. This action is now known to be mediated by the induction of RANKL (receptor activator of NFkB ligand) on the surface of osteoblasts. Furthermore, when the cultures containing the mature osteoclasts and osteoblasts

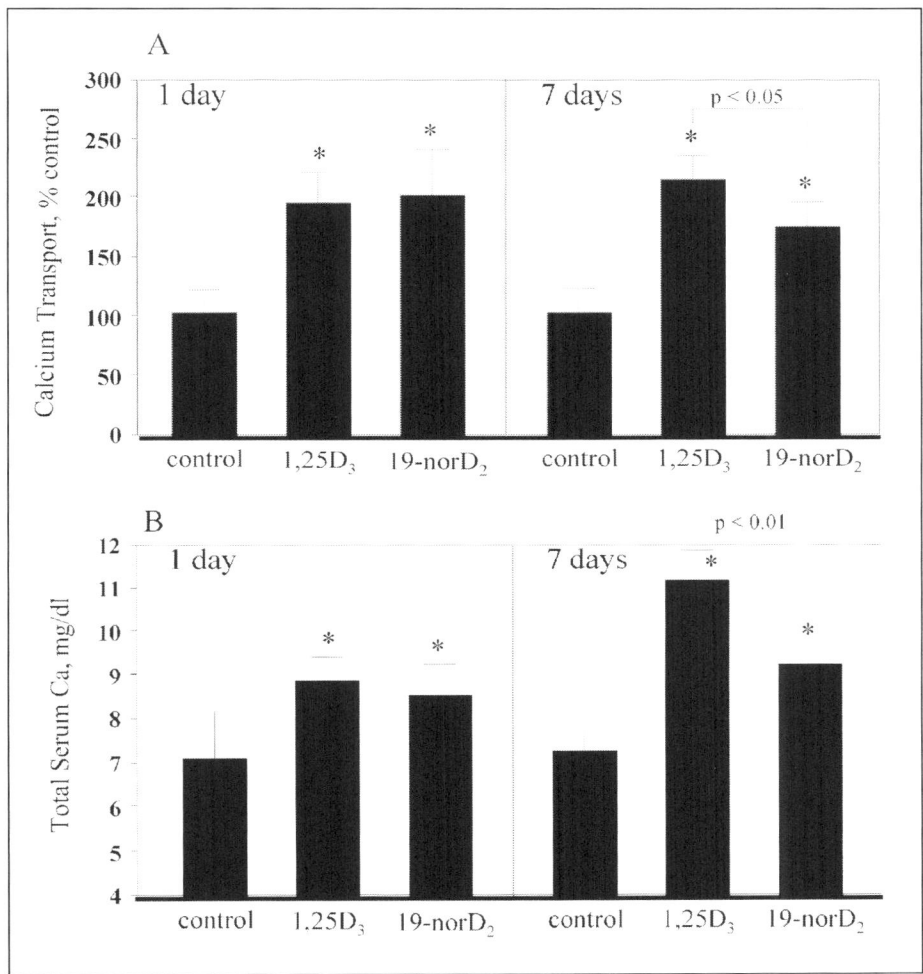

Figure 8. Effects of 19-norD$_2$ and 1,25(OH)$_2$D$_3$ on intestinal calcium transport and bone mobilization after 1 or 7 days of treatment. Vitamin D-deficient rats were fed a vitamin D- and calcium-deficient diet for 2 days and then given 1 or 7 daily injections of 600 pmol of 19-norD$_2$ or 1,25(OH)$_2$D$_3$. Calcium transport was measured by the in situ duodenal loop method; bone calcium mobilization was estimated by the increase in serum calcium since there was no dietary calcium. Data are expressed as mean ± S.E.M. (n = 12).

are plated onto bone (or dentine) slices, vitamin D compounds, via continued induction of RANKL can activate the osteoclasts to the bone surface. The mouse bone marrow model was used to assess the relative abilities of 1,25(OH)$_2$D$_3$ and 19-norD$_2$ to promote osteoclast maturation and osteoclast activity.[49] Incubation of freshly isolated bone marrow cells in the presence of various concentrations of 1,25(OH)$_2$D$_3$ and 19-norD$_2$ for 5 days revealed no differences in the potency of these compounds to induce osteoclast formation. However, when the these cultures were plated on dentine slices in the continued presence of various concentrations of the two vitamin D compounds, the amount of resorption in response to 19-norD$_2$ was less than that in the 1,25(OH)$_2$D$_3$-treated cultures. Both 19-norD$_2$ and 1,25(OH)$_2$D$_3$ produced maximum effects at 10 nM, but the number of resorption pits and the amount of bone resorption induced by 19-norD$_2$ was only 30% of that elicited by 1,25(OH)$_2$D$_3$. Further analysis revealed that the number of osteoclasts attached to the bone slices and the area resorbed per pit were the same for 19-norD$_2$ and 1,25(OH)$_2$D$_3$, indicating that the reduced bone resorption in response to 19-norD$_2$ could be attributed to a defective initiation of resorption by the osteoclasts.

In vitro bone resorption in response to 19-norD$_2$ and 1,25(OH)$_2$D$_3$ was also examined by Balint et al.[50] They found no significant difference in the potencies of the two compounds in stimulating calcium efflux from isolated neonatal calvaria. The apparent disparity may be attributed to the relatively short exposure time of 48 hours, since, as noted above,[48] differences in the in vivo bone mobilizing effects of 19-norD$_2$ and 1,25(OH)$_2$D$_3$ require more than 24 hours to become apparent.

Since the actions of vitamin D compounds on bone resorption are thought to be mediated through actions on the osteoblast, Finch et al examined the effects of 19-norD$_2$ on other vitamin D actions in the osteoblastic cell line MG63. No differences were observed in the induction of osteocalcin, alkaline phosphatase activity or VDR content, or in the suppression of cell proliferation by 19-norD$_2$ and 1,25(OH)$_2$D$_3$. Furthermore, the rate of catabolism of 19-norD$_2$ in the bone marrow cultures was not different from that of 1,25(OH)$_2$D$_3$, and VDR content was not differentially affected by 19-norD$_2$ treatment.[49] Based on our current understanding of vitamin D-mediate bone resorption, these finding would suggest that the reduced bone resorbing activity of 19-norD$_2$ may be due to a reduced ability to induce RANKL in the osteoblast, a hypothesis currently under investigation.

$1\alpha(OH)D_2$

The basis for the low calcemic activity of 1α(OH)D$_2$ is much less understood. Early studies with 1α(OH)D$_2$ showed that it was much less toxic than 1α(OH)D$_3$ when the compounds were administered chronically.[51] Paradoxically, the stimulation of calcium transport and bone mobilization by 1α(OH)D$_2$ and 1α(OH)D$_3$ were not different.[51] The reason for the lower toxicity of 1α(OH)D$_2$ remains unclear. Unlike the other vitamin D analogs, 1α(OH)D$_2$ is actually a pro-hormone that is converted, under physiologic conditions, mainly to 1,25(OH)$_2$D$_2$ which has been reported to have the same potency as 1,25(OH)$_2$D$_3$ in stimulating calcium transport in vivo and bone mobilization in vivo and in vitro.[49] However, Mawer et al[52] demonstrated that vitamin D$_2$, but not vitamin D$_3$, could be converted to 1,24(OH)$_2$D$_2$, a metabolite with potent cell differentiation activity, but much lower calcemic activity than 1,25(OH)$_2$D$_2$ or 1,25(OH)$_2$D$_3$. It has been proposed that 1α(OH)D$_2$, but not 1α(OH)D$_3$, may be 24-hydroxylated within target cells to produce an active metabolite (1,24(OH)$_2$D$_2$) that is less toxic (see Fig. 9). This hypothesis remains to be tested.

The pharmacokinetics of falecalcitriol differ from those of 1,25(OH)$_2$D$_3$. The fluorine atoms at the end of the side chain have been shown to impede catabolism of this molecule. Unlike 1,25(OH)$_2$D$_3$, which is hydroxylated at carbons 23 and 24 and then oxidatively cleaved to calcitroic acid, falecalcitriol is metabolized only partially metabolized, primarily to its 23-hydroxylated metabolite [1,23,25(OH)$_3$-26,27-F$_6$-D$_3$] in vivo[53,54] and in various cell culture systems.[55-57] This intermediate retains considerable biological activity. The fluorine atoms also affect binding to DBP; falecalcitriol has a slightly lower affinity for DBP and is cleared a bit more rapidly than 1,25(OH)$_2$D$_3$.[58] Despite the shorter circulating half-life and 3-fold lower affinity for the VDR, falecalcitriol has prolonged and potent biological activity in vivo, presumably due to its slower target tissue metabolism. Whether this analog exerts a selective action on the parathyroid glands is not known.

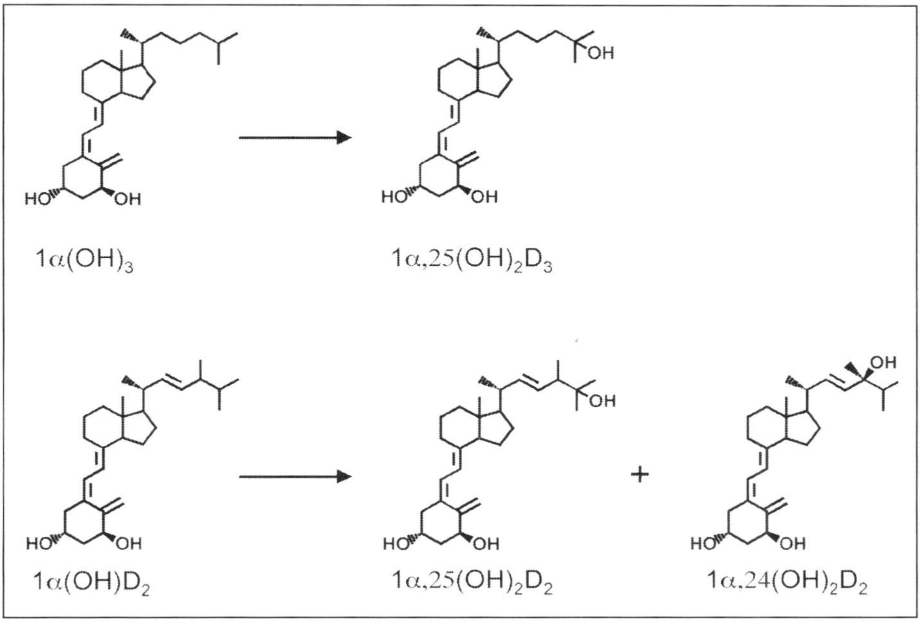

Figure 9. Differential metabolism of 1α(OH)D$_2$ and 1α(OH)D$_3$. The 24-hydroxylation of 1α(OH)D$_2$, but not 1α(OH)D$_3$, has been proposed to be a potential mechanism for the lower calcemic activity of 1α(OH)D$_2$. 1,25(OH)$_2$D$_2$ is significantly less calcemic than 1,25(OH)$_2$D$_3$ or 1,25(OH)$_2$D$_2$.

Future Perspectives

Less calcemic vitamin D analogs that can effectively suppress PTH in chronic renal failure patients offer safer alternatives to 1,25(OH)$_2$D$_3$ in the treatment of secondary hyperparathyroidism and the resulting renal osteodystrophy. The currently available analogs represent only the initial attempts at modifying the vitamin D molecule to produce more selective drugs. As we learn more about the how these modifications influence the interactions of vitamin D analogs with the VDR, DBP, the 24-hydroxylase and the membrane receptor and, in turn, the biological profile observed in vivo, it should be possible to design new analogs with greater specificity and a larger margin of safety for treatment of renal osteodystrophy and other diseases as well.

References

1. Slatopolsky E, Brown AJ, Dusso A. Pathogenesis of secondary hyperparathyroidism. Kidney Int 1999; 73:S14-9.
2. Silver J, Kilav R, Sela-Brown A et al. Molecular mechanisms of secondary hyperparathyroidism. Pediatr Nephrol 2000; 14:626-628.
3. Slatopolsky E, Dusso A, Brown AJ. The role of phosphorus in the development of secondary hyperparathyroidism and parathyroid cell proliferation in chronic renal failure. Am J Med Sci 1999; 317:370-6.
4. Roussanne MC, Duchambon P, Gogusev J et al. Parathyroid hyperplasia in renal failure: role of calcium, phosphate and calcitriol. Nephol Dial Transpl 1999; 14:S68-9.
5. Almaden Y, Canalejo A, Hernandez A et al. Direct effect of phosphorus on PTH secretion from whole rat parathyroid glands in vitro. J Bone Min Res 1996; 11:970-6.

6. Slatopolsky E, Finch J, Denda M et al. Phosphorus restriction prevents parathyroid gland growth-high phosphorus directly stimulates PTH secretion in vitro. J Clin Invest 1996; 97:2534-40.
7. Almaden Y, Canalejo A, Ballesteros E et al. Effect of high extracellular phosphate concentration on arachidonic acid production by parathyroid tissue in vitro. J Am Soc Nephol 2000; 11:1712-8.
8. Moallem E, Kilav R, Silver J et al. RNA-protein binding and post-transcriptional regulation of parathyroid hormone gene expression by calcium and phosphate. J Biol Chem 1998; 273:5253-9.
9. Sela-Brown A, Silver J, Brewer G et al. Identification of AUF1 as a parathyroid hormone mRNA 3'-untranslated region-binding protein that determines parathyroid hormone mRNA stability. J Biol Chem 2000; 275:7424-9.
10. Kilav R, Silver J, Naveh-Many T. A conserved cis-acting element in the parathyroid hormone 3'-untranslated region is sufficient for regulation of RNA stability by calcium and phosphate. J Biol Chem 2001; 276:8727-33.
11. Lopez-Hilker S, Dusso AS, Rapp NS et al. Phosphorus restriction reverses hyperparathyroidism in uremia independent of changes in calcium and calcitriol. Am J Physiol 1990; 259:F432-7.
12. Naveh-Many T, Rahamimov R, Livni N et al. Parathyroid cell proliferation in normal and chronic renal failure rats. The effects of calcium, phosphate, and vitamin D. J Clin Invest 1995; 96:1786-93.
13. Denda M, Finch J, Slatopolsky E. Phosphorus accelerates the development of parathyroid hyperplasia and secondary hyperparathyroidism in rats with renal failure. Am J Kidney Dis 1996; 28:596-602.
14. Silver J, Naveh-Many T, Mayer H et al. Regulation by vitamin D metabolites of parathyroid hormone gene transcription in vivo in the rat. J Clin Invest 1986; 78:1296-301.
15. Russell J, Lettieri D, Sherwood LM. Suppression by $1,25(OH)_2D_3$ of transcription of the pre-proparathyroid hormone gene. Endocrinology 1986; 119:2864-6.
16. Naveh-Many T, Silver J. Regulation of parathyroid hormone gene expression by hypocalcemia, hypercalcemia, and vitamin D in the rat. J Clin Invest 1990; 86:1313-9.
17. Szabo A, Merke J, Beier E et al. $1,25(OH)_2$vitamin D_3 inhibits parathyroid cell proliferation in experimental uremia. Kidney Int 1989; 35:1049-56.
18. Ritz E, Schomig M, Bommer J. Osteodystrophy in the millenium. Kidney Int 1990; 73:S94-8.
19. Drueke TB. Renal osteodystropy: management of hyperphosphatemia. Kidney Int 2000; 15:32-3.
20. Hruska K. Pathophysiology of renal osteodystrophy. Pediatr Nephrol 2000; 14:636-40.
21. Murayama E, Miyamoto K, Kubodera N et al. Synthetic studies of vitamin D_3 analogues. VIII. Synthesis of 22-oxavitamin D_3 analogues. Chem Pharm Bull 1986; 34:4410-3.
22. Brown AJ, Ritter CR, Finch JL et al. The noncalcemic analogue of vitamin D, 22-oxacalcitriol, suppresses parathyroid hormone synthesis and secretion. J Clin Invest 1989; 84:728-32.
23. Naveh-Many T, Silver J. Effects of calcitriol, 22-oxacalcitriol, and calcipotriol on serum calcium and parathyroid hormone gene expression. Endocrinology 1993; 133:2724-8.
24. Brown AJ, Finch JL, Lopez-Hilker S et al. New active analogues of vitamin D with low calcemic activity. Kidney Int 1990; 29:S22-7.
25. Hirata M, Katsumata K, Masaki T et al. 22-Oxacalcitriol ameliorates high-turnover bone and marked osteitis fibrosa in rats with slowly progressive nephritis. Kidney Int 1999; 56:2040-2047.
26. Monier-Faugere MC, Geng Z, Friedler RM et al. 22-oxacalcitriol suppresses secondary hyperparathyroidism without inducing low bone turnover in dogs with renal failure. Kidney Int 1999; 55:821-32.
27. Tsukamoto Y, Hanaoka M, Matsuo T et al. Effect of 22-oxacalcitriol on bone histology of hemodialyzed patients with severe secondary hyperparathyroidism. Am J Kidney Dis 2000; 35:458-464.
28. Slatopolsky E, Finch J, Ritter C et al. A new analog of calcitriol, 19-nor-$1,25$-$(OH)_2D_2$, suppresses parathyroid hormone secretion in uremic rats in the absence of hypercalcemia. Am J Kidney Dis 1995; 26:852-60.
29. Slatopolsky E, Lu Y, Finch J et al. The effects of 19-nor-$1,25(OH)_2D_2$ (19-nor) on PTH and bone histomorphometry in uremic rats. J Am Soc Nephrol 2000; 11:583A.
30. Takahashi F, Finch JL, Denda M et al. A new analog of $1,25$-$(OH)_2D_3$, 19-nor-$1,25$-$(OH)_2D_2$, suppresses serum PTH and parathyroid gland growth in uremic rats without elevation of intestinal vitamin D receptor content. Am J Kidney Dis 1997; 30:105-12.
31. Cozzolino M, Lu Y, Finch J et al. $1,25(OH)_2D_3$ (1,25D) and 19-nor-$1,25(OH)_2D_2$ (19-nor) prevent the phosphorus-induced parathyroid gland growth in early uremia by inducing $p21^{CIP1/WAF1}$

and reducing the expression of transforming growth factor α (TGFα) expression. J Am Soc Nephrol 2000; 11:574A.
32. Martin KJ, Gonzales EA, Gellens M et al. 19-Nor-1α,25-dihydroxyvitamin D_2 (Paricalcitol) safely and effectively reduces the levels of intact parathyroid hormone in patients on hemodialysis. J Am Soc Nephrol 1998; 9:1427-32.
33. Llach F, Keshav G, Goldblat MV et al. Suppression of parathyroid hormone secretion in hemodialysis patients by a novel vitamin D analogue: 19-nor-1,25-dihydroxyvitamin D_2. Am J Kidney Dis 1998; 32:S48-5. 4
34. Martin KJ, Gonzalez EA, Gellens ME et al. Therapy of secondary hyperparathyroidism with 19-nor-1α,25-dihydroxyvitamin D_2. Am J Kidney Dis 1998; 32:S 61-S 66.
35. Tan AU, Jr. , Levine BS, Mazess RB et al. Effective suppression of parathyroid hormone by 1_-hydroxy-vitamin D_2 in hemodialysis patients with moderate to severe secondary hyperparathyroidism. Kidney Int 1997; 51:317-23.
36. Coburn JW, Tan AU Jr, Levine BS et al. 1α-Hydroxy-vitamin D_2: a new look at an 'old' compound. Nephrol Dial Transplant 1996; 11:153-7.
37. Frazao JM, Levine BS, Tan AU et al. Efficacy and safety of intermittent oral 1α(OH)vitamin D_2 in suppressing secondary hyperparathyroidism in hemodialysis patients. Dial Transpl 1997; 26:583-595.
38. Frazao JM, Chesney RW, Coburn JW. Intermittent oral 1αhydroxyvitamin D_2 is effective and safe for the suppression of secondary hyperparathyroidism in haemodialysis patients. Nephrol Dial Transplant 1998; 3:68-72.
39. Maung HM, Elangovan L, Frazao JM et al. Efficacy and side effects of intermittent intravenous and oral doxercalciferol (1α-hydroxyvitamin D_2) in dialysis patients with secondary hyperparathyroidism:a sequential comparison. Am J Kidney Dis 2001; 37:532-543.
40. Nishizawa Y, Morii H, Ogura Y et al. Clinical trial of 26,26,26,27,27,27-hexafluoro-1,25-dihydroxyvitamin D_3 in uremic patients on hemodialysis: preliminary report. Contrib Nephrol 1991; 90:196-203.
41. Akiba T, Marumo F, Owada A et al. Controlled trial of falecalcitriol versus alfacalcidol in suppression of parathyroid hormone in hemodialysis patients with secondary hyperparathyroidism. Am J Kidney Dis 1998; 32:238-46.
42. Fan SLS, Schroeder NJ, Calverley MJ et al. Potent suppression of the parathyroid glands by hydroxylated metabolites of dihydrotachysterol(2). Nephol Dial Transpl 2000; 15:1943-1949.
43. Hruby M, Urena P, Mannstadt M et al. Effects of new vitamin D analogues on parathyroid function in chronically uraemic rats with secondary hyperparathyroidism. Nephrol Dial Transplant 1996; 11:1781-6.
44. Dusso AS, Negrea L, Gunawardhana S et al. On the mechanisms for the selective action of vitamin D analogs. Endocrinology 1991; 128:1687-92.
45. Okano T, Tsugawa N, Masuda S et al. Protein-binding properties of 22-oxa-1α,25-dihydroxyvitamin D_3, a synthetic analogue of 1α,25-dihydroxyvitamin D_3. J Nutr Sci Vitaminol 1989; 35:529-33.
46. Brown AJ, Finch J, Grieff M et al. The mechanism for the disparate actions of calcitriol and 22-oxacalcitriol in the intestine. Endocrinology 1993; 133:1158-64.
47. Kobayashi T, Tsugawa N, Okano T et al. The binding properties, with blood proteins, and tissue distribution of 22-oxa-1α,25-dihydroxyvitamin D_3, a noncalcemic analogue of 1α, 25-dihydroxyvitamin D_3, in rats. J Biochem 1994; 115:373-80.
48. Brown AJ, Finch JL, Takahashi F et al. The calcemic activity of 19-nor-1,25(OH)$_2$D$_2$ decreases with duration of treatment. J Am Soc Nephrol 2000; 11:2088-94.
49. Holliday LS, Gluck SL, Slatopolsky E et al. 1,25-dihydroxy-19-nor-vitamin D_2, a vitamin D analogs with reduced bone resorbing activity in vitro. J Am Soc Nephrol 2000; 11:1857-1864.
50. Balint E, Marshall CF, Sprague SM. Effect of the vitamin D analogues paricalcitol and calcitriol on bone mineral in vitro. Am J Kidney Dis 2000; 36:789-796.
51. Sjoden G, Smith C, Lindgren U et al. 1a-Hydroxyvitamin D_2 is less toxic than 1a-hydroxyvitamin D_3 in the rat. Proc Soc Exptl Biol Med 1985; 178:432-6.
52. Mawer EB, Jones G, Davies M et al. Unique 24-hydroxylated metabolites represent a significant pathway of metabolism of vitamin D_2 in humans: 24-hydroxyvitamin D_2 and 1,24-dihydroxyvitamin D_2 detectable in human serum. J Clin Endocrinol Metab 1998; 83:2156-66.

53. Komuro S, Kanamaru H, Nakatsuka I et al. Distribution and metabolism of F_6-1,25(OH)$_2$ vitamin D_3 and 1,25(OH)$_2$ vitamin D_3 in the bones of rats dosed with tritium-labeled compounds. Steroids 1998; 63:505-10.
54. Komuro S, Sato M, Kanamaru H et al. In vivo and in vitro pharmacokinetics and metabolism studies of 26,26,26,27,27,27-F_6-1,25(OH)$_2$ vitamin D_3 (Falecalcitriol) in rat: induction of vitamin D_3-24-hydroxylase (CYP24) responsible for 23S-hydroxylation in target tissues and the drop in serum levels. Xenobiotica 1999; 29:603-13.
55. Honda A, Nakashima N, Shida Y et al. Modification of 1α,25-dihydroxyvitamin D_3 metabolism by introduction of 26,26,26,27,27,27-hexafluoro atoms in human promyelocytic leukaemia (HL-60) cells: isolation and identification of a novel bioactive metabolite, 26,26,26,27,27,27-hexafluoro-1α,23(S),25-trihydroxyvitamin D_3. Biochem J 1993; 295:509-16.
56. Inaba M, Okuno S, Nishizawa et al. Effect of substituting fluorine for hydrogen at C-26 and C-27 on the side chain of 1,25-dihydroxyvitamin D_3. Biochem Pharmacol 1993; 45:2331-6.
57. Harada M, Miyahara T, Kajita-Kondo S et al. Differences in metabolism between 26,26,26,27,27,27-hexafluoro-1α,25-dihydroxyvitamin D_3 and 1α,25-dihydroxyvitamin D_3 in cultured neonatal mouse calvaria. Res Commun Mol Pathol Pharmacol 1994; 86:183-93.
58. Tanaka Y, DeLuca HF, Kobayashi Y et al. 26,26,26,27,27,27-hexafluoro-1,25-dihydroxyvitamin D_3: a highly potent, long-lasting analog of 1,25-dihydroxyvitamin D_3. Arch Biochem Biophy 1984; 229:348-54.

CHAPTER 9

Parathyroid Gland Hyperplasia in Renal Failure

Adriana S. Dusso, Mario Cozzolino and Eduardo Slatopolsky

Introduction

Secondary hyperparathyroidism, a frequent complication of chronic renal failure, is characterized by parathyroid hyperplasia and enhanced synthesis and secretion of parathyroid hormone (PTH).[1-3] As summarized in Figure 1, high circulating PTH levels are not only a major contributor to osteitis fibrosa and bone loss, typical features of renal osteodystrophy, but also to a variety of systemic defects including cardiovascular complications which increase mortality in renal failure patients.[4-6]

A link between the mechanisms controlling proliferation and hormonal production also exists in normal parathyroid cells, which respond to the stimulus of chronic hypocalcemia not only by an increase in PTH release but with a secondary expansion in cell mass. The mechanisms responsible for this link, however, remain poorly understood. In renal failure, hypocalcemia, hyperphosphatemia and vitamin D deficiency are the three main direct causes of hyperparathyroidism. Hyperphosphatemia and 1,25-dihydroxyvitamin D ($1,25(OH)_2D_3$) deficiency also enhance parathyroid function indirectly by lowering serum calcium (Ca).[1-3,7] The regulation of PTH synthesis by Ca, phosphate (P) and vitamin D has been extensively studied. PTH-gene transcription is tightly controlled by Ca and $1,25(OH)_2D_3$ through mechanisms that involve the calcium sensing receptor (CaSR) and the vitamin D receptor (VDR), respectively.[8,9] As renal disease progresses, a reduction in parathyroid content of both proteins renders the parathyroid glands more resistant to suppression of PTH synthesis and secretion in response to Ca and $1,25(OH)_2D_3$.[8-10] Serum Ca and P levels also control PTH synthesis through post-transcriptional mechanisms that involve the binding of cytosolic proteins to the 3'-untranslated region of the PTH mRNA, thus regulating transcript stability and consequently, PTH translation rates.[11,12]

Changes in the levels of serum Ca, P, $1,25(OH)_2D_3$ and in the parathyroid content of the CaSR and the VDR also have a dramatic impact on parathyroid tissue growth.[1-3,7] However, the lack of an appropriate parathyroid cell line has precluded a better understanding of the pathogenic mechanisms mediating the potent effects of the three main regulators on the rate of parathyroid cell proliferation already induced by uremia. This chapter summarizes the current understanding of parathyroid tissue growth and presents new insights, emerging from the 5/6 nephrectomized rat model, into the molecular mechanisms contributing to the regulation of parathyroid hyperplasia in early uremia by Ca, P and $1,25(OH)_2D_3$.

Molecular Biology of the Parathyroid, edited by Tally Naveh-Many. ©2005 Eurekah.com and Kluwer Academic / Plenum Publishers.

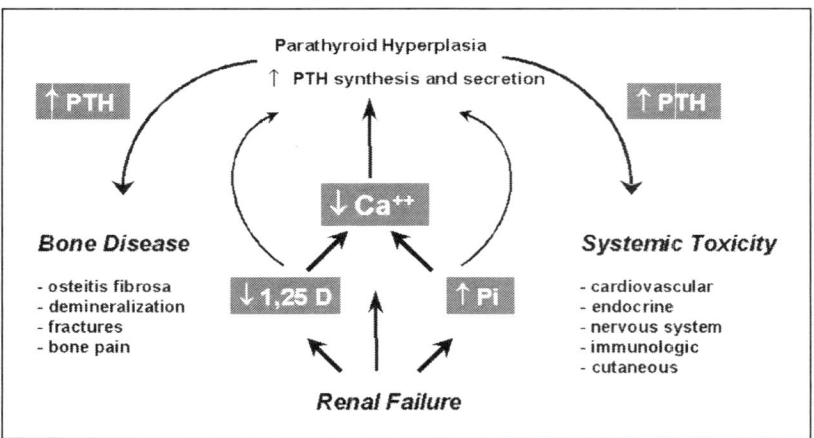

Figure 1. The pathogenesis of secondary hyperparathyroidism in renal failure: bone and systemic repercussions.

Parathyroid Tissue Growth in Normal Conditions and in Renal Failure

The parathyroid gland is a low turnover, discontinuously replicating tissue composed of cells that rarely undergo mitoses.[1-3] Quiescent parathyroid cells, however, retain the potential to proliferate in response to the growth stimuli triggered by uremia, low Ca, high P or vitamin D deficiency.[7,13]

As for most eukaryotic cells, the commitment of parathyroid cells to abandon quiescence (G0) and divide in response to growth stimuli depends on the net balance between the two opposing forces depicted in Figure 2. The forward force moving the cell through the phases of the cell cycle to complete a mitotic division is dictated by the activity of phase-specific complexes of cyclins and cyclin dependent kinases (Cdk).[14-16] The opposing force to suppress growth is exerted by a family of proteins, the Cdk- inhibitors, which bind cyclin/Cdk complexes inhibiting their activity. Mitogenic signals triggered by uremia, low Ca, high P or vitamin D deficiency induce parathyroid cells to abandon quiescence and divide by increasing the activity of cyclin/Cdk complexes, decreasing Cdk-inhibitor threshold below the levels of cyclin/Cdk complexes or a combination of both. In contrast, the antimitogenic signals elicited by high Ca, low P or vitamin D therapy could arrest cell growth by exerting the opposite effects, that is, decreasing the activity of cyclin/Cdk complexes, increasing Cdk-inhibitors to levels that exceed the amount of cyclin/Cdk complexes in the cell, or both.

The contribution of abnormalities in apoptotic rates to the enhanced parathyroid growth of uremia has also been examined.[17-19] The demonstration that the uremic state stimulates apoptosis, possibly as a compensatory reaction to the hyperproliferative activity, renders impaired apoptosis a highly improbable mechanism for the resultant hyperplasia.[17]

Different groups have demonstrated an association between parathyroid hyperplasia and changes in the content of cell cycle regulators. Overexpression of PRAD1/cyclin D1, induced by a DNA re-arrangement of the PTH gene, is one genetic disorder in primary hyperparathyroidism.[20] The importance of cyclin D1 in parathyroid cell growth was conclusively demonstrated by the recent studies of Imanishi and coworkers[21] using transgenic mice targeted to specifically overexpress cyclin D1 in the parathyroid gland. These mice slowly develop large

Parathyroid Gland Hyterplasia in Renal Failure

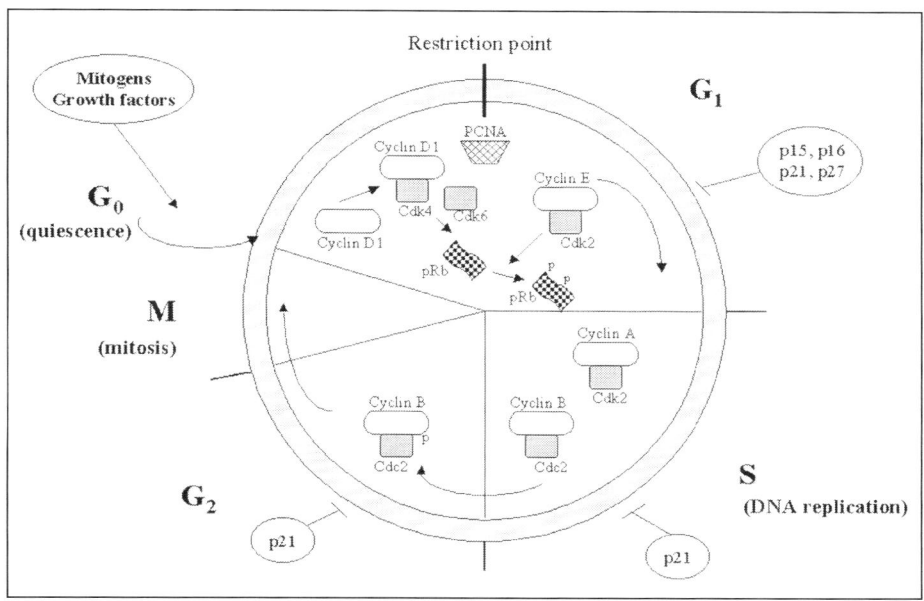

Figure 2. Schematic representation of the mammalian cell cycle. Phase-specific complexes of cyclins and cyclin dependent kinases (Cdk) facilitate progression through the cycle to complete mitosis. Phosphorylation of Rb is critical for S-phase gene activation. Phase specific Cdk-inhibitors inactivate cyclin/Cdk complexes.

hyperplastic and in some cases adenomatous glands that expressed reduced CaSR. This model mimics human parathyroid adenomas.

In renal hyperparathyroidism, parathyroid glands initially grow diffusely and polyclonally. Cells in the nodules then transform monoclonally and proliferate aggressively.[22] Neither the mechanisms triggering the initial increase in proliferative activity nor those resulting in changes in growth patterns are understood.

Changes in the expression of cell cycle components such as over-expression of cyclin D1 have been reported in patients with renal failure.[20] Studies by Tominaga and coworkers,[22] comparing diffuse versus nodular hyperplasia, also demonstrated higher cyclin D1 expression in the latter. However, different from parathyroid adenomas, there was no correlation between cyclin D1 and mitogenic activity in renal nodular hyperplasia. These findings suggest that PRAD1/cyclin D1 overexpression induced by PTH gene re-arrangement may not be the major genetic abnormality responsible for tumorigenesis. Heterogenous genetic changes appear to contribute to monoclonal proliferation of parathyroid cells induced either by the expression of PRAD1/cyclin D1 or by other mechanisms independent of the amplification of the protooncogene, such as reduction of the levels of a cell cycle repressor.[22] In fact, a reduction in the levels of the Cdk- inhibitor p27Kip1 has also been associated with hyperplastic growth.[23]

The rapid de-differentiation of hyperplastic parathyroid cells in culture[24] has precluded further assessment of the relative contribution of changes in Ca, P and vitamin D to the expression of components of the cell-cycle critical for growth control. The experimental approach in identifying molecular mechanisms, critical in the design of more effective strategies for therapy, is therefore limited to the in vivo uremic rat model.

Dietary Phosphate Regulation of Parathyroid Cell Growth in Uremia

In contrast to the slow growth rate of human parathyroid adenomas, Naveh Many and collaborators demonstrated an increased rate of parathyroid cell proliferation fifteen days after 5/6 nephrectomy in rats. In an identical time frame, uremia- induced mitotic activity was further enhanced by high dietary P but prevented by P restriction.[7] Furthermore, studies in our laboratory showed that most parathyroid growth induced by high dietary P (0.9 % P) occurred within five days of the onset of renal failure. They also confirmed that, similar to human renal hyperparathyroidism, hyperplasia rather than hypertrophy is the main contributor to the enlargement of the parathyroid glands.[13] In contrast to the mitogenic effects of high P, dietary P restriction appears to counteract the proliferative signals induced by uremia, thus preventing parathyroid cell replication and consequently, the increase in parathyroid gland size.[1-3,7,25,26] The mechanism for the growth arrest by P restriction does not involve the induction of apoptosis.[13]

Initial studies in rats were designed to identify the mechanisms mediating the opposing effects of dietary P manipulation on parathyroid cell growth in early renal failure. These studies first examined the regulation of parathyroid expression of the Cdk-inhibitor p21 by dietary P.[27] The rationale for these studies was the well known effects of high dietary P to reduce and those of P-restriction to increase serum $1,25(OH)_2D_3$ in normal individuals as well as in patients with mild and moderate renal failure.[28,29] $1,25(OH)_2D_3$, in turn, directly activates p21 gene transcription,[30] an action mediated by the vitamin D receptor (VDR) as a transcriptional enhancer. $1,25(OH)_2D_3$/VDR-induction of p21 expression is responsible for the potent antiproliferative properties of $1,25(OH)_2D_3$ in cells of the monocyte-macrophage lineage,[30] keratinocytes,[31] and prostate cancer cell lines.[32] Furthermore, in contrast to the growth arrest promoted by increases in p21, the downregulation of p21 expression accounts for growth stimulation in the human epidermoid carcinoma cell line A431,[33] human embryonic lung fibroblasts[34] and glomerular epithelial cells.[35] These reports led us to hypothesize that opposite changes in parathyroid p21 content induced by dietary P manipulations, either directly or through changes in $1,25(OH)_2D_3$, could contribute to the potent inhibition or stimulation of mitotic activity. The results of these studies demonstrated that, indeed, the low P- induction of p21 mRNA and protein content in the parathyroid glands contributes to the antiproliferative effects of P-restriction on uremia-induced parathyroid cell growth.[27] As shown in Figure 3, P-restriction decreased serum P and induced a two-fold increase in parathyroid p21 mRNA levels as soon as two days after 5/6 nephrectomy. This modest increase in p21 mRNA in P-restricted rats, remained 2-fold higher than in the high P group through day 7, and resulted in a substantially higher expression of p21 protein (See Fig, 4). The latter finding suggests the existence of an additional post-transcriptional mechanism for P restriction in the up-regulation of p21 expression in rat parathyroid glands. Although this hypothesis is difficult to test using our in vivo model, a similar induction of p21 expression through post-transcriptional regulatory mechanisms was reported both in fibroblasts and murine erythroleukemia cells.[36] Importantly, in contrast to our working hypothesis, low P induction of parathyroid p21 was not the result of an increase in serum $1,25(OH)_2D_3$ concentrations.[27] Clearly, pathways other than the $1,25(OH)_2D_3$-vitamin D receptor axis mediate both low-P activation of p21 gene transcription and the post-transcriptional enhancement of p21 protein expression.

The induction of p21 is sufficient to induce growth arrest in monocyte-macrophages,[30] keratinocytes,[31] and human cancer cells[37] and to suppress tumorigenicity in vivo.[38] In our uremic rat model, a role for low-P induction of p21 expression in the arrest of parathyroid cell growth is supported by the demonstration that temporal increases in p21 protein expression correlate inversely with parathyroid levels of the marker of mitotic activity proliferating nuclear cell antigen (PCNA) (Fig. 3). Intestinal growth as well as p21 and PCNA content remained unchanged with dietary-P manipulation,[27] thus supporting the specificity for the parathyroid glands of the antimitogenic effects of P restriction in these uremic rats.

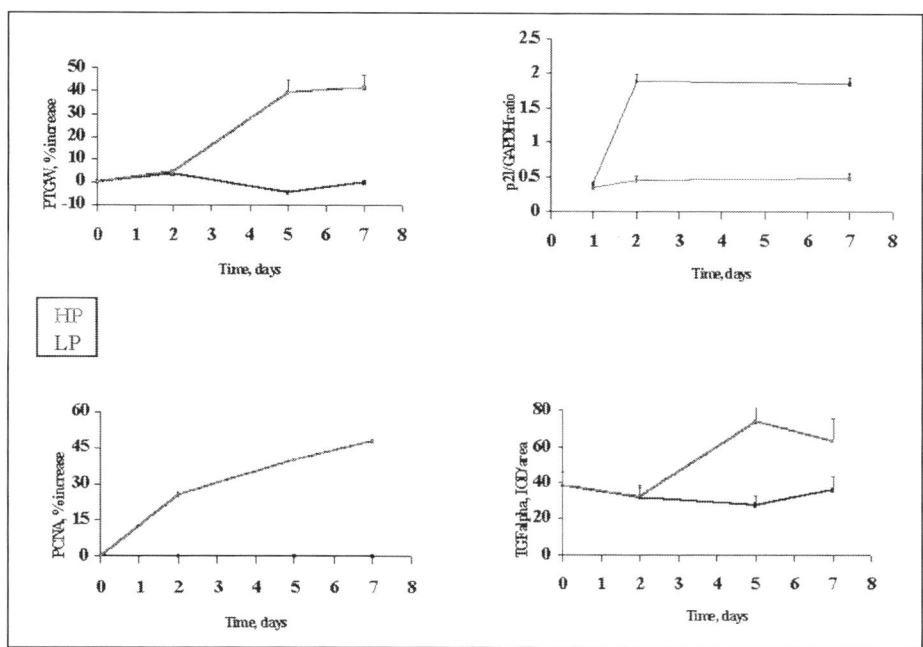

Figure 3. Dietay P control of parathyroid growth. Temporal relationship between changes in parathyroid p21 and TGFα–expression and hyperplasia in response to high (HP) and low (LP) dietary P.

The mechanisms by which p21 arrests growth are depicted in Figure 5. p21 is a member of the family of cdk-inhibitors that includes p27 and p57. p21 inhibits multiple cyclin dependent kinases including cdk1, cdk2, cdk4 and cdk6. As seen in the left panel, p21 is also a component of a multiprotein complex which includes cyclins, cdk, and PCNA. At a stoichiometry of one molecule of p21 per complex, this multicomplex phosphorylates the retinoblastoma (Rb) protein resulting in the release of E2F for autoactivation and activation of S-phase genes (DNA replication phase) to complete a mitotic division.[39] As depicted in the right panel, the induction of p21 in response to various stimuli, results in inhibition of G1-cyclin/cdk complexes and G1 growth arrest. This effect has been related, at least in part, to excess p21 blocking activation of G1-cyclin/cdk complexes by cdk7/cyclin H, thus preventing both Rb-phosphorylation and E2F activation. p21 has also been shown to inhibit replication by binding to PCNA trimers causing DNAse–polymerase to lose processivity.[39]

In contrast to our initial hypothesis, high-dietary P had no detectable effect on parathyroid expression of p21 mRNA or protein. Thus, reduction in intracellular p21 is not the cause for the mitogenic properties of high P in uremic rat parathyroid glands. In the search for mitogenic stimuli triggered by high dietary P, we next focused on transforming growth factor-α (TGFα). TGFα, known to promote growth not only in malignant transformation but also in normal tissues,[40,41] is enhanced in hyperplastic and adenomatous human parathyroid glands.[42] Immunohistochemical quantitation of TGFα content in the parathyroid glands of uremic rats showed the pattern of high TGFα expression reported for human hyperplastic glands. Furthermore, as depicted in Figure 3, parathyroid-TGFα content in these rats was slightly higher compared to their low P counterparts two days after 5/6 nephrectomy. TGFα expression peaked by day 5 and remained higher than in the low-P group through day 7. The increases in TGFα induced by high P paralleled those in PCNA expression (Figs. 3 and 5) and were specific for the

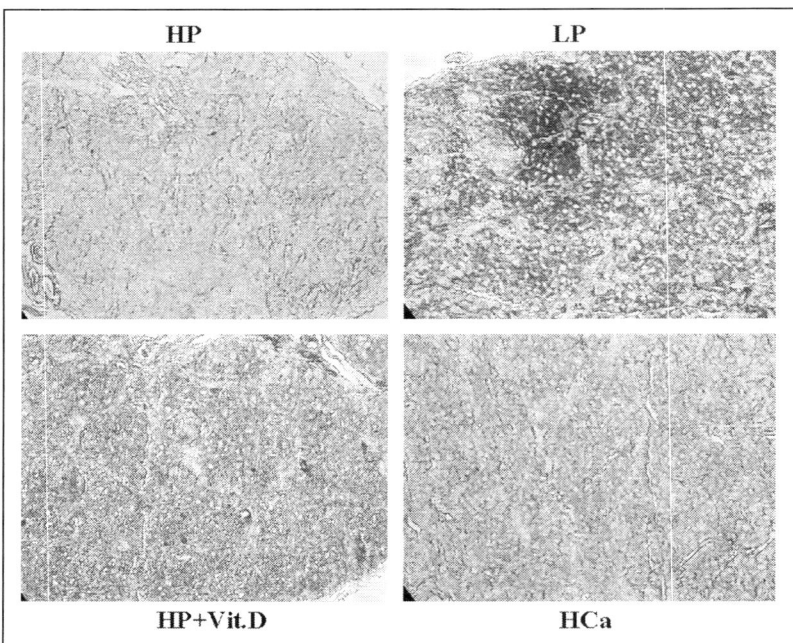

Figure 4. Regulation of parathyroid p21 expression. Induction of parathyroid p21 by P restriction (LP), vitamin D therapy (HP +Vit. D) and high dietary Ca (HCa) compared to high dietary P (HP) in rats, 7 days after 5/6 nephrectomy.

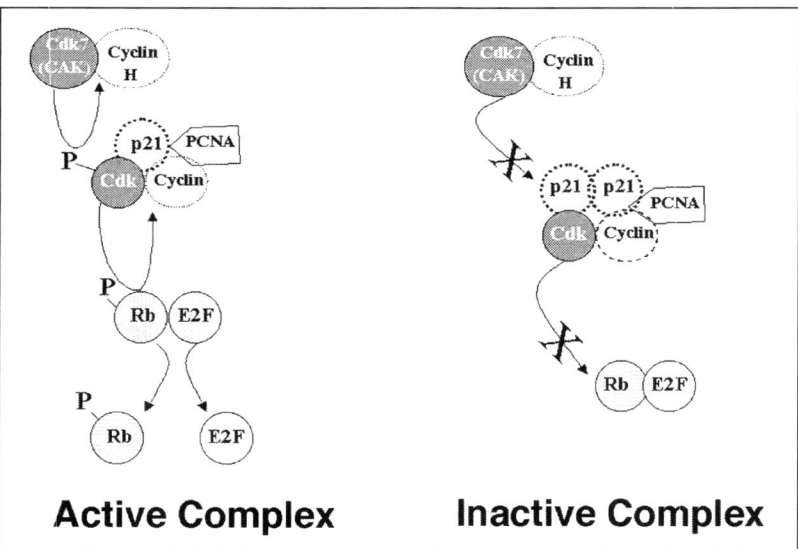

Figure 5. Mechanisms of action of p21 in normal cells. P21 is a component of the multicomplex activating Rb and E2F which promotes G_1 to S transition (Active Complex). Excess p21 inhibits Rb activation arresting growth (Inactive Complex). (Adapted from ref. 39).

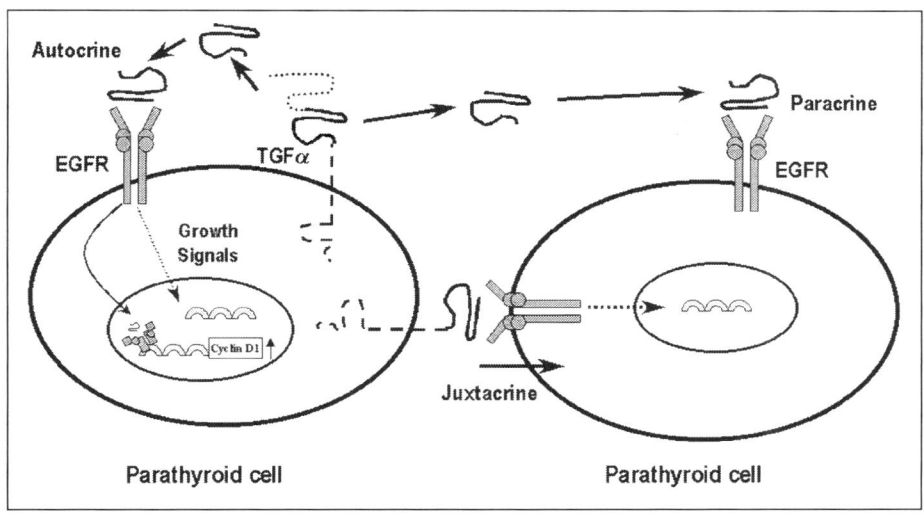

Figure 6. Mechanisms for TGFα/EGFR growth signaling. Conventional EGFR activation of growth signals upon ligand binding at the cell membrane and the TGFα/EGFR complex as a transcriptional enhancer of the cyclin D1 gene (Adapted from refs. 40 and 46).

parathyroid glands since there were no changes in intestinal growth or TGFα content.[27] Thus, high-P induction of TGFα in the uremic rat parathyroid gland may constitute an autocrine signal which further stimulates uremia-induced parathyroid cell proliferation.

Additional support for induction of parathyroid TGFα to mediate the mitogenic properties of high dietary P came from studies in rats one month after 5/6 nephrectomy.[27] The high parathyroid TGFα decreased to levels found in normal controls within three days after reducing dietary P intake from 0.9% to 0.2%. The rapid return of parathyroid TGFα content to normal levels by P restriction also suggests that low P may counteract uremia-induced parathyroid cell growth not only through induction of p21 expression, but also by preventing the enhancement of parathyroid TGFα.[27]

Figure 6 summarizes the molecular mechanisms for TGFα induced proliferating activity in the parathyroid glands. The mature, soluble form of TGFα is produced from double proteolytic cleavage of a transmembrane precursor.[40] Both membrane anchored and soluble forms of TGFα signal through the epidermal growth factor receptor (EGFR),[40,41] which is normally expressed in the parathyroid glands.[42] Increases in parathyroid TGFα could induce cell growth through autocrine and paracrine mechanisms upon activation of the EGFR by the mature TGFα isoform, and, as demonstrated in other tissues, through a less characterized juxtacrine pathway involving the transmembrane TGFα isoform from an adjacent parathyroid cell.[43,44]

The EGFR is a 170 KD membrane glycoprotein with an extracellular-ligand binding domain, a short transmembrane helix and an intracellular domain which has tyrosine kinase activity. Ligand binding induces receptor dimerization and simultaneous activation of its intrinsic tyrosine kinase. Upon activation, EGFR signals to the nucleus mainly through the Ras/mitogen-activated protein kinase (MAPK). Following MAPK activation, cyclin D1 is induced and drives the cell cycle from the G1 to the S phase.[45] In addition to this signal transduction pathway typical for cell membrane receptors, a role for nuclear EGFR as a transcription factor has recently been reported. The in vivo demonstration of nuclear EGFR associated with adenosine-thymidine-rich regions in the cyclin D1 promoter may partially explain the strong correlation between nuclear EGFR localization and high proliferating activity in several tissues.[46]

In human parathyroid glands, Gogusev and coworkers[42] demonstrated the presence of EGFR protein in 4 out of 5 adenomas, in 13 of 15 tissue samples of hyperplasia secondary to renal failure, and in most samples of normal parathyroid tissue. No differences in the expression patterns were observed among groups. However, studies in 104 human hyperplastic parathyroid glands, which failed to detect EGFR protein, showed higher EGFR mRNA expression in carcinoma and primary hyperplasia compared to adenomas and hyperplasia secondary to renal failure.[47] The strong association between high proliferative activity and nuclear EGFR localization[46] could partially account for the discrepancy between protein and mRNA expression in the latter studies.

The concept that co-expression of TGFα and EGFR could contribute to non neoplastic endocrine hyperplasia[48] led us to examine the dietary-P regulation of parathyroid EGFR expression in rat parathyroid glands. Figure 5 shows that similar to the changes in TGFα expression, high dietary P enhances parathyroid EGFR content to levels above normal, while P restriction prevented the increases in EGFR levels.

These findings indicate that the induction of parathyroid co-expression of TGFα and its receptor, EGFR, by uremia, acts as a mitogenic signal which is further enhanced by high dietary P, can be prevented by P restriction and counteracted through induction of p21. These are novel insights into the molecular mechanisms associated with the potent opposing effects of high and low P regulation of parathyroid growth. They suggest that, in addition to P restriction or the use of P binders, therapeutic maneuvers oriented to further induce p21 and inactivate TGFα/EGFR growth-promoting signals could be more effective in slowing the progression of secondary hyperparathyroidism. Despite this important step forward, the most critical target to optimize therapy, the parathyroid sensing mechanism for changes in extracellular inorganic P, remains unknown.

The assessment of the efficacy of $1,25(OH)_2D_3$ in enhancing parathyroid p21 to levels capable of counteracting the mitogenic signals triggered by uremia and high dietary P was the next step in our research.

Vitamin D Regulation of Uremia- and High Phosphate-Induced Parathyroid Cell Growth

Several laboratories have demonstrated a role for $1,25(OH)_2D_3$ in suppressing parathyroid cell proliferation in vitro,[49] and in vivo in uremia-induced PT hyperplasia in 5/6 nephrectomized rats.[50] As mentioned earlier, in several normal and transformed cell lines, the antiproliferative effects of $1,25(OH)_2D_3$ involve the induction of the Cdk-inhibitor p21, through a transcriptional mechanism that requires the vitamin D receptor. It is likely that this mechanism for $1,25(OH)_2D_3$-induction of growth arrest also operates in the parathyroid glands. An additional mechanism reported for $1,25(OH)_2D_3$ suppression of parathyroid cell proliferation is by decreasing the expression of c-myc,[49] an early replication-associated gene that regulates the progression from G1 to S phase in the cell cycle. Resistance to vitamin D action, a common occurrence in end stage renal disease, limits the efficacy of therapy. The causes for parathyroid resistance to vitamin D include a reduction in vitamin D receptor content in the parathyroid gland, and an impaired capacity of the $1,25(OH)_2D_3$/VDR complex to regulate the expression of vitamin D responsive genes. The latter persists even after correction of both serum $1,25(OH)_2D_3$ and parathyroid vitamin D receptor content with vitamin D therapy. Since hyperphosphatemia is a major determinant of the resistance to vitamin D, preliminary studies examined the efficacy of vitamin D treatment [$1,25(OH)_2D_3$ and its analog 19-Nor-$1,25(OH)_2D_2$] in counteracting the mitogenic signals induced by early uremia and high dietary P.[51] Doses of 4 ng of $1,25(OH)_2D_3$ and 30 ng of 19-nor-$1,25(OH)_2D_2$ were effective in controlling parathyroid hyperplasia with no hypercalcemic or hyperphosphatemic

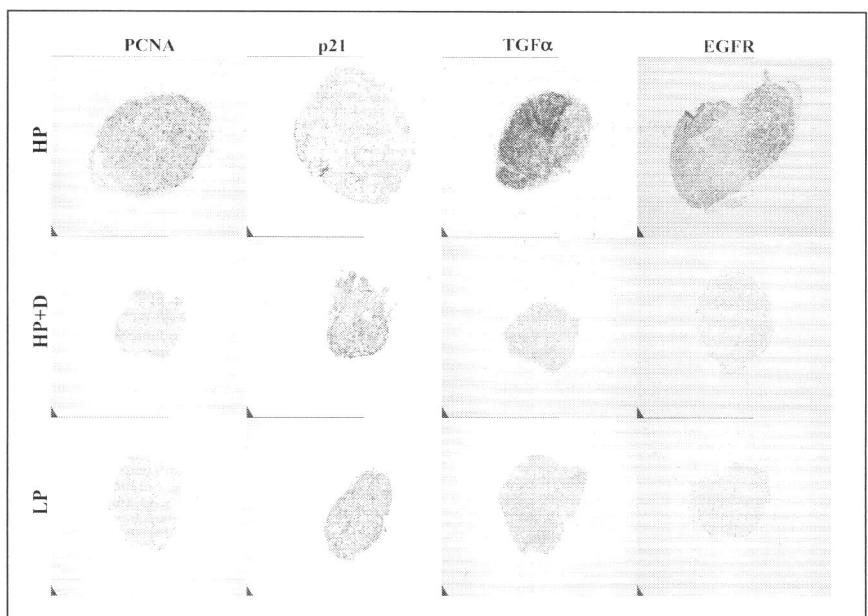

Figure 7. Dietary P- and vitamin D regulation of parathyroid growth. Regulation of parathyroid content of PCNA, p21, TGFα and EGFR by high dietary P (HP), vitamin D therapy (HP+ D) and P restriction (LP), in rats, seven days after 5/6 nephrectomy.

effects.[51] At these doses, both vitamin D compounds prevented the increase in PTH levels, mitotic activity and the enlargement of the parathyroid glands normally induced by uremia and worsened by high dietary P. In this early-uremia rat model, the efficacy of $1,25(OH)_2D_3$ and 19-nor-$1,25(OH)_2D_2$ in controlling both serum PTH and parathyroid gland growth was comparable to that described for P restriction.[27] Mechanistically, the suppression of uremic rat parathyroid cell growth by vitamin D treatment can also be partially accounted for enhanced expression of p21 (See Figs. 4 and 5). Studies in patients with secondary hyperparathyroidism[52] support our findings in uremic rats for a role of increased p21 expression in parathyroid growth arrest. The more aggressive nodular form of parathyroid hyperplasia coincided with the lowest p21 expression. Importantly, the lower the p21 content in human hyperplastic parathyroid glands, the lower the expression of the vitamin D receptor. The findings from Tokumoto and collaborators[52] and our own results[51] on induction of parathyroid p21 after vitamin D therapy suggest that $1,25(OH)_2D_3$ and 19-nor-$1,25(OH)_2D_2$ enhancement of parathyroid p21 expression is mediated by the vitamin D receptor, and could partially account not only for the antiproliferative effects of both sterols on high P-induced PT hyperplasia but for the enhanced growth associated with $1,25(OH)_2D_3$ deficiency.

More importantly, in addition to inducing p21, treatment with $1,25(OH)_2D_3$ or its analog 19-Nor-$1,25(OH)_2D_2$, prevented the increases in parathyroid TGFα and EGFR, induced by uremia and further enhanced by high dietary P (Fig. 7). This novel finding suggests that the control of parathyroid TGFα and EGFR expression by vitamin D treatment could also mediate the antiproliferative properties of $1,25(OH)_2D_3$ and 19-nor-$1,25(OH)_2D_2$ on high-P induced parathyroid hyperplasia in renal failure. Similar results were obtained when preventive vitamin D therapy (100 ng of the less calcemic vitamin D analog 19-nor-$1,25(OH)_2D_2$, three times a week, intraperitoneally), was given for two months starting after the onset of 5/6 nephrectomy, in rats fed a high P diet.[51] In these studies, the enlargement of the parathyroid glands was ameliorated by vitamin D therapy. Importantly, this reduction in parathyroid growth was paralleled by a significant reduction in TGFα content.[51] This suggests an important role for the down-regulation of TGFα/EGFR signaling in vitamin D-antiproliferative action, even with prolonged exposure to the mitogenic stimuli by renal failure and high dietary P.

Interestingly, the induction of c-myc is one of the early events following activation of the conventional EGFR signaling pathway.[53] It is possible that in addition to the complex mechanisms mediating $1,25(OH)_2D_3$/VDR inhibition of c-myc gene transcription,[54,55] prevention of the increases in the parathyroid content of TGFα and EGFR by vitamin D also contributes to the decreased c-myc levels.

In summary, our studies provide the first evidence that the inhibition of parathyroid gland growth by vitamin D therapy is associated with both the induction of parathyroid p21 and the prevention of the increases in TGFα and its receptor, EGFR.

The interruption by vitamin D of the autocrine growth pathway involving TGFα and EGFR in parathyroid cells is a very efficient one, since it simultaneously downregulates the expression of ligand and receptor. Exclusive downregulation of TGFα using antisense technology, while capable of reducing hyperproliferative activity in head and neck squamous carcinoma, was ineffective to arrest normal cell growth.[56] On the other hand, in human colon carcinoma, a marked inhibition of EGFR mRNA and protein and increased internalization of EGFR from the plasma membrane appear to mediate the antiproliferative effects of $1,25(OH)_2D_3$. Vitamin D downregulation of EGFR leads to reduced basal- and EGF mediated expression of cyclin D1 at both the mRNA and protein levels.[57] The relative contribution of this downregulation of TGFα or EGFR by vitamin D to parathyroid growth arrest is unknown. The importance of $1,25(OH)_2D_3$/vitamin D receptor–regulation of EGFR signaling is further supported by the demonstration of more aggressive growth patterns in colon carcinomas as the vitamin D receptor decreases.[58]

Additional mechanisms could also contribute to the antiproliferative properties of $1,25(OH)_2D_3$, as demonstrated in breast cancer cells.[59] In addition to inducing p21 and reducing c-myc expression, the inhibitory effects of $1,25(OH)_2D_3$ on the cell cycle machinery controlling G1 to S transition, include the prevention of the activation of cyclin D1/cdk4 by mechanisms unrelated to the increase in p21. $1,25(OH)_2D_3$ inhibits cdk2 by increased targeting of p21, decreased cyclin A and E association, and reduced phosphorylation of Rb.[59]

Taken together these results suggest that the combined effects of $1,25(OH)_2D_3$ deficiency and the resistance to vitamin D action in advanced renal failure, both worsened by hyperphosphatemia, could contribute to the switch from diffuse hyperplasia to the more aggressive nodular growth. Potential mechanisms include the simultaneous enhanced co-expression of TGFα and EGFR in the parathyroid glands, as well as the reduction of parathyroid p21and the additional $1,25(OH)_2D_3$/VDR-dependent antimitogenic signals. Identification of the molecular mechanisms mediating vitamin D downregulation of the TGFα/EGFR growth promoting activity in the hyperplastic parathyroid gland is an ongoing project in our laboratory that hopefully, will help design more effective strategies to arrest parathyroid growth.

Clinically, a more relevant question is whether vitamin D therapy is capable of suppressing parathyroid growth in established secondary hyperparathyroidism, specially in cases of concomitant hyperphosphatemia. To this end, 5/6 nephrectomized rats were fed a high P diet for two months to induce secondary hyperparathyroidism. At this point, rats were divided in two groups and received either vehicle or the vitamin D analog 19-nor $1,25(OH)_2D_2$, at a dose of 200 ng, intraperitoneally, three times a week, for two additional months.[60] While the parathyroid gland weight in the uremic rats receiving vehicle doubled between two and four months, no further increase in gland size was observed in the vitamin D treated uremic rats. Parathyroid expression of markers of mitotic activity confirmed the efficacy of vitamin D therapy in arresting growth, thus preventing parathyroid gland enlargement. The mechanisms mediating growth arrest, however, were not examined in these studies. Importantly, there were no signs of apoptosis in vitamin D treated rats indicating that there was prevention of further growth without regression of hyperplasia. The recent design of vitamin D analogs with potent pro-apoptotic capabilities in several cell types raises hope for new therapeutic approaches for parathyroid hyperplasia.[61]

Calcium Regulation of Uremia-Induced Parathyroid Growth

Further support for the pathophysiological relevance of changes in the expression of parathyroid p21, TGFα, and EGFR in controlling proliferative activity came from studies which evaluated the expression of these three proteins after the suppression of parathyroid hyperplasia by high dietary Ca or its further enhancement by low-Ca intake. High dietary Ca controlled uremia-induced parathyroid hyperplasia as demonstrated by a reduction of both parathyroid gland size and the expression of two markers of mitotic activity, Ki67 and PCNA.[51] Figures 4 and 8 show that similar to vitamin D treatment, high dietary Ca enhanced parathyroid p21 levels and prevented the increases in parathyroid content of TGFα and EGFR induced by uremia.

The mechanisms for high-Ca induction of p21 and prevention of the increases in TGFα and EGFR are unknown. Studies in vitamin D receptor-ablated mice demonstrated the ability of a Ca-enriched diet to prevent the development of parathyroid hyperplasia in both hypocalcemic and normocalcemic states.[62] This clearly demonstrates that parathyroid growth arrest occurred through mechanisms independent of vitamin D receptor- or serum Ca levels. Parathyroid p21 expression, however, was not assessed in these studies. In relation to high-Ca control of the TGFα/EGFR growth promoting signal, an association was recently reported between hypercalcemia and

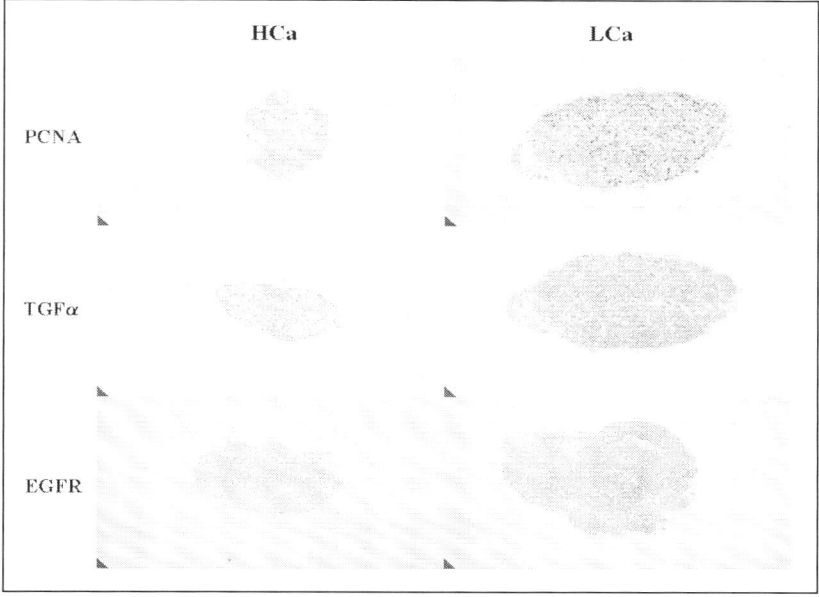

Figure 8. Dietary Ca regulation of parathyroid growth. Regulation of parathyroid expression of PCNA, TGFα and EGFR by high dietary Ca (HCa) and Ca restriction (LCa), in rats, seven days after 5/6 nephrectomy.

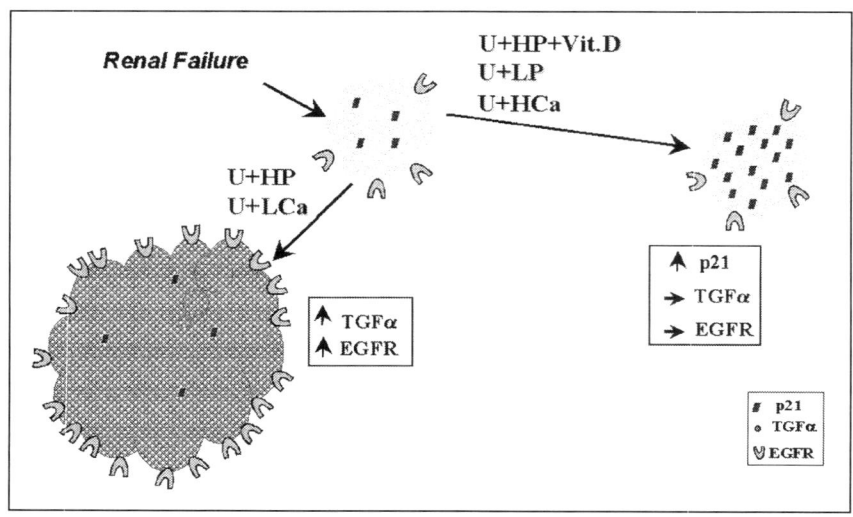

Figure 9. Molecular mechanisms regulating parathyroid hyperplasia in early uremia.

low plasma levels of TGFα in cancer patients, suggesting the possibility of systemic control of TGFα expression by Ca.[63] The persistence of such an association in renal failure patients, could partially explain the suppression of parathyroid growth by hypercalcemia.

The changes in p21, TGFα and EGFR in the parathyroid glands of uremic rats fed a high Ca diet suggest that increases in serum Ca or in intracellular Ca, induced by vitamin D therapy, could potentiate the effects of vitamin D itself in enhancing p21 expression and preventing the increases in TGFα and EGFR.

In contrast to the mechanisms for the antiproliferative effects of high dietary Ca, Figure 8 shows that, low-Ca intake exacerbated the parathyroid growth triggered by uremia by inducing TGFα and EGFR.

Conclusions

The molecular mechanisms regulating parathyroid hyperplasia in early uremia, according to our in vivo model, are depicted in Figure 9. The levels of co-expression of TGFα and EGFR, induced by uremia and further enhanced by either high dietary P or Ca restriction, correlate directly with increased proliferating activity and the enlargement of the parathyroid glands. Conversely, hypercalcemia, P restriction and vitamin D therapy, by preventing the increase in parathyroid TGFα and EGFR and simultaneous induction of p21, counteract the mitogenic signals and arrested gland growth.

The demonstration that the three main regulators of parathyroid growth, Ca, P and vitamin D modulate TGFα/EGFR signaling and p21 expression indicate the importance of these pathways in the pathogenesis of parathyroid hyperplasia and mark them as appropriate targets for more effective therapy.

Acknowledgments

This work was partially supported by Abbott Laboratories and Genzyme Int. The authors thank Ms. Yan Lu for her valuable assistance in immunohistochemical analysis, Dr. Adrian Arakaki for his supervision in computing imaging, and Ms. Jane Finch for careful proofreading of this manuscript.

References

1. Parfitt AM. The hyperparathyroidism of chronic renal failure: A disorder of growth. Kidney Int 1997; 52:3-9.
2. Slatopolsky E, Finch J, Denda M et al. Phosphorus restriction prevents parathyroid gland growth. High phosphorus directly stimulates PTH secretion in vitro. J Clin Invest 1996; 97:2534-2540.
3. Silver J, Bar Sela S, and Naveh-Many T. Regulation of parathyroid cell proliferation. Curr Op Nephrol Hyperten 1997; 6:321-326.
4. Gonzales EA, Martin KJ. Renal osteodystrophy: pathogenesis and management. Nephrol Dial Transplant 1995; 3:13-21.
5. Goodman WG, Goldin J, Kuizon BD et al. Coronary-artery calcification in young adults with end-stage renal disease who are undergoing dialysis. N Engl J Med 2000; 342:1478-1483.
6. Cozzolino M, Dusso A, E. Slatopolsky. Role of calcium x phosphate product and bone-associated proteins on vascular calcification in renal failure. J Am Soc Nephrol 2001; 12:2511-2516.
7. Naveh-Many T, Rahaminow R, Livni N, and Silver J. Parathyroid cell proliferation in normal and chronic renal failure in rats: The effects of calcium, phosphate and vitamin D. J Clin Invest 1995; 96:1786-1793.
8. Kifor O, Moore FD, Wang P et al. Reduced immunostaining for the extracellular Ca sensing receptor in primary and uremic secondary hyperparathyroidism: J Clin Endocrinol Metab 1996; 81:1598-1606.
9. Fukuda N, Tanaka H, Tominaga Y et al. Decreased 1,25-dihydroxyvitamin D3 receptor density is associated with a more severe form of parathyroid hyperplasia in chronic uremic patients. J Clin Invest 1993; 92:1436-1443.
10. Slatopolsky E. The role of calcium, phosphorus and vitamin D metabolism in the development of secondary hyperparathyroidism. Nephrol Dial Transplant 1998; 13(S3):3-8.
11. Silver J, Yalcindag C, Sela-Brown A et al. Regulation of the parathyroid hormone gene by vitamin D, calcium and phosphate. Kidney Int 1999; 73:S2-S8.
12. Sela-Brown A, Silver J, Brewer G et al. Identification of AUF1 as a parathyroid hormone mRNA-3'-untranslated region-binding protein that determines parathyroid hormone mRNA stability. J Biol Chem 2000; 275: 7424-7429.
13. Denda M, Finch J, Slatopolsky E. Phosphorus accelerated the development of parathyroid hyperplasia and secondary hyperparathyroidism in rats with renal failure. Am J Kidney Dis 1996; 28:596-602.
14. Pestell RG Albanese C, Reutens AT et al. The cyclins and cyclin-dependent kinase inhibitors in hormonal regulation of proliferation and differentiation. Endocrine Reviews 1999; 20(4):501-534.
15. Nasmyth K. Viewpoint: Putting the cell cycle in order. Science 1996; 274:1643-1645.
16. Roberts JM, Koff A, Polyak K et al. Cyclins, cdks and cyclin kinase inhibitors. Cold Spring Harbor Symp on Quantitative Biology LIX 1994:31-38.
17. Zhang P, Duchambon P, Gogusev J et al. Apoptosis in parathyroid hyperplasia of patients with primary or secondary uremic hyperparathyroidism. Kidney Int 2000; 57:437-445.
18. Jara A, Gonzales S, Felsenfeld AJ et al. Failure of high doses of calcitriol and hypercalcaemia to induce apoptosis in hyperplastic parathyroid glands of azotaemic rats: Nephrol Dial Transplant 2001; 16:506-512.
19. Canalejo A, Almaden Y, Torregrosa V et al. The in vitro effects of calcitriol on parathyroid cell proliferation and apoptosis: J Am Soc Nephrol 2000; 11:1865-1872.
20. Vasef MA, Brynes RK, Sturm M et al. Expression of cyclin D1 in parathyroid carcinomas, adenomas, and hyperplasia: a paraffin immunohistochemical study. Mod Pathol 1999; 12(4):412-416.
21. Imanishi Y, Hosokawa Y, Yoshimoto K et al. Primary hyperparathyroidism caused by parathyroid-targeted overexpression of cyclin D1 in transgenic mice: J Clin Ivest 2001; 107:1093-1101.
22. Tominaga Y, Tsuzuki T, Uchida K et al. Expression of Prad1/cyclin D1, retinoblastoma gene products, and Ki67 in parathyroid hyperplasia caused by chronic renal failure versus primary adenoma. Kidney Int 1999; 55:1375-1383.
23. Erickson LA, Jin L, Thompson GB et al. Parathyroid hyperplasia, adenomas, and carcinomas: differential expression of p27Kip1 protein. Am J Surg Pathol 1999; 23(3):288-295.

24. Brown AJ, Zhong M, RitterC et al. Loss of calcium responsiveness in cultured bovine parathyroid cells is associated with decreased calcium receptor expression: Biochem Biophys Res Commun 1995; 212:861-867.
25. Wang Q, Palnitkar S, Parfitt AM. Parathyroid cell proliferation in the rat. Effect of age, and of phosphate administration and recovery. Endocrinol 1996; 137:4558-4562.
26. Wang Q, Palnitkar S, Parffitt AM. The basal rate of cell proliferation in normal human parathyroid tissue: implications for the pathogenesis of hyperparathyroidism. Clin Endocrinol 1997; 46:343-349.
27. Dusso AS, Pavlopoulos T, Naumovich L et al. p21waf1 and TGFα mediate dietary phosphate-regulation of parathyroid cell growth. Kidney Int 2001; 59:855-865.
28. Portale AA, Halloran BP, Morris RC Jr. Physiologic regulation of serum concentration of 1,25-dihydroxyvitamin D by phosphorus in normal men. J Clin Invest 1989; 83:1494-1499.
29. Portale AA, Booth BE, Halloran BP et al. Effect of dietary phosphorus on circulating concentrations of 1,25-dihydroxyvitamin D and immunoreactive parathyroid hormone in children with moderate renal insufficiency. J Clin Invest 1984; 73:1580-1589.
30. Liu M, Lee MH, Cohen M et al. Transcriptional activation of the Cdk inhibitor p21 by vitamin D3 leads to the induced differentiation of the myelomonocytic cell line U937. Gene Develop 1996; 10:142-153.
31. DiCunto F, Topley G, Calautti E et al. Inhibitory function of p21Cip1/WAF1 in differentiation of primary mouse keratinocytes independent of cell cycle control. Science 1998; 280:1069-1072.
32. Zhuang SH, Burnstein KL. Antiproliferative effect of 1alpha,25-dihydroxyvitamin D3 in human prostate cancer cell line LNCaP involves reduction of cyclin-dependent kinase 2 activity and persistent G1 accumulation. Endocrinol 1998; 139:1197-1207.
33. Ohtsubo M, Gamou S, Shimizu N. Antisense oligonucleotide of WAF1 gene prevents EGF-induced cell cycle arrest in A431 cells. Oncogene 1998; 16:797-802.
34. Miyazaki M, Ohashi R, Tsuji T et al. Transforming growth factor-beta1 stimulates or inhibits cell growth via down-or up-regulation of p21/Waf1. Biochem Biophys Res Com 1998; 246:873-880.
35. Shankland SJ, Floege J, Thomas SE et al. Cyclin kinase inhibitors are increased during experimental membranous nephropathy: Potential role in limiting glomerular epithelial cell proliferation in vivo. Kidney Int 1997; 52:404-413.
36. Macleod KF, Sherry N, Hannon G et al. p53-dependent and independent expression of p21 during cell growth, differentiation, and DNA damage. Gene Dev 1995; 9:935-944.
37. Bonfanti M, Taverna S, Salmona M et al. P21WAF1-derived peptides linked to an internalization peptide inhibit human cancer cell growth. Cancer Res 1997; 57:1442-1446.
38. Yang ZY, Perkins ND, Ohno T et al. The p21 cyclin-dependent kinase inhibitor suppresses tumorigenicy in vivo. Nature Med 1995; 1:1052-1056.
39. Gartel AL, Serfas MS, Tyner AL. p21-Negative regulator of the cell cycle. Proc Soc Exp Biol Med 1996; 213:138-149.
40. Kumar V, Bustiin SA, McKay IA. Transforming growth factor alpha. Cell Biol Int 1995; 19:373-388.
41. Driman DK, Kobrin MS, Kudlow JE et al. Transforming growth factor-α in normal and neoplastic human endocrine tissues. Hum Pathol 1992; 23:1360-1365.
42. Gogusev J, Duchambon P, Soermann-Chopard C et al. De novo expression of transforming growth factor-α in parathyroid gland tissue of patients with primary or secondary uraemic hyperparathyroidism. Nephrol Dial Transpl 1996; 11:2155-2162.
43. Wong ST, Winchell LF, McCune BK et al. The TGF-alpha precursor expressed on the cell surface binds to the EGF receptor on adjacent cells, leading to signal transduction. Cell 1989; 56:495-500.
44. Shum L, Reeves SA, Kuo AC et al. Association of the transmembrane TGFα precursor with a protein kinase complex. J Cell Biol 1994; 125:903-916.
45. Wells A. Molecules in focus. EGF receptor. Int J Biochem Cell Biol 1999; 31:637-643.
46. Lin SY, Makino K, Xia W et al. Nuclear localization of EGF receptor and its potential new role as a transcription factor: Nature Cel Bio 2001; 3:802-808.
47. Sadler GP, Morgan JM, Jasani B et al. Epidermal growth factor receptor status in hyperparathyroidism: Immunocytochemical and in situ hybridization study. World J Surg 1996; 20:736-743.
48. Studer H, Derwahl M. Mechanisms of noneoplastic endocrine hyperplasia-A changing concept: A review focused on the thyroid gland: Endocrin Rev 1995; 16:411426.

49. Kremer R, Bolivar I, Goltzman D et al. Influence of calcium and 1,25-dihydroxycholecalciferol on proliferation and proto-oncogene expression in primary cultures of bovine parathyroid cells. Endocrinology 1989; 125(2):935-941.
50. Szabo A, Merke J, Beier E et al. 1,25(OH)2 vitamin D3 inhibits parathyroid cell proliferation in experimental uremia. Kidney Int 1989; 35:1049-1056.
51. Cozzolino M, Lu Y, Finch JL et al. p21 waf1 and TGFα mediate parathyroid growth arrest by vitamin D and high calcium. Kidney Intern 2001; 60:2109-2117.
52. Tokumoto M, Hirakawa M, Tsuruya K et al. Diminished expressions of cyclin-dependent kinase inhibitor p21 and vitamin D receptor in nodular hyperplasia and in secondary hyperparathyroidism. J Am Soc Nephrol 2000; 11:584A.
53. Tong WM, Kallay E, Hofer H et al. Growth regulation of human cancer cells by epidermal growth factor and 1,25-dihydroxyvitamin D3 is mediated by mutual modulation of receptor expression: Eur J Cancer 1998; 34:2119-2125.
54. Taoka T, CollinsED, Irino S et al. 1,25-(OH)2-vitamin D3 mediated changes in mRNA for c-myc and 1,25-(OH)2D3 receptor in HL60 cells and related subclones. Mol Cell Endocrinol 1993; 95:51-57.
55. Simpson RU, Hsu T, Wendt MD et al. 1,25-dihydroxyvitamin D3 regulation of c-myc protooncogene transcription. Possible involvement of protein kinase C. J Biol Chem 1989; 264:19710-19715.
56. Grandis JR, Chakraborty A, Zeng Q et al. Downmodulation of TGFα protein expression with antisense oligonucleotides inhibits proliferation of head and neck squamous carcinoma but not normal mucosal epithelial cells. J Cell Biochem 1998; 69:55-62.
57. Tong WM, Hofer H, Ellinger A et al. Mechanism of antimitogenic action of vitamin D in human colon carcinoma cells: Relevance for suppression of epidermal growth factor stimulated cell growth: Oncol Res 1999; 11:77-84.
58. Sheinin Y, Kaserer, Wrba F et al. In situ mRNA hybridization analysis and immunolocalization of the vitamin D receptor in normal and carcinomatous human colonic mucosa: relation to epidermal growth factor receptor expression: Virchows Arch 2000; 437:501-507.
59. Jensen SS, Madsen MW, Lukas J et al. Inhibitory effects of 1,25-dihydroxyvitamin D3 on the G1-S phase-controlling machinery. Mol Endocrinol 2001; 15:1370-1380.
60. Ritter C, Lu Y, Finch JL et al. 19-Nor-1,25(OH)2D2 (19-Nor), an analog of calcitriol, suppresses parathyroid gland growth in uremic rats with established secondary hyperparathyroidism: J Am Soc Nephrol 2000; 11:581A.
61. Nakagawa K, Kurobe M, Konno K et al. Structure-specific control of differentiation and apoptosis of human promyelocytic leukemia (HL-60) cells by A-ring diasteroisomers of 2-methyl-1, 25-dihydroxyvitamin D3 and its 20-epimer: Biochem Pharmacol 2000; 60:1937-1947.
62. Amling M, Priemel M, Holzmann T et al. Rescue of the skeletal phenotype of vitamin D receptor-ablated mice in the setting of normal mineral ion homeostasis: formal histomorphometric and biochemical analyses. Endocrinology 1999; 140:4982-4987.
63. Motellon JL, Jimenez FJ, de Miguel F et al. Relationship of plasma bone cytokines with hypercalcemia in cancer patients. Clin Chim Acta 2000; 302:59-68.

CHAPTER 10

Molecular Mechanisms in Parathyroid Tumorigenesis

Eitan Friedman

Primary hyperparathyroidism (1 HPT) is a relatively common disorder, with an estimated prevalence of 1:500-1:1000.[1] In about 80%-85% of 1 HPT cases, a single enlarged parathyroid gland with a distinct capsule that separates the tumorous from the nontumorous tissue (=adenoma or uniglandular disease) is detected.[1] In the majority of remaining individuals with 1 HPT, more than one parathyroid gland is enlarged, and at times, no distinct capsule can be detected (hyperplasia or multiglandular disease).[2] Both single adenomas and multiple hyperplastic parathyroid tumors are benign lesions, capable of local growth with no metastasizing potential. There are rare cases of parathyroid carcinoma, which account for less than 1% of all 1 HPT cases, a tumor with both local invasion and distant metastasis capabilities.[3] Uniglandular 1 HPT is usually noted in a single individual within a family, where there are no additional family members with either 1 HPT or other endocrinopathies (=sporadic 1 HPT). In about 10% of 1 HPT cases there are other family members affected with either 1 HPT (familial HPT) or other endocrinopathies, and these cases are at times associated with multiglandular disease. Several distinct inherited neoplastic disorders have 1 HPT as part of the spectrum of tumors that occur in affected families: multiple endocrine neoplasia type I (MEN1) (MIM # 131100), multiple endocrine neoplasia type 2A (MEN2A) (MIM # 171400), and familial 1 HPT with (MIM #145001) or without (MIM # 145000) jaw tumors.

The majority of human neoplasms, malignant as well as benign, are considered monoclonal tumors: they originate from a single cell, the clone precursor cell that acquires a genetic mutation which leads to an increased proliferation rate of the resulting clone, via a variety of molecular mechanisms. In order to continue and accelerate the uncontrolled proliferative activity, the clone precursor cell acquires subsequent mutations in different genes from different gene families. These subsequent genetic alterations lead to subclonal expansion of the original clone, and result in a clinically detectable tumor mass. Several gene families are intimately involved in promoting and facilitating uncontrolled cellular proliferation: oncogenes, tumor suppressor genes, telomerase genes, and DNA mismatch repair genes (reviewed in refs. 4,5).

The clonal origin of parathyroid tumors was evaluated by several investigators over the years, using a variety of technologies. Initial studies from the late 1970's and early 1980's were based on a limited number of tumors and utilized patterns of X chromosome inactivation by measurements of protein isoforms of glucose 6-phosphate dehydrogenase (G-6-PD), an X-linked gene product.[6,7] These studies surprisingly reported that parathyroid adenomas are polyclonal tumors, rather than the expected monoclonal neoplasms. In 1988, Arnold and coworkers[8] have readdressed the issue of monoclonal origin of parathyroid tumors, and applied both a DNA-based technique that measures the X-inactivation pattern in heterozygous women, and

---

Molecular Biology of the Parathyroid, edited by Tally Naveh-Many. ©2005 Eurekah.com and Kluwer Academic / Plenum Publishers.

genetic analysis (Southern blotting) to look for major gene rearrangements involving the parathyroid hormone (PTH) gene, as surrogate markers heralding the monoclonal origin of parathyroid adenomas. This breakthrough study clearly demonstrated that the majority, if not all, parathyroid tumors are in fact monoclonal. This changed concept about the clonal origin of sporadic, uniglandular parathyroid tumors, paved the way for other studies that directly or indirectly addressed the issue of clonality of parathyroid tumors encountered in different clinical settings. Indeed, MEN1-associated parathyroid tumors,[9] nonfamilial, primary parathyroid hyperplasia,[10] parathyroid hyperplasia seen in uremic patients,[10,11] and parathyroid carcinoma,[12,13] were all shown to contain a significant monoclonal component.

The well-established monoclonal origin of parathyroid tumors, prompted a search for tumor-specific genetic alterations and the precise genes that are involved in parathyroid tumorigenesis. As the tumorigenic paradigm of monoclonal tumors stipulates somatic mutations in specific genes initiate uncontrolled parathyroid cellular proliferation and somatic mutations in other genes, sustains and promotes this unregulated proliferation. Finding a coding region mutation that clearly alters the gene function (activating mutations within oncogenes or inactivating mutations in tumor suppressor genes) is the "gold standard" of defining the involvement of a gene in tumorigenesis. Yet, altered expression of the gene's product (at the mRNA or protein levels) by epigeneic mechanisms (e.g., methylation alterations, loss of imprinting, chromatin modification)[14] may also contribute to tumorigenesis.

In addition to the genes that play pivotal role in parathyroid tumorigenesis by virtue of somatic mutations or "expression altering" mechanisms, there are genes that predispose to parathyroid tumorigenesis, in the context of familial or inherited neoplastic syndromes. Individuals who harbor germline mutations within these genes have a significantly increased risk for developing parathyroid tumors during their lifetime, a risk that at times approaches 100%. This subset of genes includes the genes for MEN1,[15] MEN2A,[16] and for familial hyperparathyroidism with or without jaw tumors.[17] In addition to their role in conferring an increased lifetime risk for parathyroid tumor development, these genes are also somatically involved in parathyroid tumorigenic process.[18,19] The definition of the precise genes involved in initiation and promotion of parathyroid tumorigenesis, as well as the genes that predispose to parathyroid tumor development has gained momentum over the past few years, and is the topic of this book chapter. Noteworthy, several recent reviews of the molecular mechanism involved in parathyroid tumorigenesis are recommended for reading.[20-22]

Oncogenes Involved in Parathyroid Tumor Development

Protoncogenes are normal cellular proteins involved primarily in the signal transduction cascade initiated by binding of a ligand to a cell membrane receptor which via a series of phosphorylations of cytoplasmic proteins results in the activation of transcription factors and cellular proliferation (reviewed in refs. 23-24). When either an activating point mutation or overexpression of a nonmutated allele of a protooncogene occurs (by either DNA amplification or chromosomal translocation), the protooncogene is converted to an oncogene that promotes ligand-independent cellular proliferation. One of the oncogenes notably involved in parathyroid tumorigenesis is PRAD1 (parathyroid adenomatosis gene 1)/ cyclin D1/CCND1 (MIM# 168461). The PRAD1 gene normally localizes to the long arm of chromosome 11 (at 11q13). In a small subset of parathyroid tumors (about 5%), a major gene rearrangement occurs, involving pericentromeric inversion of one of the chromosome 11 alleles, leading to a juxtaposition of the PRAD1 gene and the 5' region of the PTH gene (localized to 11p15). This juxtaposition leads to unregulated overexpression of the PRAD1 protein, specifically in parathyroid tissue, by having the nonmutated PRAD1 gene under the control of tissue specific enhancers of the PTH gene.[25] Overexpression of PRAD1 with no demonstrable gene rearrangement has been shown by immunohistochemistry in 20-40% of parathyroid adenomas, as well as parathyroid hyperplasia and carcinoma.[26-28] The underlying mechanism of overexpression of the nonmutated, nonrearranged PRAD1 in parathyroid tumors is probably trans-acting regulatory derangement. To date, no somatic PRAD1 activating point mutations have been reported in parathyroid tumors,[29] but in a transgenic mouse model where the transgene construct mimicked the human PTH-cyclin D1 rearrangement, the resulting phenotype was that of hyperproliferative parathyroids, biochemical evidence of hypersecretion of PTH and hypercalcemia, with bone morphological changes that are similar to those of human 1 HPT.[30] The PRAD1 protein contains 295 amino acids, and belongs to a family of cyclins (hence the alternative name cyclin D1), proteins that play a pivotal role in regulation of cell cycle progression, specifically, progression through the G1 phase and G1-S transition, thus regulating whether a new cell cycle occurs (reviewed in ref. 31). The involvement of cyclins in cell cycle control is mediated by binding to and activation of cyclin dependent kinases (CDK's). In Contrast, beta catenin, a known regulator of cyclin D1 transcription, was not shown to harbor somatic mutations in exon 3 or to have an altered expression pattern in parathyroid tumors.[32]

Several oncogenes have been shown to be overexpressed in benign parathyroid tumors. The involvement of these oncogenes in parathyroid tumorigenesis was evaluated because it was either biologically plausible, or they have been isolated from parathyroid tumors and displayed transforming capability in vitro. An oncogene that was shown to be overexpressed at the protein level in sporadic uni- and multiglandular parathyroid tumors, is the Int-2 gene product, fibroblast growth factor 3 (FGF-3), a known growth promoter of parathyroid cells in culture.[33] Similarly, Epidermal growth factor receptor (EGFR) protein was overexpressed by immunostaining in 5/12 (41.6%) sporadic parathyroid tumors.[34] A novel, alternatively spliced form of the keratinocyte growth factor receptor (KGFR) isolated from parathyroid adenoma has been shown to possess oncogenic activity and transform NIH 3T3 cells.[35] In contrast, activating point mutations in Ha-Ki- and N- RAS genes, oncogenes involved in the pathogenesis of a large variety of human cancers, have not been detected in benign parathyroid tumors.[36,37] Similarly, *gsp* and *gip2* mutations, activating oncogenic mutations that have been described in other endocrine tumors, most notably in pituitary and thyroid tumors (*gsp*), adrenal and ovarian tumors (*gip2*),[38] have not been detected in benign parathyroid tumors.[39] Missense activating germline mutations of the RET proto-oncogene have been identified in the hereditary cancer syndrome MEN2A (MIM# 171400), characterized by the cooccurrence and familial clustering of medullary thyroid carcinoma, pheochromocytoma and parathyroid tumors. In addition, somatic RET proto-oncogene mutations have been identified in a subset of sporadic medullary carcinomas and pheochromocytomas (reviewed in ref. 40). Yet, no activating somatic point mutations have been detected in the RET protooncogene in sporadic parathyroid tumors.[41,42]

Indications as to the putative chromosomal regions that may contain oncogenes involved in parathyroid tumorigenesis have emerged from cytogeneic analyses. In a single parathyroid adenoma, a cytogenetically visible translocation between chromosome 1 and 5 has been reported.[43] Applying comparative genomic hybridization (CGH) to benign parathyroid tumors and targeting chromosomal regions that display an increase in DNA copy number (=amplification) may reveal novel oncogenes involved in parathyroid tumorigeneis. In the few CGH studies published,[44-46] regions on chromosomes 7, 16, and 19 have been consistently amplified. Yet, the specific genes that localize to these regions have not yet been defined.

Tumor Suppressor Genes Involved in Parathyroid Tumorigenesis

Inactivation of both alleles of a tumor suppressor gene (=bi-allelic inactivation) is a prerequisite for tumor development. Such inactivation serves to deplete the cell of the antiproliferative control that the gene product normally exerts, and uncontrolled cellular proliferation ensues. In most cases, a point mutation inactivates one allele (usually by causing a frameshift, nonsense mutation, or creating a premature stop codon), whereas the other allele is inactivated by a gross somatic deletion of the wild type bearing allele. This somatic deletion is heralded by allelic loss (Loss Of Heterozygosity—LOH), seen when the genotype of the tumor DNA is compared with the genotype of the nontumorous DNA from the same individual, using DNA markers linked to the chromosomal region that bears the tumor suppressor gene (reviewed in ref. 47). Allelotyping of parathyroid tumors revealed that there are nonrandom, chromosomal regions that display LOH, thus putatively harbor tumor suppressor genes that are relevant to parathyroid tumorigenesis.

One of the regions that most commonly display LOH in parathyroid tumors is 11q13, at the MEN1 locus.[48-50] Multiple Endocrine neoplasia type 1 (MEN1) is a rare syndrome characterized by the clustering in families (in an autosomal dominant inheritance pattern) or in individuals (sporadic cases) of benign parathyroid tumors, pancreaticoduodenal endocrine tumors (mostly benign) and pituitary adenomas. Other tumors, such as carcinoids, lipomas, nonfunctioning adrenal tumors, and angiofibromas also occur at a high rate in MEN1 patients (reviewed in refs. 51-52). Using a combined approach of allelotyping of malignant insunlinomas from MEN1 patients and subsequent family linkage analyses, Larsson and coworkers,[53] localized the MEN1 gene to the long arm of chromosome 11 (11q13). Subsequent linkage studies verified this location,[54,55] and allelotyping of MEN1-associated parathyroid tumors revealed that the majority of tumors displayed LOH using MEN1-linked markers.[9] Additionally, using the same set of genetic markers, LOH was also present in about one third of sporadic parathyroid tumors.[9] With the cloning of the MEN1 gene, somatic inactivating mutations have been detected in both MEN1-associated (in most tumors) and sporadic parathyroid tumors (in about to 20% of sporadic tumors).[8,19,56,57] The rate of LOH at 11q13 in sporadic parathyroid tumors is almost double that of the rate of MEN1 gene somatic mutations,[8,56,57] raising the possibility that there is yet an unidentified parathyroid related tumor suppressor gene on 11q13 that is targeted for inactivation by LOH.[58]

The MEN1 gene product, MENIN, is a 610 amino acid nuclear protein with no strong homologies to any known proteins, which has been shown to play a role in regulating Jun D transcription factor.[59] Protein-protein interactions with other proteins have been shown: Smad3 (a protein that acts in the TGF-β pathway) and NF-κB transcription factors.[60-62] These latter proteins have been shown to bind to the Cyclin D1 promotor and increase its transcriptional activity, whereas binding to MENIN inhibits their activity. Thus, parenthetically and speculatively, loss of MENIN in parathyroid tissue (by biallelic inactivation) may result in more binding of NF-kβ to Cyclin D1 promotor, which leads to an increase in its parathyroid pro-proliferative activity.

In support of the tumor suppressor activity of the *MEN1* gene, a mouse model was generated, where the mouse homologue of the *Men1* gene was heterozygously knocked out. The resulting phenotype included parathyroid hyperplasia (without hypercalcemia), as well as pancreatic endocrine, pituitary and adrenal tumor formation,[63] all part of the tumor spectrum of human MEN1 phenotype.

Allelotyping studies revealed other regions in the genome that are nonrandomly, clonally deleted in parathyroid tumors: 1p,[64,65] 1q,[66] 6q, 9p, and 15q.[67,68] These regions putatively contain tumor suppressor genes involved in parathyroid tumorigenic process. Applying CGH to parathyroid tumors, 20q12-13 displayed a nonrandom decrease in DNA copy numbers, indicative of the existence of tumor suppressor gene(s) relevant to parathyroid tumor development in that chromosomal region, in addition to 11q, 1p and the other regions displaying LOH detected by allelotyping.[69]

Of the chromosomal loci showing LOH, the most common region is at 1p, a region displaying LOH in about 40% of sporadic parathyroid tumors.[12,64] The region that defined the minimally deleted area localizes to 1p36, a region that is also deleted in a subset of medullary thyroid tumors, and pheochromocytomas,[70,71] tumors that form the clinical spectrum of MEN2A. Thus, it would appear, that tumor suppressor gene(s) that locate to 1p36 are important in parathyroid tumorigenesis as well as the pathogenesis of other endocrine tumors. There are several candidate genes that localize to that region, two of whom, the p18, cyclin dependent kinase inhibitor,[66] and the p73 tumor suppressor gene[72] were excluded as involved in parathyroid tumorigenesis, as no somatic inactivating mutations were detected in either gene in the analyzed parathyroid tumors. Intriguingly, the region on 1p shown to be deleted in benign parathyroid adenomas is distinct from a more proximal region on 1p that displays allelic loss in parathyroid carcinomas.[12] This latter finding may indicate that different tumor suppressor genes that localize to 1p are involved in benign and malignant parathyroid tumor development. Two genes that localize to 9p, p15 p16, both cyclin kinase inhibitors, and the PPP2R1B gene that localizes to 11q23, were also excluded as contributors to parathyroid tumorigenesis.[66,73] Another candidate tumor suppressor gene whose involvement was invoked in parathyroid tumorigenesis is the *Smad3* gene: it localizes to 15q, a region displaying nonrandom LOH, and the encoded protein is a TGF-β signaling molecule, a known binding partner of MENIN.[60] Yet, no somatic inactivating mutations could be shown in this gene as well.[74]

Another class of tumor suppressor genes that were evaluated as potential contributors to parathyroid tumorigenesis, are "classical tumors genes": tumor suppressor genes that have been shown to be involved in a variety of tumors, including some of endocrine origin: p53 (MIM# 191170), Rb (MIM# 180200). While no somatic p53 inactivating mutations or positive immunostaining of mutant p53 protein could be shown in most parathyroid tumors analyzed,[75-78] inactivation of the Rb gene seems to significantly contribute to parathyroid carcinoma, but not to benign parathyroid tumors.[79,80] Additionally, loss of Rb expression seems to be limited to aggressive parathyroid tumors.[81] Given this seemingly intimate involvement of the Rb gene in malignant parathyroid tumorigenesis, it was surprising that no somatic inactivating mutations could be shown in the coding region of the Rb gene in parathyroid carcinomas.[13]

There is a well-established association between external ionizing irradiation exposure and the subsequent development of head and neck tumors, including parathyroid neoplasms (reviewed in ref. 82). Thus, genes that are involved in mediating radiation damage repair and meiotic recombination have become candidate genes to be involved in parathyroid tumorigenesis. Three genes, *RAD51* (MIM# 179617), *RAD54* (MIM# 603615) and *BRCA2* (MIM# 600185) are involved in DNA damage repair and localize to 15q, 1p, and 13q, respectively, regions that display nonrandom allelic loss in parathyroid adenomas. Thus, these genes were considered candidate parathyroid tumor suppressor genes. However, there is currently no evidence that any of these genes is somatically inactivated in benign or malignant parathyroid tumors.[13,83,84]

Other Molecular Pathways Involved in Parathyroid Tumorigenesis

A number of genes have been suggested as contributors to parathyroid tumorigenesis by virtue of their known biological function, disease-associated and normal tissue expression patterns, chromosomal locations, and/or involvement in familial syndrome relevant to parathyroid pathological states. Some of these genes were evaluated for harboring somatic mutations in parathyroid tumors, but none proved to harbor such pathogenic mutations. Thus, while these genes do not have an established role in parathyroid tumorigenesis, they may still be important indirect contributors to hyperparathyroidism.

Microsatellite Instability in Parathyroid Tumors

The hallmark of involvement of DNA mismatch repair (DNA-MMR) genes in the tumorigenic pathway, is the demonstration of somatic genomic instability, heralded by microsatellite instability (MSI) (reviewed in ref. 85): comparison of tumor and nontumorous DNA from the same individual shows the existence of novel, different size alleles in the tumor tissue. When the DNA mismatch repair system is nonfunctional by mutational inactivation of one or more of the participating genes or by epigenetic mechanisms, the rate of random accumulation of DNA replication errors increases by two to three orders of magnitude (reviewed in ref. 85). This increased rate of base mismatching usually does not result in a tumorous phenotype, unless an activation of an oncogene or inactivation of a tumor suppressor gene, occurs. The involvement of an abnormal DNA-MMR as determined by the demonstration of MSI is most notable in colon cancer, especially in the context of hereditary nonpolyposis colon cancer (HNPCC) (MIM# 114500),[86] but a variety of other tumors have also shown to display MSI.[87] Only a handful of reports that document MSI in parathyroid tumors exist. In a single large parathyroid adenoma from a young Brazilian girl, MSI was shown in 9/23 markers from chromosomes 1, 10 and 11.[88] MSI was also documented in 6/14 single parathyroid adenomas, especially with chromosome 11 markers, and in 5/6 multiglandular hyperplastic lesions, especially with 17p markers.[89] No mutation detection was attempted in either study, so the precise gene(s) underlying this apparent MSI in parathyroid tumorigenesis are presently unknown. Furthermore, the rarity of the reports on MSI in parathyroid tumors, combined with the extensive allelotyping studies reported in the same tumor types,[67,68] probably indicate that DNA-MMR gene involvement in parathyroid tumorigenesis is a rare event.

Telomerase Activity in Parathyroid Tumors

A teleomere is composed of a large tract of a short (6 bases) repeat—TTAGGG—located at the ends of chromosomes.[90] The size of the telomeres in human chromosomes ranges from 5,000 to 15,000,[91] and their putative function is to protect against degeneration by exonucleases and ligases, protect against chromosomal end fusion, protect from activation of DNA damage check points, and also to play a role in homologous pairing.[92,93] Normal dividing somatic cells lose telomeric sequences progressively with each cell division. When the telomeres have been shortened to a critical length, the cell recognizes the DNA damage and enters a senescence phase.[94] Thus, normal cells have a finite number of cell replications in vitro (about 40-70) before senescence occurs, a feature that is related to the decrease in telomere length.[95] The length of the telomere is in part controlled by the enzyme telomerase, a ribonucleoprotein capable of maintaining telomere length. In normal adult tissues, its levels are negligible, but it is overexpressed in about 90% of human cancers (reviewed in ref. 96). The activity of telomerase can be measured in vivo by the Telomeric Repeat Amplification Protocol (TRAP),[97] whereas the consequences of its activity can be quantified by measuring and comparing telomere lengths in tumorous and nontumorous tissue. Few studies have looked at telomeres as contributors to parathyroid tumor development. Falchetti and coworkers[98] reported that a single metastasis from a parathyroid carcinoma and primary cell culture derived from the primary tumor exhibit a high telomerase activity. Subsequently, Kammori et al[99] demonstrated that telomere length was significantly shorter in benign (uni and polyglandular disease) and malignant parathyroid tumors compared with normal parathyroid tissue, but that telomerase activity was only observed in the malignant parathyroid tumors and in none of the benign parathyroid adenomas. These results may indicate that telomerase activity may distinguish benign from malignant parathyroid tumors, but that telomere shortening is observed in all parathyroid neoplasms, regardless of biological behavior and specific histological features.

Calcium Set Point and the Regulatory Genes Involved

Familial hypocalciuric hypercalcemia (FHH) (MIM# 145980) is an autosomal dominant disorder characterized by parathyroid hyperfunction resulting from reduced sensitivity to extracellular calcium.[100,101] In most cases, a heterozygous inactivating mutation in the calcium sensing receptor (CaSR) gene (localized to 3q13.3-21) (MIM# 601199) can be found.[102-104] Although this disorder is usually not accompanied by parathyroid cellular proliferation, a specific heterozygous mutation was associated with an FHH variant characterized by parathyroid adenomas and hypercalciuria,[105] and a homozygous mutation in CaSR causes severe neonatal hyperparathyroidism (MIM# 239200), a disorder associated with parathyroid hypercellularity.[106] These observations pertaining to inherited disorders associated with parathyroid proliferation and perturbations of calcium metabolism, made it plausible that acquired, biallelic, somatic inactivation of the CaSR gene may contribute to parathyroid tumorigenesis. However, several investigators analyzed sporadic parathyroid tumors for such mutations, with negative results.[107-109] These findings are indicative that inactivating mutations in the CaSR gene do not contribute to parathyroid tumorigenesis in a classical tumor suppressor gene manner, and probably do not offer growth advantage to parathyroid cells. However, decreased expression in CaSR protein assessed by immunohistochemistry was shown in about half of parathyroid adenomas tested.[110-112] Combined with the lack of demonstrable somatic mutations, one plausible interpretation of these results is that mutations in genes involved in the regulation of the calcium sensing pathway (or even epigenetic mechanisms affecting CaSR expression levels) may contribute to parathyroid tumorigenesis, and perhaps play a pivotal role in the well established insensitivity of parathyroid tumors to extracellular calcium.

Vitamin D Receptor

1, 25 dihydroxyvitamin D3, the ligand of the vitamin D receptor (VDR) is capable of inhibiting parathyroid proliferation in vitro.[113,114] This observation, prompted a search for inactivating mutations in the VDR gene in sporadic parathyroid tumors, with negative results in both sporadic primary hyperparathyroidism and in secondary hyperparathyroidism of uremia.[115] Thus, akin to the CaSR gene, the VDR gene does not function as a classical tumor suppressor gene in parathyroid tissue, and its inactivation is noncontributory to parathyroid proliferation. Yet, its involvement in parathyroid tumorigenesis is inferred from the altered expression pattern at the mRNA and protein levels,[116] and the known effects of the ligand, 1, 25 OH VitD3, on PTH secretion, as well as the known effects of vitamin D deficiency on parathyroid cellular proliferation.[117]

Summary

Over the past decade a more comprehensive understanding of the molecular mechanisms involved in initiation and progression of benign and malignant parathyroid tumors has been achieved. These new insights provide better fundamental understanding of the biology and interaction of endocrine tumor related genes and have the potential of providing more accurate, biologically rational diagnostic and therapeutic tools. The application of novel technologies such as microarrays for concomitant analyses of thousands of genes at the RNA and protein levels as well as proteomic technology over the next few years will hopefully lead to new and exciting revelations in parathyroid tumorigenesis.

References

1. Bilezikian JP. Primer in the metabolic bone diseases and disorders of mineral metabiolism. In: Favus MJ, ed. New York: Raven Press, 1993:155-159.
2. Grimelius L, Johansson H. Pathology of parathyroid tumors. Semin Surg Oncol 1997; 13:142-54.
3. Rubello D, Casara D, Dwamena BA et al. Parathyroid carcinoma. A concise review. Minerva Endocrinol 2001; 26:59-64.
4. Knudson AG. Cancer genetics. Am J Med Genet 2002; 111:96-102.
5. Balmain A, Gray J, Ponder B. The genetics and genomics of cancer. Nat Genet 2003; 33(Suppl):238-244.
6. Fialkow PJ, Jackson CE, Block MA et al. Multicellular origin of parathyroid "adenomas". N Engl J Med 1977; 297:696-698.
7. Jackson CE, Cerny JC, Block MA et al. Probable clonal origin of aldosteronomas versus multicellular origin of parathyroid "adenomas". Surgery 1982; 92:875-879.
8. Arnold A, Staunton CE, Kim HG et al. Monoclonality and abnormal parathyroid hormone genes in parathyroid adenomas. N Engl J Med 1988; 318:658-662.
9. Friedman E, Sakaguchi K, Bale AE et al. Clonality of parathyroid tumors in familial multiple endocrine neoplasia type 1. N Engl J Med 1989; 321:213-218.
10. Arnold A, Brown MF, Urena P et al. Monoclonality of parathyroid tumors in chronic renal failure and in primary parathyroid hyperplasia. J Clin Invest 1995; 95:2047-2053.
11. Farnebo F, Farnebo LO, Nordenstrom J et al. Allelic loss on chromosome 11 is uncommon in parathyroid glands of patients with hypercalcaemic secondary hyperparathyroidism. Eur J Surg 1997; 163:331-337.
12. Valimaki S, Forsberg L, Farnebo LO et al. Distinct target regions for chromosome 1p deletions in parathyroid adenomas and carcinomas. Int J Oncol 2002; 21:727-735.
13. Shattuck TM, Kim TS, Costa J et al. Mutational analyses of RB and BRCA2 as candidate tumour suppressor genes in parathyroid carcinoma. Clin Endocrinol (Oxf) 2003; 59:180-189.
14. Lee MP. Genome-wide analysis of epigenetics in cancer. Ann N Y Acad Sci 2003; 983:101-109.
15. Chandrasekharappa SC, Guru SC, Manickam P et al. Positional cloning of the gene for multiple endocrine neoplasia-type 1. Science 1997; 276:404-406.
16. Mulligan LM, Kwok JBJ, Healey CS et al. Germ-line mutations of the RET proto-oncogene in multiple endocrine neoplasia type 2A. Nature 1993; 363:458-460.
17. Carpten JD, Robbins CM, Villablanca A et al. HRPT2, encoding parafibromin, is mutated in hyperparathyroidism-jaw tumor syndrome. Nature Genet 2002; 32:584-588.
18. Heppner C, Kester MB, Agarwal SK et al. Somatic mutation of the MEN1 gene in parathyroid tumours. Nature Genet 1997; 16:375-378.
19. Marx SJ, Agarwal SK, Kester MB et al. Germline and somatic mutation of the gene for multiple endocrine neoplasia type 1 (MEN1). J Intern Med 1998; 243(6):447-453.
20. Hendy GN. Molecular mechanisms of primary hyperparathyroidism. Rev Endocr Metab Disord 2000; 1(4):297-305.
21. Arnold A, Shattuck TM, Mallya SM et al. Molecular pathogenesis of primary hyperparathyroidism. J Bone Miner Res 2002; 17(Suppl 2):N30-36.
22. Imanishi Y. Molecular pathogenesis of tumorigenesis in sporadic parathyroid adenomas. J Bone Miner Metab 2002; 20(4):190-195.
23. Wahl G, Vafa O. Genetic instability, oncogenes, and the p53 pathway. Cold Spring Harb Symp Quant Biol 2000; 65:511-520.
24. Felsher, DW. Cancer revoked: Oncogenes as therapeutic targets. Nature Reviews Cancer 2003; 3:375-379.
25. Rosenberg CL, Kim HG, Shows TB et al. Rearrangement and overexpression of D11S287E, a candidate oncogene on chromosome 11q13 in benign parathyroid tumors. Oncogene 1991; 6:449-453.
26. Hsi ED, Zukerberg LR, Yang WI et al. Cyclin D1/PRAD1 expression in parathyroid adenomas: An immunohistochemical study. J Clin Endocrinol Metab 1996; 81(5):1736-1739.

27. Tominaga Y, Tsuzuki T, Uchida K et al. Expression of PRAD1/cyclin D1, retinoblastoma gene products, and Ki67 in parathyroid hyperplasia caused by chronic renal failure versus primary adenoma. Kidney Int 1999; 55(4):1375-1383.
28. Vasef MA, Brynes RK, Sturm M et al. Expression of cyclin D1 in parathyroid carcinomas, adenomas and hyperplasias: A paraffin immunohistochemical study. Mod Patholm 1999; 12(4):412-416.
29. Hosokawa Y, Tu T, Tahara H et al. Absence of cyclin D1/PRAD1 point mutations in human breast cancers and parathyroid adenomas and identification of a new cyclin D1 gene polymorphism. Cancer Lett 1995; 93(2):165-170.
30. Imanishi Y, Hosokawa Y, Yoshimoto K et al. Primary hyperparathyroidism caused by parathyroid-targeted overexpression of cyclin D1 in transgenic mice. J Clin Invest 2001; 107(9):1093-1102.
31. Sherr CJ. The Pezcoller lecture: Cancer cell cycles revisited. Cancer Res 2000; 60(14):3689-3695.
32. Ikeda S, Ishizaki Y, Shimizu Y et al. Immunohistochemistry of cyclin D1 and beta-catenin and mutational analysis of exon 3 of beta-catenin gene in parathyroid adenomas. Int J Oncol 2002; 20(3):463-466.
33. Tseleni-Balafouta S, Thomopoulou G, Lazaris ACh et al. A comparative study of the int-2 gene product in primary and secondary parathyroid lesions. Eur J Endocrinol 2002; 146(1):57-60.
34. Gulkesen KH, Kilicarslan B, Altunbas HA et al. EGFR and p53 expression and proliferative activity in parathyroid adenomas; an immunohistochemical study. APMIS 2001; 109(12):870-874.
35. Sakaguchi K, Lorenzi MV, Matsushita H et al. Identification of a novel activated form of the keratinocyte growth factor receptor by expression cloning from parathyroid adenoma tissue. Oncogene 1999; 18(40):5497-5505.
36. Friedman E, Bale AE, Marx SJ et al. Genetic abnormalities in sporadic parathyroid adenomas. J Clin Endocrinol Metab 1990; 71(2):293-297.
37. Yoshimoto K, Iwahana H, Fukuda A et al. *Ras* mutations in endocrine tumors: Mutation detection by polymerase chain reaction-single strand conformation polymorphism. Jpn J Cancer Res 1992; 83(10):1057-1062.
38. Lyons J, Landis CA, Harsh G et al. Two G protein oncogenes in human endocrine tumors. Science 1990; 249(4969):655-659.
39. Vessey SJ, Jones PM, Wallis SC et al. Absence of mutations in the Gs alpha and Gi2 alpha genes in sporadic parathyroid adenomas and insulinomas. Clin Sci (Lond) 1994; 87(5):493-497.
40. Marsh DJ, Mulligan LM, Eng C. RET proto-oncogene mutations in multiple endocrine neoplasia type 2 and medullary thyroid carcinoma. Horm Res 1997; 47(4-6):168-178.
41. Padberg BC, Schroder S, Jochum W et al. Absence of RET proto-oncogene point mutations in sporadic hyperplastic and neoplastic lesions of the parathyroid gland. Am J Pathol 1995; 147(6):1600-1607.
42. Williams GH, Rooney S, Carss A et al. Analysis of the RET proto-oncogene in sporadic parathyroid adenomas. J Pathol 1996; 180(2):138-141.
43. Orndal C, Johansson M, Heim S et al. Parathyroid adenoma with t(1;5)(p22;q32) as the sole clonal chromosome abnormality. Cancer Genet Cytogenet 1990; 48(2):225-228.
44. Palanisamy N, Imanishi Y, Rao PH et al. Novel chromosomal abnormalities identified by comparative genomic hybridization in parathyroid adenomas. J Clin Endocrinol Metab 1998; 83(5):1766-1770.
45. Agarwal SK, Schrock E, Kester MB et al. Comparative genomic hybridization analysis of human parathyroid tumors. Cancer Genet Cytogenet 1998; 106(1):30-36.
46. Farnebo F, Kytola S, Teh BT et al. Alternative genetic pathways in parathyroid tumorigenesis. J Clin Endocrinol Metab 1999; 84(10):3775-3780.
47. Osada H, Takahashi T. Genetic alterations of multiple tumor suppressors and oncogenes in the carcinogenesis and progression of lung cancer. Oncogene 2002; 21(48):7421-7434.
48. Friedman E, De Marco L, Gejman PV et al. Allelic loss from chromosome 11 in parathyroid tumors. Cancer Res 1992; 52(24):6804-6809.
49. Imanishi Y, Tahara H. Putative parathyroid tumor suppressor on 1p: Independent molecular mechanisms of tumorigenesis from 11q allelic loss. Am J Kidney Dis 2001; 38(4 Suppl 1):S165-167.

50. Correa P, Juhlin C, Rastad J et al. Allelic loss in clinically and screening-detected primary hyperparathyroidism. Clin Endocrinol (Oxf) 2002; 56(1):113-117.
51. Marx S, Spiegel AM, Skarulis MC et al. Multiple endocrine neoplasia type 1: Clinical and genetic topics. Ann Intern Med 1998; 129(6):484-494.
52. Thakker RV. Multiple endocrine neoplasia. Horm Res 2001; 56(Suppl 1):67-72.
53. Larsson C, Skogseid B, Oberg K et al. Multiple endocrine neoplasia type 1 gene maps to chromosome 11 and is lost in insulinoma. Nature 1988; 332(6159):85-87.
54. Bale SJ, Bale AE, Stewart K et al. Linkage analysis of multiple endocrine neoplasia type 1 with INT2 and other markers on chromosome 11. Genomics 1989; 4(3):320-322.
55. Thakker RV, Wooding C, Pang JT et al. Linkage analysis of 7 polymorphic markers at chromosome 11p11.2-11q13 in 27 multiple endocrine neoplasia type 1 families. Ann Hum Genet 1993; 57 (Pt 1):17-25.
56. Carling T, Correa P, Hessman O et al. Parathyroid MEN1 gene mutations in relation to clinical characteristics of nonfamilial primary hyperparathyroidism. J Clin Endocrinol Metab 1998; 83(8):2960-2963.
57. Farnebo F, Teh BT, Kytola S et al. Alterations of the MEN1 gene in sporadic parathyroid tumors. J Clin Endocrinol Metab 1998; 83(8):2627-2630.
58. Chakrabarti R, Srivatsan ES, Wood TF et al. Deletion mapping of endocrine tumors localizes a second tumor suppressor gene on chromosome band 11q13. Genes Chromosomes Cancer 1998; 22(2):130-137.
59. Agarwal SK, Guru SC, Heppner C et al. Menin interacts with the AP1 transcription factor JunD and represses JunD-activated transcription. Cell 1999; 96(1):143-152.
60. Kaji H, Canaff L, Lebrun JJ et al. Inactivation of menin, a Smad3-interacting protein, blocks transforming growth factor type beta signaling. Proc Natl Acad Sci USA 2001; 98(7):3837-3842.
61. Heppner C, Bilimoria KY, Agarwal SK et al. The tumor suppressor protein menin interacts with NF-kappaB proteins and inhibits NF-kappaB-mediated transactivation. Oncogene 2001; 20(36):4917-4925.
62. Ohkura N, Kishi M, Tsukada T et al. Menin, a gene product responsible for multiple endocrine neoplasia type 1, interacts with the putative tumor metastasis suppressor nm23. Biochem Biophys Res Commun 2001; 282(5):1206-1210.
63. Crabtree JS, Scacheri PC, Ward JM et al. A mouse model of multiple endocrine neoplasia, type 1, develops multiple endocrine tumors. Proc Natl Acad Sci USA 2001; 98(3):1118-1123.
64. Cryns VL, Yi SM, Tahara H et al. Frequent loss of chromosome arm 1p DNA in parathyroid adenomas. Genes Chromosomes Cancer 1995; 13(1):9-17.
65. Williamson C, Pannett AA, Pang JT et al. Localisation of a gene causing endocrine neoplasia to a 4 cM region on chromosome 1p35-p36. J Med Genet 1997; 34(8):617-619.
66. Tahara H, Smith AP, Gaz RD et al. Parathyroid tumor suppressor on 1p: Analysis of the p18 cyclin-dependent kinase inhibitor gene as a candidate. J Bone Miner Res 1997; 12(9):1330-1334.
67. Tahara H, Smith AP, Gaz RD et al. Loss of chromosome arm 9p DNA and analysis of the p16 and p15 cyclin-dependent kinase inhibitor genes in human parathyroid adenomas. J Clin Endocrinol Metab 1996; 81(10):3663-3667.
68. Tahara H, Smith AP, Gaz RD et al. Genomic localization of novel candidate tumor suppressor gene loci in human parathyroid adenomas. Cancer Res 1996; 56(3):599-605.
69. Garcia JL, Tardio JC, Gutierrez NC et al. Chromosomal imbalances identified by comparative genomic hybridization in sporadic parathyroid adenomas. Eur J Endocrinol 2002; 146(2):209-213.
70. Yang KP, Nguyen CV, Castillo SG et al. Deletion mapping on the distal third region of chromosome 1p in multiple endocrine neoplasia type IIA. Anticancer Res 1990; 10(2B):527-533.
71. Benn DE, Dwight T, Richardson AL et al. Sporadic and familial pheochromocytomas are associated with loss of at least two discrete intervals on chromosome 1p. Cancer Res 2000; 60(24):7048-7051.
72. Shan L, Yang Q, Nakamura Y et al. Frequent loss of heterozygosity at 1p36.3 and p73 abnormality in parathyroid adenomas. Mod Pathol 2001; 14(4):273-278.
73. Hemmer S, Wasenius VM, Haglund C et al. Alterations in the suppressor gene PPP2R1B in parathyroid hyperplasias and adenomas. Cancer Genet Cytogenet 2002; 134(1):13-17.

74. Shattuck TM, Costa J, Bernstein M et al. Mutational analysis of Smad3, a candidate tumor suppressor implicated in TGF-beta and menin pathways, in parathyroid adenomas and enteropancreatic endocrine tumors. J Clin Endocrinol Metab 2002; 87(8):3911-3914.
75. Yoshimoto K, Iwahana H, Fukuda A et al. Role of p53 mutations in endocrine tumorigenesis: Mutation detection by polymerase chain reaction-single strand conformation polymorphism. Cancer Res 1992; 52(18):5061-5064.
76. Cryns VL, Rubio MP, Thor AD et al. p53 abnormalities in human parathyroid carcinoma. J Clin Endocrinol Metab 1994; 78(6):1320-1324.
77. Hakim JP, Levine MA. Absence of p53 point mutations in parathyroid adenoma and carcinoma. J Clin Endocrinol Metab 1994; 78(1):103-106.
78. Kishikawa S, Shan L, Ogihara K et al. Overexpression and genetic abnormality of p53 in parathyroid adenomas. Pathol Int 1999; 49(10):853-857.
79. Cryns VL, Thor A, Xu HJ et al. Loss of the retinoblastoma tumor-suppressor gene in parathyroid carcinoma. N Engl J Med 1994; 330(11):757-761.
80. Dotzenrath C, Teh BT, Farnebo F et al. Allelic loss of the retinoblastoma tumor suppressor gene: A marker for aggressive parathyroid tumors? J Clin Endocrinol Metab 1996; 81(9):3194-3196.
81. Szijan I, Orlow I, Dalamon V et al. Alterations in the retinoblastoma pathway of cell cycle control in parathyroid tumors. Oncol Rep 2000; 7(2):421-425.
82. Ron E, Saftlas AF. Head and neck radiation carcinogenesis: Epidemiologic evidence. Otolaryngol Head Neck Surg 1996; 115(5):403-408.
83. Carling T, Imanishi Y, Gaz RD et al. RAD51 as a candidate parathyroid tumour suppressor gene on chromosome 15q: Absence of somatic mutations. Clin Endocrinol (Oxf) 1999; 51(4):403-407.
84. Carling T, Imanishi Y, Gaz RD et al. Analysis of the RAD54 gene on chromosome 1p as a potential tumor-suppressor gene in parathyroid adenomas. Int J Cancer 1999; 83(1):80-82.
85. Duval A, Hamelin R. Genetic instability in human mismatch repair deficient cancers. Ann Genet 2002; 45(2):71-75.
86. Chung DC, Rustgi AK. The hereditary nonpolyposis colorectal cancer syndrome: Genetics and clinical implications. Ann Intern Med 2003; 138(7):560-570.
87. Hussein MR, Wood GS. Microsatellite instability and its relevance to cutaneous tumorigenesis. J Cutan Pathol 2002; 29(5):257-267.
88. Sarquis M, Friedman E, Boson WL et al. Microsatellite instability in sporadic parathyroid adenoma. J Clin Endocrinol Metab 2000; 85(1):250-252.
89. Koshiishi N, Chong JM, Fukasawa T et al. Microsatellite instability and loss of heterozygosity in primary and secondary proliferative lesions of the parathyroid gland. Lab Invest 1999; 79(9):1051-1058.
90. Shampay J, Blackburn EH. Tetrahymena micronuclear sequences that function as telomeres in yeast. Nucleic Acids Res 1989; 17(8):3247-3260.
91. Blackburn EH. Structure and function of telomores. Nature 1991; 350:569-573.
92. Moyzis RK, Buckingham JM, Cram LS et al. A highly conserved repetitive DNA sequence, (TTAGGG)n, present at the telomeres of human chromosomes. Proc Natl Acad Sci USA 1988; 85(18):6622-6626.
93. Levy MZ, Allsopp RC, Futcher AB et al. Telomere end-replication problem and cell aging. J Mol Biol 1992; 225(4):951-960.
94. Holt SE, Shay JW. Role of telomerase in cellular proliferation and cancer. J Cell Physiol 1999; 180(1):10-18.
95. Artandi SE, DePinho RA. Mice without telomerase: What can they teach us about human cancer? Nat Med 2000; 6(8):852-855.
96. Buys CH. Telomeres, telomerase, and cancer. N Engl J Med 2000; 342(17):1282-1283.
97. Kim NW, Piatyszek MA, Prowse KR et al. Specific association of human telomerase activity with immortal cells and cancer. Science 1994; 266(5193):2011-2015.
98. Falchetti A, Becherini L, Martineti V et al. Telomerase repeat amplification protocol (TRAP): A new molecular marker for parathyroid carcinoma. Biochem Biophys Res Commun 1999; 265(1):252-255.

99. Kammori M, Nakamura K, Kanauchi H et al. Consistent decrease in telomere length in parathyroid tumors but alteration in telomerase activity limited to malignancies: Preliminary report. World J Surg 2002; 26(9):1083-1087.
100. Law Jr WM, Heath 3rd H. Familial benign hypercalcemia (hypocalciuric hypercalcemia). Clinical and pathogenetic studies in 21 families. Ann Intern Med. 1985; 102(4):511-519.
101. Fuleihan Gel-H. Familial benign hypocalciuric hypercalcemia. Bone Miner Res 2002; 17 (Suppl 2):N51-56.
102. Pollak MR, Brown EM, Chou YH et al. Mutations in the human Ca(2+)-sensing receptor gene cause familial hypocalciuric hypercalcemia and neonatal severe hyperparathyroidism. Cell 1993; 75(7):1297-1303.
103. Pearce SH, Trump D, Wooding C et al. Calcium-sensing receptor mutations in familial benign hypercalcemia and neonatal hyperparathyroidism. J Clin Invest 1995; 96(6):2683-2692.
104. Heath 3rd H, Odelberg S, Jackson CE et al. Clustered inactivating mutations and benign polymorphisms of the calcium receptor gene in familial benign hypocalciuric hypercalcemia suggest receptor functional domains. J Clin Endocrinol Metab 1996; 81(4):1312-1317.
105. Carling T, Szabo E, Bai M et al. Familial hypercalcemia and hypercalciuria caused by a novel mutation in the cytoplasmic tail of the calcium receptor. J Clin Endocrinol Metab 2000; 85(5):2042-2047.
106. Pollak MR, Chou YH, Marx SJ et al. Familial hypocalciuric hypercalcemia and neonatal severe hyperparathyroidism. Effects of mutant gene dosage on phenotype. J Clin Invest 1994; 93(3):1108-1112.
107. Thompson DB, Samowitz WS, Odelberg S et al. Genetic abnormalities in sporadic parathyroid adenomas: Loss of heterozygosity for chromosome 3q markers flanking the calcium receptor locus. J Clin Endocrinol Metab 1995; 80(11):3377-3380.
108. Hosokawa Y, Pollak MR, Brown EM et al. Mutational analysis of the extracellular Ca(2+)-sensing receptor gene in human parathyroid tumors. J Clin Endocrinol Metab 1995; 80(11):3107-3110.
109. Cetani F, Pinchera A, Pardi E et al. No evidence for mutations in the calcium-sensing receptor gene in sporadic parathyroid adenomas. J Bone Miner Res 1999; 14(6):878-882.
110. Kifor O, Moore Jr FD, Wang P et al. Reduced immunostaining for the extracellular Ca2+-sensing receptor in primary and uremic secondary hyperparathyroidism. J Clin Endocrinol Metab 1996; 81(4):1598-1606.
111. Farnebo F, Enberg U, Grimelius L et al. Tumor-specific decreased expression of calcium sensing receptor messenger ribonucleic acid in sporadic primary hyperparathyroidism. J Clin Endocrinol Metab 1997; 82(10):3481-3486.
112. Gogusev J, Duchambon P, Hory B et al. Depressed expression of calcium receptor in parathyroid gland tissue of patients with hyperparathyroidism. Kidney Int 1997; 51(1):328-336.
113. Nygren P, Larsson R, Johansson H et al. 1,25(OH)2D3 inhibits hormone secretion and proliferation but not functional dedifferentiation of cultured bovine parathyroid cells. Calcif Tissue Int 1988; 43(4):213-218.
114. Kremer R, Bolivar I, Goltzman D et al. Influence of calcium and 1,25-dihydroxycholecalciferol on proliferation and proto-oncogene expression in primary cultures of bovine parathyroid cells. Endocrinology 1989; 125(2):935-941.
115. Brown SB, Brierley TT, Palanisamy N et al. Vitamin D receptor as a candidate tumor-suppressor gene in severe hyperparathyroidism of uremia. J Clin Endocrinol Metab 2000; 85(2):868-872.
116. Carling T, Rastad J, Szabo E et al. Reduced parathyroid vitamin D receptor messenger ribonucleic acid levels in primary and secondary hyperparathyroidism. J Clin Endocrinol Metab 2000; 85(5):2000-2003.
117. Rao DS, Honasoge M, Divine GW et al. Effect of vitamin D nutrition on parathyroid adenoma weight: Pathogenetic and clinical implications. J Clin Endocrinol Metab 2000; 85(3):1054-1058.

CHAPTER 11

Molecular Genetic Abnormalities in Sporadic Hyperparathyroidism

Trisha M. Shattuck, Sanjay M. Mallya and Andrew Arnold

Abstract

The biochemical state of primary hyperparathyroidism is generally caused by hypercellular parathyroid glands categorized as multigland hyperplasia, benign adenoma or malignant carcinoma. Most, and probably all, adenomas and carcinomas are monoclonal in origin, and specific clonal genetic lesions have been identified in most of these tumors. Only two genes have been definitively proven to be important players in the pathogenesis of typical sporadic parathyroid tumors, an oncogene, cyclin D1 and a tumor suppressor gene, *MEN1*. The cyclin D1 oncoprotein is overexpressed in 20-40% of parathyroid adenomas and the identification of clonal rearrangements which activate the cyclin D1 gene in a subset of tumors indicates that such activation is a primary genetic driver of parathyroid neoplasia. Cyclin D1 plays an important role in regulation of the cell cycle and may have non-cell cycle effects which contribute to tumorigenesis as well. The central role that cyclin D1 plays in parathyroid tumorigenesis has been confirmed in a mouse model where cyclin D1 is overexpressed specifically in the parathyroids and in which many features of human hyperparathyroidism are reproduced. Germline mutations of *MEN1* cause multiple endocrine neoplasia type 1, a genetic syndrome in which patients develop tumors of multiple endocrine (and some nonendocrine) tissues including the parathyroid glands. Acquired (somatic) mutations in *MEN1* have also been identified in 12-17% of sporadic parathyroid adenomas. The function of *MEN1* remains elusive, but the discovery of proteins that interact with the *MEN1* protein product, menin, and the development of a mouse model of MEN1 syndrome may help to shed light on menin's function. The *HRPT2* gene has recently been identified as a major contributor to the development of sporadic parathyroid carcinoma. Identification of acquired chromosomal aberrations in parathyroid adenomas and carcinomas using techniques such as molecular allelotyping and comparative genomic hybridization has highlighted several areas of the genome that may harbor other important parathyroid tumor suppressor genes and oncogenes. The eventual identification of the full spectrum of genes involved in parathyroid tumorigenesis will be important for developing a complete understanding of the molecular basis of primary hyperparathyroidism.

Introduction

Primary hyperparathyroidism (HPT), a disorder characterized by hypercalcemia due to the excessive secretion of parathyroid hormone (PTH), affects about 1:1000 people.[1] In these patients, the parathyroid gland(s) increases in mass, and the control of PTH secretion from the

---

Molecular Biology of the Parathyroid, edited by Tally Naveh-Many. ©2005 Eurekah.com and Kluwer Academic / Plenum Publishers.

parathyroid cell by the ambient calcium concentration (represented by the calcium set-point) is reset. One or more enlarged parathyroid gland(s) secretes excessive quantities of PTH due to a defect within the gland itself rather than resulting from other physiological disturbances as occurs in secondary hyperparathyroidism. In the majority of cases (85%) a single, benign adenoma is identified, while multiple hypercellular glands are found in about 15%.[2] Malignant parathyroid carcinoma is responsible for less than 1% of cases of primary hyperparathyroidism. The classical symptoms of primary hyperparathyroidism result from metabolic abnormalities rather than the tumor mass per se, and include kidney stones, gastrointestinal disruption, neuropsychiatric symptoms and a prototypical bone disorder osteitis fibrosa cystica.[3] However, in many parts of the world it is now more common for this disorder to be diagnosed in asymptomatic or minimally symptomatic patients, with abnormal calcium levels discovered in routine blood tests or blood tests for other conditions (reviewed in ref. 4). In many such patients the serum calcium and PTH levels may remain stable for many years.[5] Concordant with this clinical observation, the rates of parathyroid tumor cell proliferation, once hyperparathyroidism is fully established, often appear to be quite low.[6] In addition to the mode of patient ascertainment, a number of factors may bear upon the severity of the proliferative and biochemical defects in hyperparathyroidism, including the patient's vitamin D status.[7] The molecular basis of the relationship between the proliferative defect and the PTH regulatory abnormality characteristic of such tumors is a fundamental issue in endocrine neoplasia.

Implications of the Monoclonality of Parathyroid Tumors

Parathyroid adenomas have been differentiated from primary parathyroid hyperplasia on the basis of the number of abnormal, hypercellular parathyroid glands found in the patient. A hypercellular gland is defined as an adenoma when only one gland is affected. It is, however, impossible to determine histologically if a hypercellular gland is a solitary tumor or one of several.[2] Multiple glands may be enlarged in a nonuniform fashion with one gland being larger than the others. Such asymmetric hyperplasia may be confused with a solitary adenoma, or with true "double adenomas".[8-10] To better define the pathogenic mechanisms underlying the development of adenomas and hyperplasias, the clonality of these entities was examined. A method frequently used to study the clonality of a tumor involves assessment of the X-chromosome inactivation pattern of the cells in a tumor. According to the Lyon phenomenon, in females, one X-chromosome is randomly inactivated in each cell early in embryonic development.[11,12] This X-inactivation pattern is stably passed on to all of the progeny of a cell. A monoclonal collection of cells will therefore all have the same X-chromosome inactivated, whereas a polyclonal group of cells should have inactivated either of the X-chromosomes in a random distribution. Initially, X-chromosome inactivation patterns of parathyroid adenomas were examined using the glucose-6-phosphate dehydrogenase (G6PD) protein polymorphism. These studies indicated that the adenomas were polyclonal rather than monoclonal growths,[13,14] suggesting that a parathyroid adenoma is actually a form of asymmetric multigland hyperplasia and that no clonal DNA changes that are characteristic of monoclonal tumors should be found in these growths. Subsequently, X-chromosome inactivation analysis using a DNA polymorphism-based method was used to reexamine the clonality of parathyroid adenomas. These DNA based approaches are informative in a much higher percentage of tumors and avoid many of the pitfalls of the G6PD protein polymorphism technique (reviewed in ref. 15). These studies concluded that many (and probably all) parathyroid adenomas are monoclonal in origin.[16] The findings of clonal DNA rearrangements, mutations and areas of loss of heterozygosity, which will be described below, have overwhelmingly confirmed the monoclonality of parathyroid adenomas. Similar discoveries have established that parathyroid carcinomas are also monoclonal lesions.[17-19]

The implications of monoclonality in the context of modern concepts of tumor genetics are considerable. Clonal DNA alterations in key growth-regulating genes, generally categorized as protooncogenes and tumor suppressor genes, are key contributors in the conversion of normal cells to neoplastic cells. Protooncogenes are often involved in the physiological control of cellular growth, proliferation, or differentiation. Protooncogenes are converted into oncogenes by DNA damage of various sorts including chromosomal translocations or inversions, point mutations, proviral insertions, or gene amplifications. These result in a deregulation of the expression of the protein product or formation of an intrinsically abnormal product. Such an "activated" oncogene then can contribute to the development or growth of the tumor. In contrast, tumor-suppressor genes normally serve to restrain cellular proliferation or activate cell death, directly or indirectly (for example through maintenance of genomic stability). Inactivation of these genes leads to a growth advantage for a cell. This inactivation can occur by mutation or deletion of the gene or by regulatory derangements such as abnormal methylation, which causes loss of transcription of the gene. Inactivation of both alleles of a "classic" tumor suppressor gene is necessary to completely deplete the protein product and promote neoplastic transformation.

Cells gain a selective advantage when a sufficient number of key changes occur in protooncogenes and/or tumor suppressor genes. Progeny of these cells grow and accumulate more genetic changes in a process known as clonal evolution, ultimately forming a clinically apparent mass of cells. As all cells in the tumor arose from the same single parent cell that had acquired the initial rare genetic changes, they comprise a clonal population. For certain genes, the same pattern of DNA damage can be found in each cell of a clonal tumor and the damaged regions represent and can help to define the important genetic events that led to the clonal expansion. Subpopulations of a clone may develop due to the acquisition of additional DNA damage after the initial clone is established. Therefore, there are some genes involved in the initial clonal expansion of the tumor and others that are important for the clonal evolution of the tumors. Those genes that are involved in each process have not been clearly defined in most tumor types, but the importance of selection as the driving process cannot be overemphasized.

It is thought that it is necessary for several different genes, within the same cell, to be damaged for the cell to become neoplastic. While some genes are implicated in one or a few types of tumors, other genes have been implicated in many different tumor types. The emergence of a particular tumor type, for example parathyroid adenoma vs. carcinoma, may reflect the specific biochemical pathways that are disrupted. However, it is unlikely that a disruption of one specific gene is both a necessary and sufficient cause of a specific tumor type. Furthermore, different combinations of mutated genes may lead to similar pathological and clinical results. More studies are necessary to better understand how specific genetic changes in a cell determine the pathological outcome.

Molecular Genetics of Parathyroid Adenomas

Oncogenes

Cyclin D1

As discussed, the early studies of parathyroid adenomas indicated that parathyroid adenomas are indeed monoclonal neoplasms, suggesting that these tumors are caused by mutations that alter the growth regulation of parathyroid cells.[16] Subsequent molecular analyses revealed the presence of tumor-specific DNA rearrangements in a subset of adenomas—in such rearrangements the 5' regulatory region of the *PTH* gene became separated from its coding exons and was shown to recombine with a novel DNA locus, D11S287.[20] This rearrangement was both clonal, suggesting its importance in tumor cell selection, and remarkably simi-

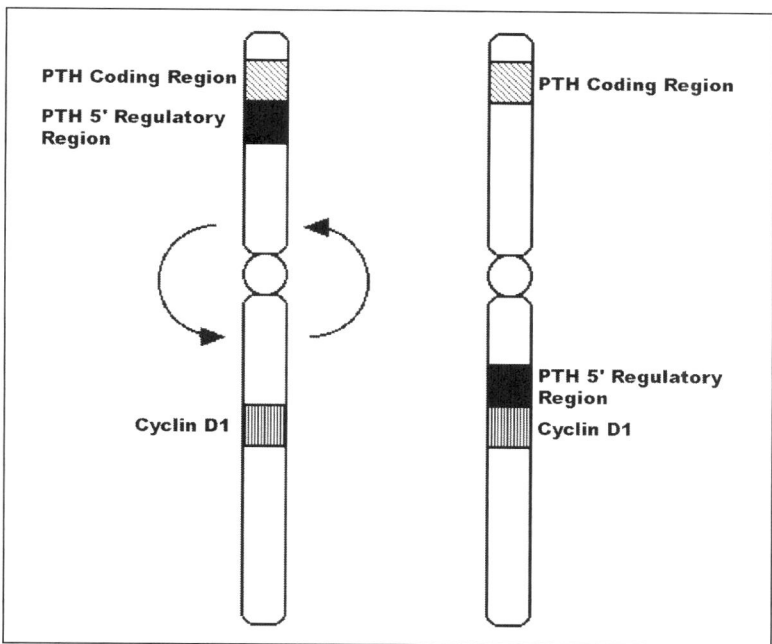

Figure 1. Schematic representation of a pericentromeric inversion that activates the *cyclin D1* oncogene. Chromosomal breaks occur at the *PTH* and *cyclin D1* loci and the resulting rearrangement places the 5' regulatory elements of the *PTH* gene upstream of the *cyclin D1* gene leading to specific overexpression of cyclin D1 in parathyroid cells.

lar to specific chromosomal translocations observed in various B-cell lymphomas, wherein the tissue-specific regulatory sequences of the immunoglobulin heavy chain gene are juxtaposed with oncogenes like *BCL-2* or *C-MYC*, causing their overexpression in the B-lineage cells. Thus, by analogy, it was hypothesized that DNA from the non-*PTH* side of the breakpoint in the parathyroid tumors would harbor an oncogene whose deregulation provided the host parathyroid cell with a selective growth advantage. An mRNA transcript from this breakpoint-adjacent gene was identified and was found to be dramatically overexpressed in these parathyroid adenomas,[21] providing evidence that the tissue-specific enhancer elements of the *PTH* gene were indeed deregulating the expression of this putative oncogene. Subsequent cloning of this candidate oncogene led to the identification of a novel gene (*PRAD1/Cyclin D1*) with sequence similarities to cyclins.[22] The *PTH* gene is localized to chromosomal region 11p15, whereas the *cyclin D1* oncogene maps to 11q13.[20] Thus, the simplest explanation for this rearrangement is a pericentromeric inversion, inv(11)(p15;q13), that positions the 5' PTH regulatory region upstream of the *cyclin D1* gene (Fig. 1). To date, *cyclin D1* is the only established parathyroid tumor oncogene.

Cyclin D1 was also cloned independently by two other groups—as a murine gene that was induced by growth factor exposure in a macrophage cell line[23] and as a human cDNA that rescued yeast with mutant G1-phase cyclins.[24] The *cyclin D1* gene encodes a 35 kDa protein that shares structural homology and some functional properties with other cyclins.[22,24,25] The cyclin D1 protein contains the conserved 'cyclin box' and a retinoblastoma (pRB)-binding domain.[26] During the G1 phase of the cell cycle, cyclin D1 complexes with and activates its kinase partners CDK4 or CDK6. The activated kinases participate in the phosphorylation and

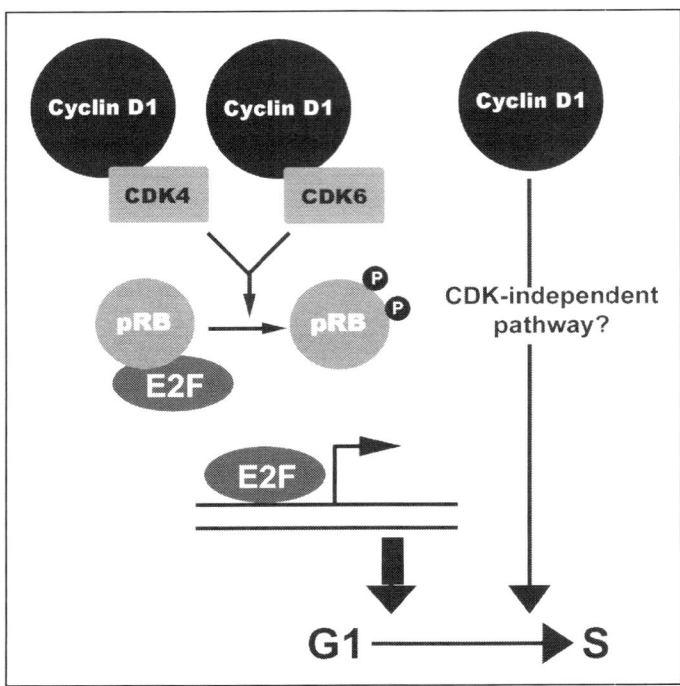

Figure 2. Schematic representation of the oncogenicity of *cyclin D1* overexpression in parathyroid tumorigenesis. Cyclin D1 is known to function in cell-cycle regulation. Cyclin D1 binds to and activates its kinase partners cdk4 or cdk6, which phosphorylate pRb. Phosphorylated pRb releases transcription factors in the E2F family, permitting transcription of genes involved in transition of the cell from G1 to S phase of the cell cycle. Alternatively, it is also plausible that cyclin D1 may regulate parathyroid cell proliferation through non-cdk-dependent mechanisms.

inactivation of pRb, effecting entry into S-phase (Fig. 2). Thus, one plausible mechanism for the oncogenic activity of cyclin D1 is via inactivation of the growth inhibitory effects of pRb, a well-established tumor suppressor gene product. Innumerable studies have demonstrated an important role for the cyclin D1 pathway beyond its role in parathyroid neoplasia. The *cyclin D1* gene is amplified in multiple human malignancies, including breast cancers[27-29] and head and neck carcinomas,[30] and is activated by gene rearrangements (analogous to those in parathyroid tumors) in mantle cell lymphomas and multiple myeloma.[31-34] Moreover, activating mutations of *CDK4* have been detected in human melanomas[35] and *CDK4* is amplified in sarcomas and gliomas.[36,37] Also, the cyclin D1/cdk4-inhibitor p16 (p16^{INK4a}) has been well established as a tumor suppressor. Thus, the cyclin D1 pathway is a central target in oncogenesis in many tissue types.

Cyclin D1 in Parathyroid Tumorigenesis

As noted above, a subset of parathyroid adenomas contains clonal rearrangements that juxtapose the *PTH* 5' regulatory region with the *cyclin D1* gene. The 11q13 chromosomal breakpoint may be found close to *cyclin D1* or may be located as much as 300 kb centromeric of *cyclin D1* (Y. Hosokawa et al, unpublished data). This variability in the location of the breakpoint complicates standard approaches for detection of the rearrangement such as

Southern blotting. Thus, there are no accurate estimates to date of the percentage of parathyroid adenomas that harbor such *cyclin D1*-activating rearrangements. However, as determined by immunohistochemistry, 20-40% of parathyroid adenomas overexpress cyclin D1.[38-41] It is highly plausible that in addition to the described chromosomal rearrangement, other molecular mechanism such as gene amplification, rearrangement with other enhancer/promoters active in parathyroid cells, or transcriptional activation can serve as alternative routes toward the cell's acquisition of the selective advantage inherent in cyclin D1 overexpression. For cyclin D1, overexpression of the normal gene product is oncogenic, as no internal activating mutations have been found in parathyroid adenomas[42] or other human tumors. Other than the still unproven possibility that pRb may be the tumor suppressor target of 13q deletions in parathyroid carcinomas,[18] molecular analyses of parathyroid tumors have failed to detect mutations in other genes involved in the cyclin D1 pathway. Specifically, inactivating mutations or homozygous deletions of the p16 and p15 cdk inhibitor genes occur uncommonly, if ever, in parathyroid adenomas.[43] Also, it is not yet known whether activating mutations of β-catenin, which is a reported transcriptional activator of *cyclin D1*, play a role in parathyroid tumorigenesis. Interestingly, there are data suggesting that the existing paradigm that cyclin D1 promotes tumorigenesis only through its effects on pRb and cdks may be too simplistic. In breast cancer cells, for example, cyclin D1 was reported to complex with the estrogen receptor and activate estrogen receptor-mediated transcription, independent of a cdk partner.[44,45] Certainly, the possibility that cyclin D1 may regulate parathyroid-cell growth via yet unknown mechanisms must be seriously considered in the future.

An Animal Model of Primary Hyperparathyroidism

Our laboratory has generated a transgenic mouse model for parathyroid neoplasia, by targeting cyclin D1 overexpression to the parathyroid glands.[46] These mice (PTH-cyclin D1 mice) harbor a transgene in which the *cyclin D1* gene is placed under the control of a 5.2 kb fragment of the *PTH* regulatory region, thereby mimicking the rearrangement and resultant cyclin D1 overexpression observed in the human tumors. The resulting phenotype in these animals is remarkably similar to the abnormalities that develop in patients with primary parathyroid neoplasia. By the age of 6-10 months, PTH-cyclin D1 mice develop chronic biochemical HPT, as evidenced by increased serum calcium and PTH levels, and develop bone abnormalities characteristic of HPT. The parathyroid glands in PTH-cyclin D1 mice are hypercellular (Fig. 3), and the relative PTH-calcium setpoint, as estimated in vivo by the concentration of serum calcium needed to half-maximally suppress PTH levels, is increased (see below). Expression of the calcium sensing receptor protein (CaSR) is decreased in the parathyroid glands of the HPT animals to approximately the same extent as occurs in human parathyroid adenomas. Furthermore, assessments of parathyroid cell proliferation in HPT animals discovered increased uptake of 5'-bromo-2-deoxyuridine and increased levels of proliferating cell nuclear antigen, as detected by immunohistochemistry. Thus, tissue-specific overexpression of cyclin D1 does induce parathyroid cell proliferation resulting in HPT, substantiating the role of cyclin D1 as a driver of parathyroid cell growth. The development of this animal model provides an attractive system to study parathyroid biology and endocrine neoplasia on several fronts.

Parathyroid neoplasia is a complex process that involves abnormal cell proliferation coupled with an aberrant control of hormonal secretion. This link between proliferation and hormonal production/secretion is also evident in normal parathyroid cells which respond to the stimulus of chronic hypocalcemia not only by increasing hormonal secretion but also by a secondary expansion of parathyroid cell mass. Likewise, in addition to a proliferative defect leading to parathyroid hypercellularity, in vivo studies on parathyroid neoplasms in humans have shown

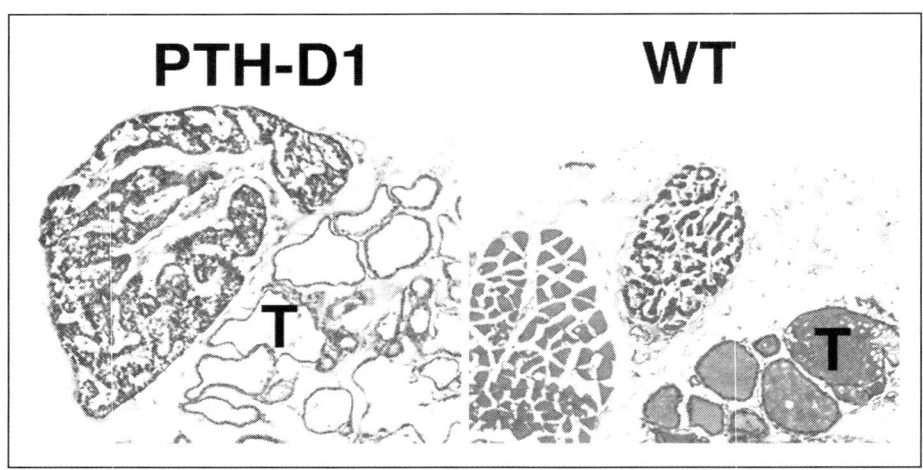

Figure 3. Hematoxylin and eosin stained sections of parathyroid glands from 14-month-old PTH-cyclin D1 transgenic (left) and wild-type (right) mice. The parathyroid glands from the PTH-cyclin D1 mice exhibit marked hypercellularity and enlargement, compared with the parathyroid gland from wild-type mice. T: thyroid gland.

an abnormality in the feedback system through which extracellular calcium regulates PTH secretion—an apparent resetting of the 'set-point' mechanism that normally tightly couples PTH secretion with ambient calcium levels.[47,48] It has been hypothesized that most parathyroid adenomas are caused by acquired mutation(s) in the genes of the set-point pathway, with the abnormal PTH response to the ambient calcium level being the initial driving force for parathyroid tumor cell proliferation.[49,50] However, while severe germline deficiency of the CaSR can cause the parathyroid hyperproliferation of neonatal severe hyperparathyroidism, somatic mutations of the Ca^{++}-sensing receptor gene have not been found in sporadic parathyroid tumors.[51] The PTH-cyclin D1 mouse model has shed light on the issue of whether abnormal cell proliferation is the result or cause of parathyroid cell hormonal setpoint dysregulation. In wild-type mice, as in humans, the inverse relationship between PTH and serum calcium levels takes the shape of a sigmoidal curve. In the PTH-cyclin D1 mice, this curve is shifted upward and to the right, resulting in an abnormally high relative set-point (Fig. 4), similar to findings in patients with primary or severe secondary hyperparathyroidism. Thus, it is now clear that the hormonal regulatory defect need not be primary, but can result secondary to primary growth-control disturbances in the gland. Similarly, the model shows that a primary growth disturbance, at least as driven by cyclin D1, can cause a secondary diminution in expression of the parathyroid calcium-sensing receptor. In more general terms, the PTH-cyclin D1 mouse model provides an experimental system in which the molecular mechanisms that deregulate set-point control can be dissected. For example, the status of the vitamin D receptor, an important regulator of PTH gene expression and parathyroid cell proliferation, could be examined in a relevant in vivo context. These studies will further elucidate the critical links between proliferation and functional abnormalities in parathyroid neoplasia. Finally, the development of larger adenoma-like lesions in some animals suggests that they may be monoclonal expansions, and the screening of these lesions for acquired genetic alterations could aid in the identification of additional genes involved in human parathyroid neoplasia.

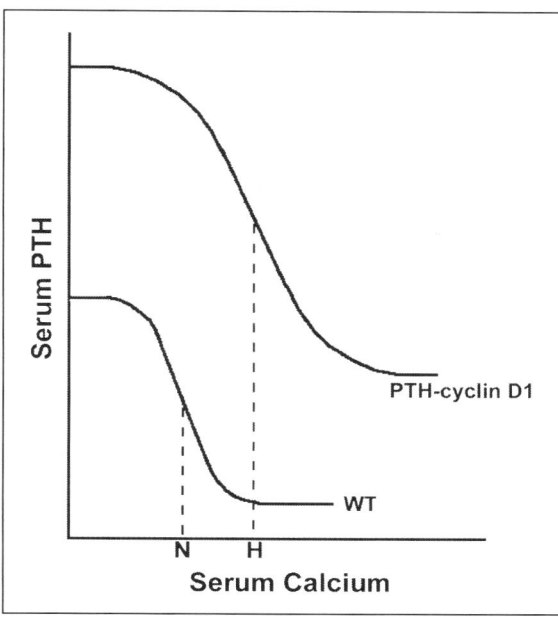

Figure 4. Schematic representation of alteration of the relative calcium-PTH relationship in the PTH-cyclin D1 mice. The Ca-PTH relationship is represented by a sigmoidal curve. The setpoint (N) is the calcium level at which there is half-maximal PTH secretion. In PTH-cyclin D1 mice and patients with HPT, this curve is shifted upward and to the right, resulting in an abnormally high setpoint (H).

Tumor Suppressor Genes: MEN1

The Multiple Endocrine Neoplasia Type 1 Syndrome

The multiple endocrine neoplasia type 1 (MEN1) syndrome, an autosomal dominant inherited disease, is characterized by multiple tumors of the parathyroids, enteropancreatic tissues and anterior pituitary (see Chapter 10 by E. Friedman). Linkage studies in families affected by MEN1 discovered that the gene responsible for MEN1 is located on chromosome 11q13 near *PYGM*, a muscle phosphorylase gene.[52] The *MEN1* gene, like the genes in other previously characterized familial tumor syndromes, was thought to act as a classic tumor suppressor gene, as acquired inactivating alterations at 11q13 uniformly involved the allele inherited from the unaffected parent in MEN1-associated tumors. These DNA alterations were seen, in many such tumors including those of the parathyroid, as loss of polymorphic DNA markers from this region of chromosome 11.[52-55] Thus, it was hypothesized that an inherited mutation in one allele in combination with an acquired deletion of the remaining, normal *MEN1* allele would result in the absence of functional protein product and a selective growth advantage to the transformed cell. The *MEN1* gene was eventually identified by positional cloning[56,57] and a classic tumor suppressor role for *MEN1* was confirmed. Germ-line mutations were found throughout the protein-coding region of *MEN1* in patients with MEN1 syndrome (reviewed in refs. 58,59). Nonsense or frameshift mutations that result in a truncated protein make up 70% of germline mutations while missense mutations or in-frame deletions make up the remaining 30%. No correlation between mutation and phenotype of the syndrome has been found, and the functional consequences of many of the subtler mutations have not yet been precisely elucidated.

Clues to the Action of Menin

Menin, the protein product of the 10 exon *MEN1* gene, is a 610-amino acid protein with no homology to any known proteins. Its sequence was initially reported to contain no common-motifs that provide clues to its function. Subsequently, two nuclear localization signal sequences were identified at the carboxyl-terminus of the menin protein, and as expected, menin seems to be targeted to the nucleus.[60] Interestingly, *MEN1* is expressed at the mRNA and protein level in nearly all tissue types examined, rather than being limited to the tissue types susceptible to MEN1-associated tumors.[61] Menin levels vary as cells in culture proliferate. Pituitary cells synchronized at the G1-S phase boundary express menin at a lower level than G0-G1-synchronized cells,[62] and the expression of menin increases as the cell enters S phase. Cells synchronized at the G2-M phase express lower levels of menin.[62] The stable overexpression of menin in NIH3T3 cells transformed with *ras* was able to inhibit cell growth and tumor formation by these cells in xenografts.[63] This finding supports the hypothesis that menin can act as a growth inhibitor, but it is still unclear how it does so.

Proteins that interact with menin have been, and continue to be, identified in hopes that this information may provide further clues to its function. Menin was shown to interact with the activator protein 1 factor JunD in a yeast two hybrid system, and can repress JunD-activated transcription.[64] Mutants of JunD have been characterized that fail to interact with menin. These studies indicate that JunD binds menin at JunD's amino terminus in a region that shares little homology with other jun proteins.[65] Unlike other jun family members, JunD appears to be antimitogenic. Therefore, it is far from clear how menin could act as a tumor suppressor by repressing JunD activity. It also has been reported that menin can directly interact with Smad3, a downstream component of the TGF-β signaling pathway.[66] When TGF-β receptors are activated, Smad3 becomes phosphorylated and can subsequently enter the nucleus to alter transcription (reviewed in refs. 67,68). TGF-β signals usually inhibit cell growth. However, when menin levels are decreased by antisense *MEN1* gene transcripts, cultured pituitary cells are less susceptible to the growth inhibitory effects of TGF-β. Antisense menin also inhibits the transcription of TGF-β target genes.[66] These results indicate that menin may play a role in the growth inhibitory actions of the TGF-β pathway, and loss of menin may decrease the growth restriction exerted by this pathway. This explanation of menin's role in normal cell biology seems more consistent with its tumor suppressor activity than its paradoxical effects on JunD. However, more work needs to be done to better understand menin's role in the TGF-β pathway, JunD transcription, as well as perhaps in other cellular processes. A better understanding of menin's function may also help to explain why menin is expressed in most tissues, but only a limited number of tissues are affected by tumors in the MEN1 syndrome.

Menin in Sporadic Parathyroid Adenomas

Studies of parathyroid tumors have found that allelic loss of chromosome 11 markers occurs in 25-40% of sporadic parathyroid adenomas.[55,69,70] This allelic loss usually includes the chromosomal region to which the *MEN1* gene has been mapped. Somatic mutation and/or deletion of *MEN1* resulting in its complete (biallelic) inactivation have been documented in 12-17% of sporadic parathyroid adenomas.[71-73] This 12-17% accounts for approximately half of the 25-40% of parathyroid adenomas with losses of 11q13. Noncoding mutations in *MEN1* such as mutations within promoter or enhancer regions, and/or epigenetic inactivation by promoter hypermethylation, which would have been missed in these analyses, might explain some of this discrepancy. However, there is also a real possibility that a second tumor suppressor gene on chromosome 11 might be important in the pathogenesis of parathyroid adenomas and is disrupted in those tumors with 11q13 loss and no *MEN1* mutations. Consistent with this possibility, studies in another endocrine tumor, follicular thyroid tumors, have also suggested that a tumor suppressor gene in addition to *MEN1* may be located on chromosome 11q13.[74] Somatic *MEN1* mutation is not limited to sporadic parathyroid tumors as it has also been reported in sporadic gastrinomas, insulinomas, lung carcinoids and angiofibromas.[75-77]

A Mouse Model of MEN1

A mouse model of MEN1 has been developed.[78] Mouse *Men1* shares 97% identity with its human counterpart at the amino acid level and, like the human version, is expressed in many tissues.[79] A mouse with a floxed neomycin cassette in intron 2 and a third lox P site in intron 8 on the *MEN1* gene was created. Mice homozygous for this targeted disruption were not viable. Mice heterozygous for the disrupted *men1* allele developed tumors characteristic of MEN1 syndrome including pancreatic islet cell tumors, pituitary adenomas, adrenal cortical tumors, and parathyroid tumors. Pituitary tumors that developed in the mouse secreted prolactin as is commonly seen in the pituitary tumors which arise in humans with the MEN1 syndrome. Pancreatic hyperplasias and tumors produced insulin leading to elevated serum insulin levels that correlated in severity with the abnormalities found in the pancreas. However, mice with parathyroid tumors did not develop elevated PTH or calcium levels. This suggests that despite their increased cell mass, these parathyroids may retain the ability to appropriately regulate PTH production in each cell in order to maintain normal calcium and PTH levels. One might therefore hypothesize that these cells maintain normal expression of the calcium sensing receptor. It will be important to determine this in order to better understand the parathyroid pathophysiology in this mouse model and illuminate the basis of this interesting discrepancy in its physiology as compared with the cognate human condition. Alternatively, unlike human tumors, the mouse parathyroid tumors may be nonfunctional. Tumors that were analyzed for loss of heterozygosity showed acquired loss of the normal allele, indicating that tumors are developing in this mouse model in a manner similar to the way tumors develop in human MEN1.[78] Further studies of this model may help to understand the role of *MEN1* in tumor formation, although the lack of hormonal activity of tumors in this model appears to represent a significant limitation in its potential utility for investigating some of the hormonal syndromes of MEN1.

RET and Calcium Sensing Receptor

Rare inherited conditions such as MEN1 with predisposition to a particular tumor type have yielded important information about the pathogenesis of the more common, sporadic type of the same tumor. Several such disorders in addition to MEN1 are accompanied by altered parathyroid function. Benign parathyroid tumors are found in 10-20% of MEN2A patients. Inherited mutations in the *RET* gene, encoding a receptor tyrosine kinase, have been found in nearly all patients with MEN2A, but in contrast with *MEN1* no somatic *RET* mutations have been observed in sporadic parathyroid adenomas.[80-83] Thus, acquired activating mutations in *RET* either do not occur in normal individuals' parathyroid cells with any appreciable frequency, or such mutations do not appear to confer a clinically significant selective advantage upon a parathyroid cell when they do occur.

Familial hypocalciuric hypercalcemia (FHH) can result from a germline inactivating mutation in one allele of the calcium-sensing receptor (*CASR*), located on chromosome 3q21.1. Although the parathyroid cells of typical FHH patients do not properly sense the level of blood calcium, parathyroid cell proliferation is not significantly increased.[84,85] However, inheritance of two defective *CASR* genes results in neonatal severe hyperparathyroidism, a disease in which the parathyroids grow excessively at a very young age. This link between inherited severe calcium sensing defects and parathyroid cell proliferation raised the possibility that acquired severe (i.e., biallelic inactivating) alterations in the *CASR* gene might also increase growth in parathyroid cells and result in sporadic parathyroid tumors. However, no inactivating mutations in *CASR* have been discovered in sporadic tumors[51,86] indicating that *CASR* mutations do not appear to provide a growth advantage when somatically acquired in a parathyroid cell. It remains possible that the decreased expression of the CaSR observed in parathyroid adenomas, apparently a secondary result of distinct primary tumorigenic alterations,[46] might still play a role in their pathogenesis. Nonetheless, among the genes most directly responsible for the rare inherited forms of hyperparathyroidism, only *MEN1* has been solidly linked to the development of the common sporadic form of the disease.

Vitamin D Receptor

Vitamin D, more specifically 1,25 dihydroxyvitamin D3, is known to inhibit parathyroid proliferation.[87,88] The *vitamin D receptor* (VDR) gene has been investigated as a possible target for acquired inactivation in parathyroid tumors but no specific clonal mutations have been found, either in parathyroid adenomas or in severe secondary/tertiary HPT of uremia.[89,90] Thus, VDR does not function as a classic tumor suppressor gene in parathyroid tissue, and its inactivation does not appear to be a primary driving force in parathyroid tumorigenesis. Studies have found a reduced level of VDR messenger RNA and protein levels in parathyroid adenomas and have correlated specific VDR germline polymorphisms with an increased susceptibility to parathyroid tumor formation, but the exact significance and mechanism of these associations remain to be defined.[7,91,92]

Other Genetic Abnormalities in Sporadic Parathyroid Adenomas

Studies of sporadic parathyroid adenomas have revealed highly recurrent clonal allelic losses that may indicate the genomic locations of key parathyroid tumor suppressor genes. Molecular allelotyping studies have shown frequent loss of heterozygosity on chromosomes 1, 6, 11 and 15 as well as less frequent losses on 9 and 13.[70,93] Studies of sporadic adenomas using comparative genomic hybridization (CGH), a molecular cytogenetic method which detects regions of chromosomal gains/losses in the tumor cell genome, have confirmed these common areas of deletion. CGH has also identified areas of chromosomal gain on chromosome 7, 16 and 19 that may signify the locations of new parathyroid oncogenes.[17,93,94] Unfortunately, traditional cytogenetic studies have not yielded significant insight into the locations of important genes involved in the pathogenesis of parathyroid tumors. While a cytogenetic translocation between chromosomes 1 and 5 has been reported in a single parathyroid adenoma,[95] the significance of this finding remains unclear.

Several candidate tumor suppressor genes that map to these regions of loss have been examined for mutations in parathyroid adenomas. *RAD51* and *RAD54*, located on 15q and 1p respectively, served as good candidate genes as they play an important role in DNA repair and recombination mechanisms that are important following radiation.[96,97] The incidence of parathyroid adenomas is increased following ionizing radiation,[98,99] and possibly, a defective DNA repair machinery may contribute to the development of these tumors as well as tumors not associated with radiation. CDK inhibitors, p15, p16 (both on 9p) and p18 (on 1p), which are involved in the negative regulation of the cell cycle, are located in known regions of chromosomal loss.[43,100] No tumor-specific mutations were found in either the CDK inhibitors or the RAD genes in parathyroid adenomas, indicating that these genes do not appear to play an important role in the development of sporadic adenomas. No proto-oncogenes in candidate regions have been investigated for activating mutations. However, the *ras* gene was examined, and no tumor-specific mutations were discovered.[101] Knowledge of the patterns of acquired and recurrent chromosomal aberrations will hopefully prove useful for the identification of the full set of oncogenes and tumor suppressor genes that contribute to the development of parathyroid adenomas.

Microsatellite Instability

Studies of several tumor types have demonstrated that cancer cells lack regulation of genomic stability. One type of such instability is microsatellite instability (MSI), caused by a defective mismatch repair mechanism. Microsatellites, or short tandem repeats are composed of di-, tri-, tetra and pentanucleotide repeat sequences. MSI is defined as a change in the length of repeats within a tumor, when compared with normal tissue from the same individual.[102] This tumor-specific allelic change, either due to insertion or deletion of repeating units, is reflective of a defective mismatch repair (MMR) system. The defective MMR fails to recognize and repair DNA replication errors, and thus, enhances the accumulation of single nucleotide mutations and alterations in the length of simple, repetitive microsatellite sequences that occur ubiquitously throughout the genome. MSI is observed in most hereditary nonpolyposis colorectal cancers (HNPCC), and is associated with inherited mutations in the MMR genes *hMSH2*, *hMLH1*, *hPMS1*, *hPMS2* and *hMSH3*. In addition to HNPCC, MSI is also observed in nearly 20% of sporadic colorectal tumors,[103] and extracolonic tumors including breast, endometrial, gastric and ovarian cancer.

It has been hypothesized that the growth rates of parathyroid tumors may be too low to account for the number of clonal mutational events that have been detected in these neoplasms,[6] although this argument may not adequately recognize the power of selection over an extended time period. In any case, a mutational event(s) that results in an increase the rate of genomic instability could potentially be operational in the parathyroid. Indeed, two studies have suggested that MSI may play a role in parathyroid tumorigenesis.[104,105] However, a limitation of these studies was their small sample size and a lack of systematic analysis of a genome-wide state of microsatellite instability. Thus, the contribution of this type of instability to the genesis of parathyroid tumors needs to be examined more conclusively.

Molecular Genetics of Parathyroid Carcinoma

While parathyroid carcinomas are very rare, they are associated with significant morbidity and mortality and are important to consider in the differential diagnosis of patients with primary HPT. Carcinomas often present in younger patients than do adenomas, and are more frequently symptomatic with more severe hypercalcemia. Nevertheless, it remains difficult to distinguish between adenomas and carcinomas on clinical and histological grounds, in the absence of distant metastases. Therefore, the discovery of genetic changes unique to carcinomas may aid in the discrimination between the two types of tumors. Such molecular diagnostic information might be especially valuable in the "atypical adenoma", which has certain histologic features that suggest, but are not specific for, an aggressive phenotype.[106] [Note added in proof: Mutation of the *HRPT2* tumor suppressor gene has recently been identified as a major factor in the pathogenesis of parathyroid carcinoma (Shattuck et al, N Engl J Med 2003; Howell et al, J Med Genet 2003).] Observed patterns of acquired clonal chromosomal changes appear to hold promise for their identification and for diagnostic use in their own right.

Molecular allelotyping and CGH have been used to identify areas in the genome where oncogenes and tumor suppressor genes involved in parathyroid carcinomas might be located.[17,107] Losses on 1p, 3q, 4q, 13q and 21q indicate possible areas where tumor suppressor genes involved in the development of carcinomas may lie.[17-19,107] CGH studies have described nonrandom chromosome gains, suggesting the possible genomic locations of oncogenes involved in parathyroid carcinogenesis.[17,19,107] However, most gains have not been seen consistently in multiple studies. A subset of these genetic abnormalities appears to occur with greater frequency in parathyroid carcinomas than in adenomas. These special regions may harbor genes that contribute to the invasive or metastatic behavior of the carcinomas. Interestingly, carcinomas also tend to lack most of the chromosomal changes that are commonly present in adenomas. This finding suggests that carcinomas do not generally originate from typical benign adenomas, but instead that carcinomas and adenomas develop along separate pathways driven by distinct genetic changes.

Strong evidence indicates that one key tumor suppressor gene important for the development of malignant parathyroid tumors is located on chromosome 13. Several groups have demonstrated that definite carcinomas and other clinically aggressive parathyroid tumors frequently have loss of heterozygosity on 13q.[108-110] The region of loss was shown to include the *RB1* gene[18] and *BRCA2*.[110] In tumors with 13q loss, protein levels of pRB, as detected by immunohistochemistry, were decreased in accord with a potential role for RB; analogous expression evidence is not available for BRCA2. That said, there are many genes on 13q, and it is quite conceivable that the true target of these acquired deletions, i.e., a classic 13q tumor suppressor whose biallelic inactivation is a driving force in selection of parathyroid cancers, will prove to be neither RB nor BRCA2. Resolution of this issue for any given 13q candidate gene awaits analysis of its sequence for specific internal inactivating mutations, the key evidence required to prove involvement as a parathyroid tumor suppressor gene. 13q loss has also been reported in a smaller percentage of parathyroid adenomas.[93] Future research will determine if the relevant 13q tumor suppressor genes in parathyroid carcinomas vs. adenomas are identical.

Loss of the *p53* tumor suppressor gene has been found occasionally in parathyroid carcinomas, but direct mutations have not been described. Therefore, *p53* is unlikely to play a major role as a classic tumor suppressor in the development of parathyroid carcinomas.[111,112]

Immunohistochemical studies have detected overexpression of the cyclin D1 oncogene in 50-91% of parathyroid carcinomas,[41] apparently even higher than the 20-40% overexpression observed in parathyroid adenomas.[39] These findings raise the possibility that cyclin D1 may also play a major role in the development of parathyroid cancers. In addition to its established role in adenomas, it will be important to determine the effects of cyclin D1 overexpression in the context of parathyroid malignancy, and whether cyclin D1 can act as a primary driver of parathyroid carcinomas.

Molecular Genetics of Secondary and Tertiary Hyperparathyroidism

The molecular basis of severe secondary or tertiary HPT is poorly understood. Because of multigland involvement, it was previously assumed that this condition predominantly involves polyclonal (non-neoplastic) cellular proliferations. Although this is likely to be the case in the initial proliferative phase, it is now clear that monoclonal parathyroid expansion does occur in most patients with severe secondary or tertiary HPT.[113,114] Emergence of such monoclonal expansions may well be a major factor in the acquisition of an increasingly autonomous PTH regulation and the refractoriness to conventional medical therapy found in severe secondary or tertiary HPT. Immunohistochemical studies have not detected overexpression of cyclin D1 in parathyroid glands from patients with uremic HPT,[40] suggesting an infrequent role for cyclin D1 in the development of this disease. Interestingly, acquired inactivation of the *MEN1* gene, also relatively common in sporadic primary HPT, seems to play only a negligible role in the clonal expansion of uremia-associated lesions. Only 2-3% of this form of sporadic parathyroid tumors exhibit allelic loss at 11q13, and *MEN1* mutation was found in just a subset of these.[114,117] Thus, the molecular genetic basis for the development of monoclonal parathyroid tumors in uremic patients appears to differ markedly from that in primary HPT. These findings lend further support to the hypothesis that different forms of parathyroid neoplasia develop through unique pathogenic mechanisms (Fig. 5).

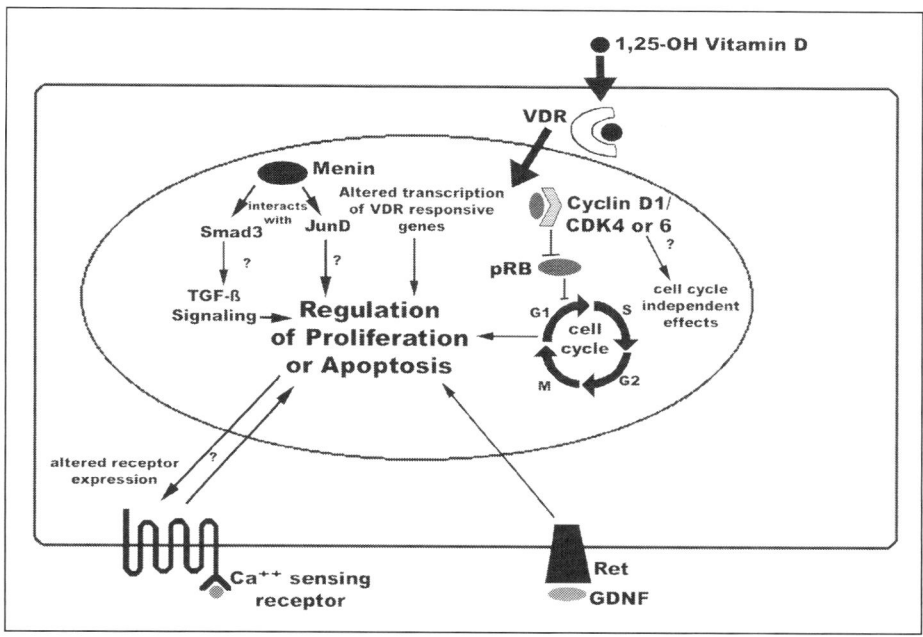

Figure 5. Mediators of parathyroid growth regulation. Many molecules that regulate parathyroid growth have been identified. The calcium-sensing receptor acts, at least in the context of severe germ-line defects, as a growth suppressor. Complete loss of functional calcium receptor in familial syndromes leads to abnormal proliferation of the parathyroid gland. Moreover, proliferative changes may alter expression of the CaSR. Loss of menin leads to abnormal proliferation. Protein partners that bind menin have been identified and include JunD and Smad3, but the mechanisms by which these interactions regulate growth remain unclear. Overexpression of cyclin D1 increases proliferation of parathyroid cells. Cyclin D1 binds CDK4 or 6 and can promote G1 to S transition by releasing the negative regulation of the cell cycle progression by pRb. This allows progression of a cell through the cell cycle and proliferation. It is also possible that cyclin D1 may have cell cycle independent effects as well. Overexpression of cyclin D1 is frequently seen in sporadic parathyroid tumors. Inherited mutations in RET, a tyrosine kinase receptor, lead to parathyroid tumors as is seen in MEN2A. Neither RET nor one of its ligands, Glial Derived Neurotrophic Factor (GDNF) seems to play a significant role in sporadic tumorigenesis. The vitamin D receptor (VDR), a nuclear receptor, binds 1, 25 dihydroxy-vitamin D and can then alter transcription and have been seen to decrease in parathyroid cell proliferation. No mutations in VDR have been identified in sporadic parathyroid tumors.

References

1. Melton LJ. Epidemiology of primary hyperparathyroidism. J Bone Miner Res 1991; 6(Suppl 2):S25-30; discussion S31-2.
2. Black WC 3rd, Utley JR. The differential diagnosis of parathyroid adenoma and chief cell hyperplasia. Am J Clin Pathol 1968; 49(6):761-75.
3. Silverberg SJ, Bilezikian JP. Clinical Presentation of Primary Hyperparathyroidism in the United States. In: Bilezikian JP, Marcus R, Levine MA, eds. The Parathyroids: Basic and Clinical Concepts. Second ed. San Diego: Academic Press, 2001:349-360.
4. Silverberg SJ, Bilezikian JP. Primary hyperparathyroidism: still evolving? J Bone Miner Res 1997; 12(5):856-62.
5. Silverberg SJ, Bilezikian JP. Clinical Course of Primary Hyperparathyroidism. In: Bilezikian JP, Marcus R, Levine MA, eds. The Parathyroids: Basic and Clinical Concepts. San Diego: Academic Press, 2001:387-398.

6. Parfitt AM, Wang Q, Palnitkar S. Rates of cell proliferation in adenomatous, suppressed, and normal parathyroid tissue: implications for pathogenesis. J Clin Endocrinol Metab 1998; 83(3):863-9.
7. Rao DS, Honasoge M, Divine GW et al. Effect of vitamin D nutrition on parathyroid adenoma weight: pathogenetic and clinical implications. J Clin Endocrinol Metab 2000; 85(3):1054-8.
8. Attie JN, Bock G, Auguste LJ. Multiple parathyroid adenomas: report of thirty-three cases. Surgery 1990; 108(6):1014-9; discussion 1019-20.
9. Verdonk CA, Edis AJ. Parathyroid "double adenomas": fact of fiction? Surgery 1981; 90(3):523-6.
10. Tezelman S, Shen W, Shaver JK et al. Double parathyroid adenomas. Clinical and biochemical characteristics before and after parathyroidectomy. Ann Surg 1993; 218(3):300-7; discussion 307-9.
11. Lyon MF. Gene action in the X-chromosome of the mouse (Mus musculus L.). Nature 1961; 290:372-373.
12. Migeon BR, Kennedy JF. Evidence for the inactivation of an X-chromosome early in the development of the human female. Am J Hum Genet 1975; 27(2):233-9.
13. Jackson CE, Cerny JC, Block MA et al. Probable clonal origin of aldosteronomas versus multicellular origin of parathyroid "adenomas". Surgery 1982; 92(5):875-9.
14. Fialkow PJ, Jackson CE, Block MA et al. Multicellular origin of parathyroid "adenomas". N Engl J Med 1977; 297(13):696-8.
15. Williams GT, Wynford-Thomas D. How may clonality be assessed in human tumours? Histopathology 1994; 24(3):287-92.
16. Arnold A, Staunton CE, Kim HG et al. Monoclonality and abnormal parathyroid hormone genes in parathyroid adenomas. N Engl J Med 1988; 318(11):658-62.
17. Agarwal SK, Schrock E, Kester MB et al. Comparative genomic hybridization analysis of human parathyroid tumors. Cancer Genet Cytogenet 1998; 106(1):30-6.
18. Cryns VL, Thor A, Xu HJ et al. Loss of the retinoblastoma tumor-suppressor gene in parathyroid carcinoma. N Engl J Med 1994; 330(11):757-61.
19. Imanishi Y, Palanisamy N, Tahara H et al. Molecular pathogenetic analysis of parathyroid carcinoma. J Bone Miner Res 1999; 14 (Suppl 1):S421.
20. Arnold A, Kim HG, Gaz RD et al. Molecular cloning and chromosomal mapping of DNA rearranged with the parathyroid hormone gene in a parathyroid adenoma. J Clin Invest 1989; 83(6):2034-40.
21. Rosenberg CL, Kim HG, Shows TB et al. Rearrangement and overexpression of D11S287E, a candidate oncogene on chromosome 11q13 in benign parathyroid tumors. Oncogene 1991; 6:449-53.
22. Motokura T, Bloom T, Kim HG et al. A novel cyclin encoded by a bcl1-linked candidate oncogene. Nature 1991; 350:512-5.
23. Matsushime H, Roussel MF, Ashmun RA et al. Colony-stimulating factor 1 regulates novel cyclins during the G1 phase of the cell cycle. Cell 1991; 65(4):701-13.
24. Xiong Y, Connolly T, Futcher B et al. Human D-type cyclin. Cell 1991; 65(4):691-9.
25. Lew DJ, Dulic V, Reed SI. Isolation of three novel human cyclins by rescue of G1 cyclin (Cln) function in yeast. Cell 1991; 66(6):1197-206.
26. Dowdy SF, Hinds PW, Louie K et al. Physical interaction of the retinoblastoma protein with human D cyclins. Cell 1993; 73(3):499-511.
27. Arnold A. The cyclin D1/PRAD1 oncogene in human neoplasia. J Investig Med 1995; 43(6):543-9.
28. Lammie GA, Fantl V, Smith R et al. D11S287, a putative oncogene on chromosome 11q13, is amplified and expressed in squamous cell and mammary carcinomas and linked to BCL-1. Oncogene 1991; 6(3):439-44.
29. Peters G, Fantl V, Smith R et al. Chromosome 11q13 markers and D-type cyclins in breast cancer. Breast Cancer Res Treat 1995; 33(2):125-35.
30. Izzo JG, Papadimitrakopoulou VA, Li XQ et al. Dysregulated cyclin D1 expression early in head and neck tumorigenesis: in vivo evidence for an association with subsequent gene amplification. Oncogene 1998; 17(18):2313-22.
31. Chesi M, Bergsagel PL, Brents LA et al. Dysregulation of cyclin D1 by translocation into an IgH gamma switch region in two multiple myeloma cell lines. Blood 1996; 88(2):674-81.
32. Swerdlow SH, Yang WI, Zukerberg LR et al. Expression of cyclin D1 protein in centrocytic/ mantle cell lymphomas with and without rearrangement of the BCL1/cyclin D1 gene. Hum Pathol 1995; 26(9):999-1004.

33. Rosenberg CL, Wong E, Petty EM et al. PRAD1, a candidate BCL1 oncogene: mapping and expression in centrocytic lymphoma. Proc Natl Acad Sci USA 1991; 88:9638-42.
34. Williams ME, Swerdlow SH, Rosenberg CL et al. Chromosome 11 translocation breakpoints at the PRAD1/cyclin D1 gene locus in centrocytic lymphoma. Leukemia 1993; 7:241-5.
35. Wolfel T, Hauer M, Schneider J et al. A p16INK4a-insensitive CDK4 mutant targeted by cytolytic T lymphocytes in a human melanoma. Science 1995; 269(5228):1281-4.
36. He J, Allen JR, Collins VP et al. CDK4 amplification is an alternative mechanism to p16 gene homozygous deletion in glioma cell lines. Cancer Res 1994; 54(22):5804-7.
37. Khatib ZA, Matsushime H, Valentine M et al. Coamplification of the CDK4 gene with MDM2 and GLI in human sarcomas. Cancer Res 1993; 53(22):5535-41.
38. Hemmer S, Wasenius VM, Haglund C et al. Deletion of 11q23 and cyclin D1 overexpression are frequent aberrations in parathyroid adenomas. Am J Pathol 2001; 158(4):1355-62.
39. Hsi ED, Zukerberg LR, Yang WI et al. Cyclin D1/PRAD1 expression in parathyroid adenomas: an immunohistochemical study. J Clin Endocrinol Metab 1996; 81(5):1736-9.
40. Tominaga Y, Tsuzuki T, Uchida K et al. Expression of PRAD1/cyclin D1, retinoblastoma gene products, and Ki67 in parathyroid hyperplasia caused by chronic renal failure versus primary adenoma. Kidney Int 1999; 55(4):1375-83.
41. Vasef MA, Brynes RK, Sturm M et al. Expression of cyclin D1 in parathyroid carcinomas, adenomas, and hyperplasias: a paraffin immunohistochemical study. Mod Pathol 1999; 12(4):412-6.
42. Hosokawa Y, Tu T, Tahara H et al. Absence of cyclin D1/PRAD1 point mutations in human breast cancers and parathyroid adenomas and identification of a new cyclin D1 gene polymorphism. Cancer Lett 1995; 93(2):165-70.
43. Tahara H, Smith AP, Gaz RD et al. Loss of chromosome arm 9p DNA and analysis of the p16 and p15 cyclin- dependent kinase inhibitor genes in human parathyroid adenomas. J Clin Endocrinol Metab 1996; 81(10):3663-7.
44. Neuman E, Ladha MH, Lin N et al. Cyclin D1 stimulation of estrogen receptor transcriptional activity independent of cdk4. Mol Cell Biol 1997; 17(9):5338-47.
45. Zwijsen RM, Wientjens E, Klompmaker R et al. CDK-independent activation of estrogen receptor by cyclin D1. Cell 1997; 88(3):405-15.
46. Imanishi Y, Hosokawa Y, Yoshimoto K et al. Primary hyperparathyroidism caused by parathyroid-targeted overexpression of cyclin D1 in transgenic mice. J Clin Invest 2001; 107(9):1093-1102.
47. Brown EM, Wilson RE, Thatcher JG et al. Abnormal calcium-regulated PTH release in normal parathyroid tissue from patients with adenoma. Am J Med 1981; 71(4):565-70.
48. Khosla S, Ebeling PR, Firek AF et al. Calcium infusion suggests a "set-point" abnormality of parathyroid gland function in familial benign hypercalcemia and more complex disturbances in primary hyperparathyroidism. J Clin Endocrinol Metab 1993; 76(3):715-20.
49. Parfitt AM, Fyhrie DP. Gompertzian growth curves in parathyroid tumours: further evidence for the set-point hypothesis. Cell Prolif 1997; 30(8-9):341-9.
50. Parfitt AM. Parathyroid growth: normal and abnormal. In: Bilezikian JPM R, Levine MA, ed. The Parathyroids: Basic and Clinical Concepts. Second ed. San Diego: Academic Press, 2001:293-329.
51. Hosokawa Y, Pollak MR, Brown EM et al. Mutational analysis of the extracellular Ca(2+)-sensing receptor gene in human parathyroid tumors. J Clin Endocrinol Metab 1995; 80(11):3107-10.
52. Larsson C, Skogseid B, Oberg K et al. Multiple endocrine neoplasia type 1 gene maps to chromosome 11 and is lost in insulinoma. Nature 1988; 332(6159):85-7.
53. Friedman E, Sakaguchi K, Bale AE et al. Clonality of parathyroid tumors in familial multiple endocrine neoplasia type 1. N Engl J Med 1989; 321(4):213-8.
54. Thakker RV, Bouloux P, Wooding C et al. Association of parathyroid tumors in multiple endocrine neoplasia type 1 with loss of alleles on chromosome 11. N Engl J Med 1989; 321(4):218-24.
55. Friedman E, De Marco L, Gejman PV et al. Allelic loss from chromosome 11 in parathyroid tumors. Cancer Res 1992; 52(24):6804-9.
56. Chandrasekharappa SC, Guru SC, Manickam P et al. Positional cloning of the gene for multiple endocrine neoplasia-type 1. Science 1997; 276(5311):404-7.
57. European Consortium on MEN1. Identification of the multiple endocrine neoplasia type 1 (MEN1) gene. Hum Mol Genet 1997; 6:1177-83.

58. Marx SJ, Agarwal SK, Kester MB et al. Multiple endocrine neoplasia type 1: clinical and genetic features of the hereditary endocrine neoplasias. Recent Prog Horm Res 1999; 54:397-438.
59. Schussheim DH, Skarulis MC, Agarwal SK et al. Multiple endocrine neoplasia type 1: new clinical and basic findings. Trends Endocrinol Metab 2001; 12(4):173-8.
60. Guru SC, Goldsmith PK, Burns AL et al. Menin, the product of the MEN1 gene, is a nuclear protein. Proc Natl Acad Sci USA 1998; 95(4):1630-4.
61. Bassett JH, Rashbass P, Harding B et al. Studies of the murine homolog of the multiple endocrine neoplasia type 1 (MEN1) gene, men1. J Bone Miner Res 1999; 14(1):3-10.
62. Kaji H, Canaff L, Goltzman D et al. Cell cycle regulation of menin expression. Cancer Res 1999; 59(20):5097-101.
63. Kim YS, Burns AL, Goldsmith PK et al. Stable overexpression of MEN1 suppresses tumorigenicity of RAS. Oncogene 1999; 18(43):5936-42.
64. Agarwal SK, Guru SC, Heppner C et al. Menin interacts with the AP1 transcription factor JunD and represses JunD-activated transcription. Cell 1999; 96(1):143-52.
65. Knapp JI, Heppner C, Hickman AB et al. Identification and characterization of JunD missense mutants that lack menin binding. Oncogene 2000; 19(41):4706-12.
66. Kaji H, Canaff L, Lebrun JJ et al. Inactivation of menin, a Smad3-interacting protein, blocks transforming growth factor type beta signaling. Proc Natl Acad Sci USA 2001; 98(7):3837-42.
67. Attisano L, Wrana JL. Smads as transcriptional co-modulators. Curr Opin Cell Biol 2000; 12(2):235-43.
68. Itoh S, Itoh F, Goumans MJ et al. Signaling of transforming growth factor-beta family members through Smad proteins. Eur J Biochem 2000; 267(24):6954-67.
69. Bystrom C, Larsson C, Blomberg C et al. Localization of the MEN1 gene to a small region within chromosome 11q13 by deletion mapping in tumors. Proc Natl Acad Sci USA 1990; 87(5):1968-72.
70. Tahara H, Smith AP, Gas RD et al. Genomic localization of novel candidate tumor suppressor gene loci in human parathyroid adenomas. Cancer Res 1996; 56(3):599-605.
71. Heppner C, Kester MB, Agarwal SK et al. Somatic mutation of the MEN1 gene in parathyroid tumours. Nat Genet 1997; 16(4):375-8.
72. Carling T, Correa P, Hessman O et al. Parathyroid MEN1 gene mutations in relation to clinical characteristics of nonfamilial primary hyperparathyroidism. J Clin Endocrinol Metab 1998; 83(8):2960-3.
73. Farnebo F, Teh BT, Kytola S et al. Alterations of the MEN1 gene in sporadic parathyroid tumors. J Clin Endocrinol Metab 1998; 83(8):2627-30.
74. Nord B, Larsson C, Wong FK et al. Sporadic follicular thyroid tumors show loss of a 200-kb region in 11q13 without evidence for mutations in the MEN1 gene. Genes Chromosomes Cancer 1999; 26(1):35-9.
75. Debelenko LV, Brambilla E, Agarwal SK et al. Identification of MEN1 gene mutations in sporadic carcinoid tumors of the lung. Hum Mol Genet 1997; 6(13):2285-90.
76. Zhuang Z, Vortmeyer AO, Pack S et al. Somatic mutations of the MEN1 tumor suppressor gene in sporadic gastrinomas and insulinomas. Cancer Res 1997; 57(21):4682-6.
77. Boni R, Vortmeyer AO, Pack S et al. Somatic mutations of the MEN1 tumor suppressor gene detected in sporadic angiofibromas. J Invest Dermatol 1998; 111(3):539-40.
78. Crabtree JS, Scacheri PC, Ward JM et al. A mouse model of multiple endocrine neoplasia, type 1, develops multiple endocrine tumors. Proc Natl Acad Sci USA 2001; 98(3):1118-23.
79. Stewart C, Parente F, Piehl F et al. Characterization of the mouse Men1 gene and its expression during development. Oncogene 1998; 17(19):2485-93.
80. Pausova Z, Soliman E, Amizuka N et al. Role of the RET proto-oncogene in sporadic hyperparathyroidism and in hyperparathyroidism of multiple endocrine neoplasia type 2. J Clin Endocrinol Metab 1996; 81(7):2711-8.
81. Padberg BC, Schroder S, Jochum W et al. Absence of RET proto-oncogene point mutations in sporadic hyperplastic and neoplastic lesions of the parathyroid gland. Am J Pathol 1995; 147(6):1600-7.
82. Kimura T, Yoshimoto K, Tanaka C et al. Obvious mRNA and protein expression but absence of mutations of the RET proto-oncogene in parathyroid tumors. Eur J Endocrinol 1996; 134(3):314-9.

83. Williams GH, Rooney S, Carss A et al. Analysis of the RET proto-oncogene in sporadic parathyroid adenomas. J Pathol 1996; 180(2):138-41.
84. Law WM Jr, Carney JA, Heath H 3rd. Parathyroid glands in familial benign hypercalcemia (familial hypocalciuric hypercalcemia). Am J Med 1984; 76(6):1021-6.
85. Law WM Jr, James EM, Charboneau JW et al. High-resolution parathyroid ultrasonography in familial benign hypercalcemia (familial hypocalciuric hypercalcemia). Mayo Clin Proc 1984; 59(3):153-5.
86. Cetani F, Pinchera A, Pardi E et al. No evidence for mutations in the calcium-sensing receptor gene in sporadic parathyroid adenomas. J Bone Miner Res 1999; 14(6):878-82.
87. Kremer R, Bolivar I, Goltzman D et al. Influence of calcium and 1,25-dihydroxycholecalciferol on proliferation and proto-oncogene expression in primary cultures of bovine parathyroid cells. Endocrinology 1989; 125(2):935-41.
88. Nygren P, Larsson R, Johansson H et al. 1,25(OH)2D3 inhibits hormone secretion and proliferation but not functional dedifferentiation of cultured bovine parathyroid cells. Calcif Tissue Int 1988; 43(4):213-8.
89. Wu HI, Arnold A. Vitamin D receptor gene as a candidate tumor suppressor gene in parathyroid adenomas. J Bone Mineral Res 1996; 11 (Suppl 1):S488.
90. Brown SB, Brierley TT, Palanisamy N et al. Vitamin D receptor as a candidate tumor-suppressor gene in severe hyperparathyroidism of uremia. J Clin Endocrinol Metab 2000; 85:868-72.
91. Carling T, Kindmark A, Hellman P et al. Vitamin D receptor genotypes in primary hyperparathyroidism. Nat Med 1995; 1(12):1309-11.
92. Carling T, Rastad J, Szabo E et al. Reduced parathyroid vitamin D receptor messenger ribonucleic acid levels in primary and secondary hyperparathyroidism. J Clin Endocrinol Metab 2000; 85(5):2000-3.
93. Palanisamy N, Imanishi Y, Rao PH et al. Novel chromosomal abnormalities identified by comparative genomic hybridization in parathyroid adenomas. J Clin Endocrinol Metab 1998; 83(5):1766-70.
94. Farnebo F, Kytola S, Teh BT et al. Alternative genetic pathways in parathyroid tumorigenesis. J Clin Endocrinol Metab 1999; 84(10):3775-80.
95. Orndal C, Johansson M, Heim S et al. Parathyroid adenoma with t(1; 5)(p22; q32) as the sole clonal chromosome abnormality. Cancer Genet Cytogenet 1990; 48(2):225-8.
96. Carling T, Imanishi Y, Gaz RD et al. Analysis of the RAD54 gene on chromosome 1p as a potential tumor- suppressor gene in parathyroid adenomas. Int J Cancer 1999; 83(1):80-2.
97. Carling T, Imanishi Y, Gaz RD et al. RAD51 as a candidate parathyroid tumour suppressor gene on chromosome 15q: absence of somatic mutations. Clin Endocrinol (Oxf) 1999; 51(4):403-7.
98. Fujiwara S, Sposto R, Ezaki H et al. Hyperparathyroidism among atomic bomb survivors in Hiroshima. Radiat Res 1992; 130(3):372-8.
99. Schneider AB, Gierlowski TC, Shore-Freedman E et al. Dose-response relationships for radiation-induced hyperparathyroidism. J Clin Endocrinol Metab 1995; 80(1):254-7.
100. Tahara H, Smith AP, Gaz RD et al. Parathyroid tumor suppressor on 1p: analysis of the p18 cyclin- dependent kinase inhibitor gene as a candidate. J Bone Miner Res 1997; 12:1330-4.
101. Friedman E, Bale AE, Marx SJ et al. Genetic abnormalities in sporadic parathyroid adenomas. J Clin Endocrinol Metab 1990; 71(2):293-7.
102. Boland CR, Thibodeau SN, Hamilton SR et al. A National Cancer Institute Workshop on Microsatellite Instability for cancer detection and familial predisposition: development of international criteria for the determination of microsatellite instability in colorectal cancer. Cancer Res 1998; 58(22):5248-57.
103. Liu B, Parsons R, Papadopoulos N et al. Analysis of mismatch repair genes in hereditary non-polyposis colorectal cancer patients. Nat Med 1996; 2(2):169-74.
104. Sarquis M, Friedman E, Boson WL et al. Microsatellite instability in sporadic parathyroid adenoma. J Clin Endocrinol Metab 2000; 85(1):250-2.
105. Koshiishi N, Chong JM, Fukasawa T et al. Microsatellite instability and loss of heterozygosity in primary and secondary proliferative lesions of the parathyroid gland. Lab Invest 1999; 79(9):1051-8.
106. Shane E. Parathyroid carcinoma. J Clin Endocrinol Metab 2001; 86(2):485-493.

107. Kytola S, Farnebo F, Obara T et al. Patterns of chromosomal imbalances in parathyroid carcinomas. Am J Pathol 2000; 157(2):579-86.
108. Cryns VL, Thor A, Xu HJ et al. Loss of the retinoblastoma tumor-suppressor gene in parathyroid carcinoma [see comments]. N Engl J Med 1994; 330(11):757-61.
109. Dotzenrath C, Teh BT, Farnebo F et al. Allelic loss of the retinoblastoma tumor suppressor gene: a marker for aggressive parathyroid tumors? J Clin Endocrinol Metab 1996; 81(9):3194-6.
110. Pearce SH, Trump D, Wooding C et al. Loss of heterozygosity studies at the retinoblastoma and breast cancer susceptibility (BRCA2) loci in pituitary, parathyroid, pancreatic and carcinoid tumours. Clin Endocrinol (Oxf) 1996; 45(2):195-200.
111. Cryns VL, Rubio MP, Thor AD et al. p53 abnormalities in human parathyroid carcinoma. J Clin Endocrinol Metab 1994; 78(6):1320-4.
112. Hakim JP, Levine MA. Absence of p53 point mutations in parathyroid adenoma and carcinoma. J Clin Endocrinol Metab 1994; 78(1):103-6.
113. Arnold A, Brown MF, Urena P et al. Monoclonality of parathyroid tumors in chronic renal failure and in primary parathyroid hyperplasia. J Clin Invest 1995; 95(5):2047-53.
114. Shan L, Nakamura Y, Murakami M et al. Clonal emergence in uremic parathyroid hyperplasia is not related to MEN1 gene abnormality. Jpn J Cancer Res 1999; 90:965-9.
115. Imanishi Y, Tahara H, Salusky I et al. MEN1 gene mutations in refractory hyperparathyroidism of uremia. J Bone Miner Res 1999; 14 (Suppl 1):S446.
116. Tahara H, Imanishi Y, Yamada T et al. Rare somatic inactivation of the multiple endocrine neoplasia type 1 gene in secondary hyperparathyroidism of uremia. J Clin Endocrinol Metab 2000; 85:4113-7.
117. Farnebo F, Teh BT, Dotzenrath C et al. Differential loss of heterozygosity in familial, sporadic, and uremic hyperparathyroidism. Hum Genet 1997; 99(3):342-9.

Chapter 12

Genetic Causes of Hypoparathyroidism

Rachel I. Gafni and Michael A. Levine

Abstract

Hypoparathyroidism is characterized clinically by the presence of hypocalcemia and hyperphosphatemia due to inadequate supply or effectiveness of circulating parathyroid hormone (PTH). It may be present either as an isolated finding or as a component of a more complex developmental, metabolic, or endocrinologic syndrome. While the most common cause of hypoparathyroidism continues to be surgical destruction,[1] several genetic etiologies have been identified that help define the molecular basis for less common causes. These genetic disorders can result in impaired embryologic development of the parathyroids, disordered synthesis or secretion of PTH, autoimmune destruction of the parathyroid gland, or inappropriate end-organ response to PTH (Table 1).

Disorders of Parathyroid Gland Formation

The parathyroid glands are derived from the epithelial endodermal lining of the third and fourth pharyngeal pouches (Fig. 1). The differentiation of these pouches into parathyroid tissue begins at approximately the 5th week of gestation.[2] The ventral portion of the 3rd pharyngeal pouch also gives rise to the thymus. The fetal parathyroids produce not only PTH but also parathyroid hormone-related peptide (PTHrP), a distinct polyhormone also produced in several other tissues.[3] The calcium homeostasis of a fetus differs from a post-term infant in that maternal calcium is actively transported across the placenta, under the regulation of PTHrP.[4] Although these observations had led to the notion that PTH is not necessary in fetal calcium metabolism, recent studies in transgenic animals lacking PTH, PTHrP, or both have demonstrated that PTH does play an important role in regulating serum concentrations of calcium in the fetus.[5] It is clear, however, that PTH secreted from parathyroid tissue after birth is the primary regulator of calcium and phosphate homeostasis in the growing child and adult.

DiGeorge Syndrome/Catch-22

The DiGeorge syndrome (DGS) is the most well-described condition in which neonatal hypoparathyroidism is associated with other developmental anomalies. In its most classical form, maldevelopment of the third and fourth pharyngeal pouches results not only in parathyroid hypoplasia but also in thymic hypoplasia and, consequently, impaired T-cell mediated immunity. In addition, these patients often manifest conotruncal cardiac abnormalities, cleft palate and dysmorphic facies (Fig. 2). Molecular mapping has attributed this syndrome to hemizygous microdeletions within 22q11.21-q11.23 (Fig. 3), most within a critical 250 kb region.[6] DGS most commonly arises from de novo deletion or translocation of 22q11, but autosomal dominant inheritance can occur. Large deletions of 22q11 are associated with several continuous gene

Table 1. Genetic disorders associated with hypoparathyroidism

	Inheritance	Locus	Gene	OMIM	Associated Abnormalities
Disorders of Parathyroid Gland Formation					
DiGeorge Syndrome/CATCH-22	Sporadic or AD	22q11.21-q11.23 10p13-p14 (DGS II)	?	*188400 *601362	Thymic hypoplasia w/ immune deficiency, conotruncal defects, cleft palate dysmorphic facies
Hypoparathyroidism, sensorineural deafness, renal dysplasia syndrome	AD	10p14-10pter	GATA3	*131320 #146255	Deafness, renal dysplasia
Hypoparathyroidsim-retardation-dysmorphism syndrome	AR	1q42-43	?	*241410	Growth retardation, developmental delay, microcephaly, microphthalmia, small hands/feet
Sanjad-Sakati syndrome	AR	1q42-43		*241410	
Kenny-Caffey Syndrome	AR or AD	1q42-43 (in AR)		*244460	Medullary stenosis, otic abnormalities
Mitochondrial Disease	variable				
Kearns-Sayre		mtDNA		#530000	Encephalopathy, eye disease, heart block
Pearson Marrow-Pancreas syndrome		mtDNA		#557000	Anemia, pancreatic dysfunction
MELAS		mt tRNA		#590050	Myopathy, encephalopathy, acidosis, stroke
LCHAD		2p23	MTP	*600890	Hypoglycemia, hypotonia
Familial isolated hypoparathyroidism	AR	6p23-24	GCMB	*603716	
	XR	Xq26-27	?	*307700	

continued on next page

Table 1. Continued

	Inheritance	Locus	Gene	OMIM	Associated Abnormalities
Disorder of Parathyroid Gland Secretion					
PTH gene mutations	AR or AD	11p15.3-15.1		*168450	
Calcium-sensing receptor mutations-activating	AD Sporadic	3q13.3-21	CaSR	*601199	
Parathyroid Gland Destruction					
Autoimmune polyendocrinopathy-candidiasis-ectodermal dystrophy syndrome	AR	21q22.3	AIRE	*240300	Adrenal insufficiency, mucocutaneous candidiasis, malbsorption, vitiligo, alopecia, hepatitis, pernicious anemia
Resistance to Parathyroid Hormone					
Pseudohypoparathyroidism Ia	AD	20q13.2-13.3 (maternal)	GNAS1	#103580 *139320	Albright's Hereditary Osteodystrophy
Pseudopseudohypoparathyroidism	AD	20q13.2-13.3 (paterna;)	GNAS1	#103580 *139320	Lacks biochemical hypoparathyroidsim
Pseudohypoparathyroidism Ib	Sporadic	20q13.3 (pat)	GNAS1	#603233	Methylation defects upstream of $G_s\alpha$ promoter

OMIM = Online Mendelian Inheritance in Man, http://www.ncbi.nlm.nih.gov/Omim/
AD, autosomal dominant; AR, autosomal recessive; XR, X-linked recessive

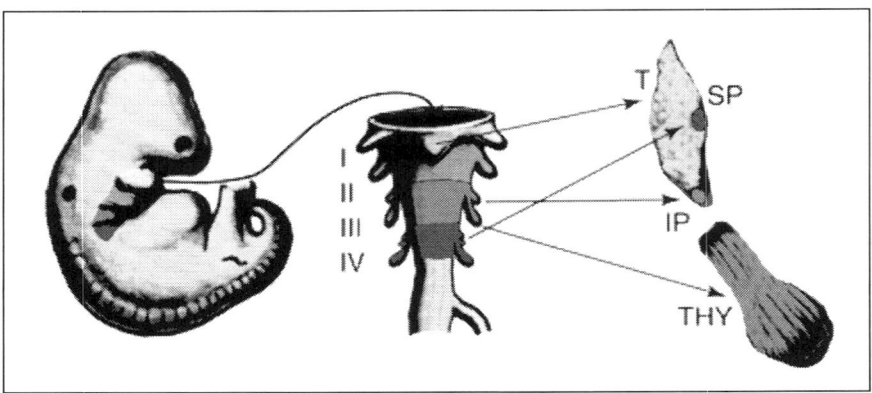

Figure 1. Parathyroid gland formation. The four endodermal pharyngeal pouches (I-IV) differentiate into the inferior parathyroid (IP), the superior parathyroid (SP), the thyroid (T), and the thymus (THY). Modified with permission from Wegener et al, Trends Genet 2001; 17:286-290.

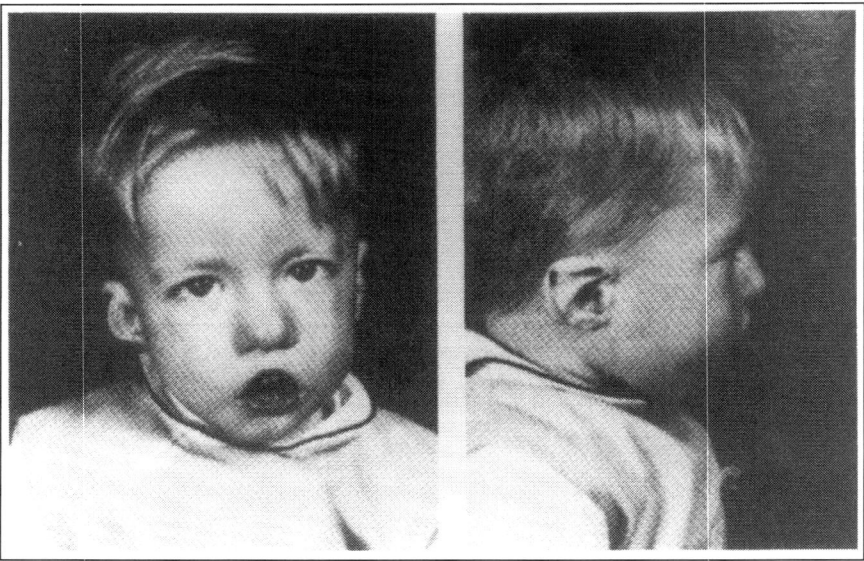

Figure 2. Characteristic facial features of DiGeorge Syndrome including micrognathia, hypertelorism, short philtrum and ear malformations. Reproduced with permission from Kretschmer et al, NEJM 1968; 279:1295.

deletion syndromes, including conotruncal anomaly face syndrome and velocardiofacial syndrome (VCFS).[6,7] VCFS is typically diagnosed later in childhood, and recent series have shown hypocalcemia to be present in up to 20% of cases.[8] Patients with deletions of 22q11 may exhibit partial or complete resolution of neonatal hypocalcemia[8,9] or may manifest latent hypoparathyroidism.[10] Because of the phenotypic variability of the various overlapping syndromes, it has been proposed that these conditions all be included within the acronym "CATCH-22", representing a syndrome of Cardiac abnormality, Abnormal facies, Thymic hypoplasia, Cleft palate, and Hypocalcemia with deletion of chromosome 22q11.

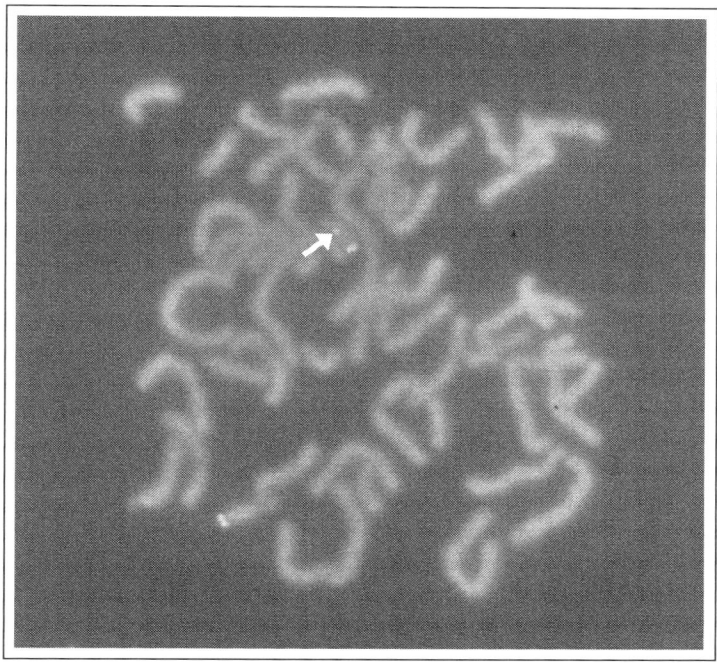

Figure 3. Metaphase spread of a patient with DiGeorge syndrome after fluorescence in situ hybridization (FISH). Green signals represent a chromosome-22-specific probe. The pink signal identifies the DiGeorge critical region on one chromosome 22. Absence of pink signal on the other allele indicates a microdeletion within this region.

Recent investigations have brought geneticists closer to identifying the specific genes within the critical region responsible for the phenotypic aberrations that occur in the CATCH-22 spectrum. One candidate is *CKRL*, which encodes an adapter protein implicated in growth factor and focal adhesion signaling.[11] Mice lacking the murine homologue, *Crkrol*, normally expressed in neural crest cells, demonstrate multiple defects in neural crest derivatives, including cranial ganglia, cardiac outflow tracts, thymus, parathyroids, and craniofacial structures.[12] Interestingly, only the homozygotes, which die in utero, display these abnormalities, while heterozygotes are phenotypically normal, suggesting a species-specific difference in sensitivity to gene dosage.[12] Similarly, mice with a homozygous deficiency of *Tbx1*, which encodes a T-box transcription factor expressed in non-neural crest cells, cranial mesenchyme, and pharyngeal pouches, demonstrate a phenotype similar to *Crkrol-/-* mice, while *Tbx1* heterozygotes display only cardiac anomalies.[13,14] Haploinsufficiency of *TBX1*, the human homologue, has emerged as the likely explanation for the developmental defects of DGS1, as heterozygous *TBX1* mutations have been identified in five patients (three of whom were from the same family) with DGS who lack the 22q11 microdeletion.[15] Most of these patients had a typical *del22q11DS*/DiGeorge syndrome phenotype (including heart defects) but did not have learning disabilities. Hence, consistent with mouse genetic results, human *TBX1* mutation (presumably leading to haploinsufficiency, but this is not known yet) is sufficient to cause most of the abnormalities observed in *del22q11DS* or DiGeorge syndrome.[16] Other genes that may be involved include *UDF1L* (ubiquitin fusion degradation gene), in which a hemizygous mutation is associated with some of the cardiac and craniofacial anomalies.[17] Taken together, these findings suggest that multiple genes may be involved in generating the broad and highly variable phenotypic spectrum seen in this disorder.

Hypoparathyroidism, Sensorineural Deafness and Renal Dysplasia Syndrome

In addition to the well-described mutations in chromosome 22q11, deletions within two nonoverlapping regions of 10p have been found to contribute to a phenotype similar to DiGeorge syndrome. Mutations within the DiGeorge critical region II, located at 10p13-14, are associated with DGS/VCFS.[18] Located more telomeric at 10p14-10pter is the locus for the hypoparathyroidism, sensorineural deafness, and renal dysplasia syndrome (HDR).[19] Unlike CATCH-22, individuals with HDR do not exhibit cardiac, palatal, or immunologic abnormalities. This disorder, first described as a syndromic form of autosomal dominant familial hypoparathyroidism in 1992,[20] was recently found to be due to haplo-insufficiency of the GATA binding protein-3 (*GATA3*) gene (Fig. 4).[21] The *GATA3* gene, which is located within a 200 kb critical HDR deletion region on 10p14-10pter, encodes a carboxy-terminal zinc-finger protein essential for DNA binding. It is expressed in the developing vertebrate kidney, otic vesicle, and parathyroids, as well as the central nervous system and organs of T-cell development.[21] Heterozygous mice do not demonstrate the HDR phenotype, and homozygous *GATA3* knockout mice develop multiple nonendocrine anomalies and immune deficiency, suggesting both tissue and species-specific differences.[21] Because immune disorders may be present in patients with other 10p deletions, it is likely that immunodeficiency is related to mutations within other genes on 10p genes, including those within the DGS II region.

Hypoparathyroidism-Retardation-Dysmorphism Syndrome/Sanjad-Sakati/Kenny-Caffey

The hypoparathyroidism-retardation-dysmorphism syndrome (HRD), also known as the Sanjad-Sakati (SS) syndrome, is a rare form of autosomal recessive hypoparathyroidism associated with other developmental anomalies. In addition to parathyroid dysgenesis, affected patients have severe growth and mental retardation, microcephaly, microphthalmia, small hands and feet and abnormal teeth.[22,23] This disorder, seen almost exclusively in individuals of Arab descent, has been mapped to a 1-cM interval on chromosome 1q42-43.[24,25] The Kenny-Caffey Syndrome is a clinically similar syndrome that includes hypocalcemia, dwarfism, medullary stenosis of the long bones and eye abnormalities.[26] This syndrome may be inherited as an autosomal recessive or autosomal dominant condition. The autosomal recessive variant, more prevalent among Arab kindreds, also includes growth retardation, dysmophic features and mental retardation, similar to the HRD/SS syndrome. Despite clinical variability, both of these autosomal recessive disorders are linked to the same 1.0 cM region on chromosome 1q42-q43,[26] and are both are associated with mutations of the *TBCE* gene, which encodes one of several chaperone proteins required for the proper folding of α-tubulin subunits and the formation of α-β-tubulin heterodimers.[27]

Mitochondrial Disease

Several syndromes due to deletions in mitochondrial DNA have been associated with endocrinopathies, including hypoparathyroidism. These include the Kearns-Sayre syndrome (encephalomyopathy, ophthalmoplegia, retinitis pigmentosa, heart block),[28] the Pearson Marrow-Pancreas syndrome (sideroblastic anemia, neutropenia, thrombocytopenia, pancreatic dysfunction),[29] and the maternally inherited diabetes and deafness syndrome.[30] Hypoparathyroidism has also been described in the MELAS (Mitochondrial myopathy, Encephalopathy, Lactic Acidosis, and Stroke-like episodes) syndrome, due to point mutations in mitochondrial tRNA.[31] Finally, mutations in the mitochondrial trifunctional protein (MTP), resulting in

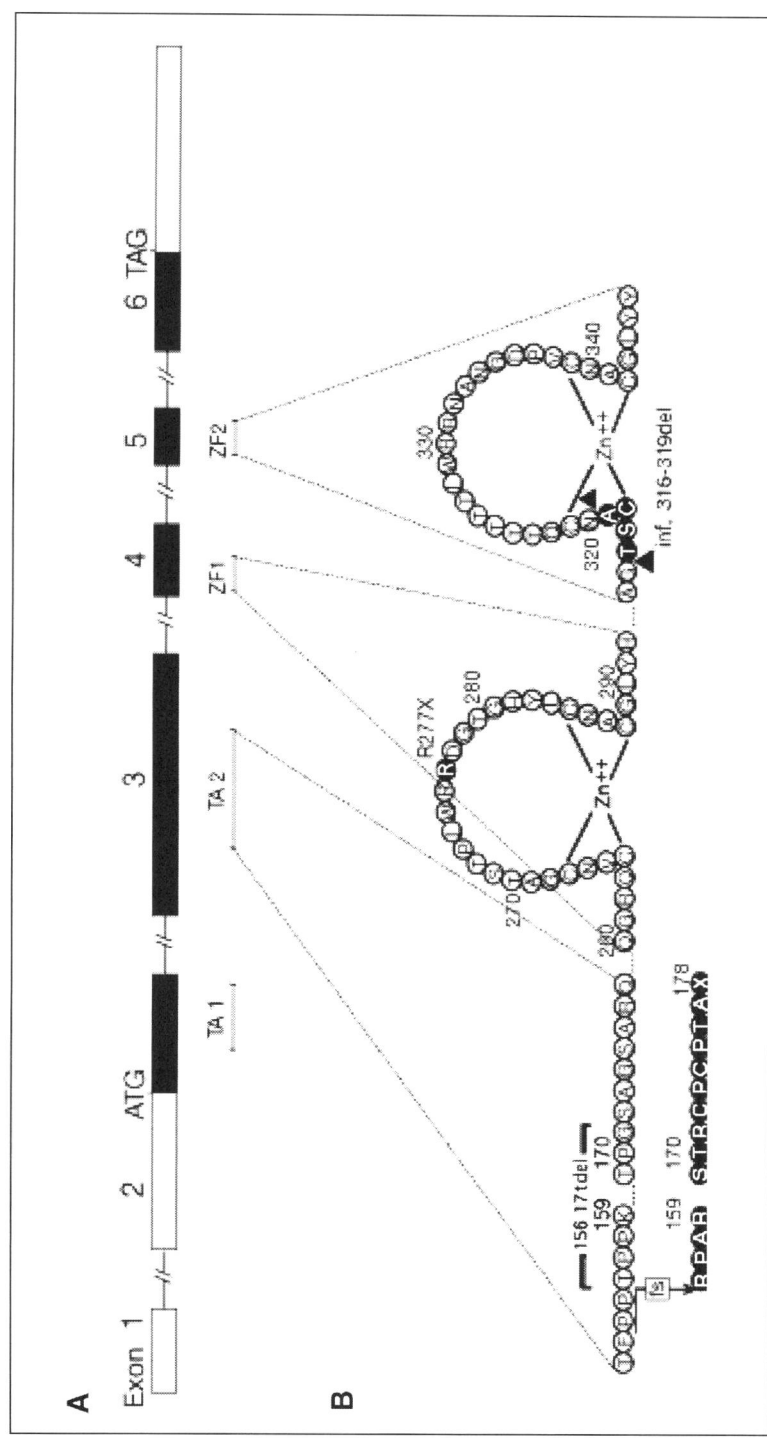

Figure 4. A) Organization of the 6-exon *GATA3* gene consisting of 2 transactivating domains (TA1 & TA2) and two zinc finger domains (ZF1 & ZF2). ATG, start codon; TAG, stop codon. B) Locations of *GATA3* mutations identified in the human HDR syndrome. fs, frameshift mutation resulting in missense peptide and premature stop at codon 178; R277X, transition (C → T) resulting in a premature termination codon at 277; inf, in-frame deletion (del) of codons 316-319 flanked by arrowheads. The altered amino acids are highlighted in black. Reproduced with permission from Van Esch et al.[21]

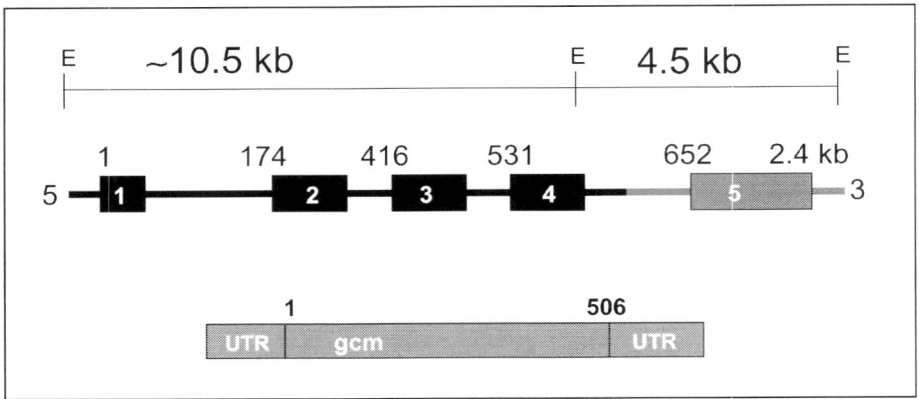

Figure 5. *GCMB* Gene Reproduced with permission from Ding et al.[38]

long-chain 3-hydroxy-acyl-coenzyme A dehydrogenase (LCHAD) deficiency or combined MTP deficiency have been associated with hypoparathyroidism in two unrelated patients.[32,33] Although a patient with LCHAD deficiency was noted to have absent parathyroids at autopsy,[32] the mechanisms by which these mitochondrial defects in fatty acid β-oxidation and respiratory chain function affect parathyroid development or function have yet to be elucidated.

Autosomal Recessive Hypoparathyroidism

Hypoparathyrodism may also occur as an isolated condition with an autosomal recessive inheritance pattern. Animal studies have identified the family of GCM (Glial Cell Missing) transcription factors as a possible molecular etiology for this disorder. The *GCMA* and *GCMB* genes are the human orthologues of the Drosophila *glial cell missing (gcm)* gene, which encodes a protein that determines glial versus neuronal differentiation of pluripotent neural stem cells in the fly. Accordingly, inactivating mutations of *gcm* result in preferential differentiation of stem cells to neurons.[34] Mice that are deficient in *Gcm2* (murine homologue of *GCMB*), which is expressed predominantly, if not uniquely, in developing parathyroid cells, lack parathyroid glands yet exhibit only mild hypocalcemia and normal levels of circulating parathyroid hormone.[35] This paradox is explained by the observation that the murine thymus is able to serve as an auxiliary source of parathyroid hormone, under control of the related *Gcm1* gene, expressed by thymic cells.[35]

The two human CGM homologs, encoded by the *GCMA* and *GCMB* genes, possess several characteristics unique to the GCM family of nuclear transcription factors.[36,37] Unlike the Drosophila *gcm* protein, the mammalian proteins are found primarily in nonneural tissues. *GCMA* is expressed predominantly in the placenta while *GCMB*, like *Gcm2*, is expressed predominantly in the developing and mature parathyroid gland.[36] Recently, a large intragenic mutation in the *GCMB* genes (Fig. 5), located on chromosome 6p23-24, was described in the proband of an extended kindred with autosomal recessive hypoparathyroidism. Both parents, as well as several other unaffected family members were heterozygous for the mutation.[38] Despite the absence of known consanquinity, microsatellite analysis suggested that both of the proband's mutant alleles were derived from a common ancestor.[38] Therefore, although the prevalence of *GCMB* mutations in autosomal recessive isolated hypoparathyroidism is as yet unknown, it is evident that *GCMB*, like *Gcm2* in the mouse, is a master regulator of parathyroid development.

X-Linked Hypoparathyroidism

Isolated familial hypoparathyroidism may also be inherited as an X-linked recessive trait. Affected males present with neonatal or infantile hypocalcemic seizures while hemizygous females are unaffected. Autopsy of an affected individual revealed complete agenesis of the parathyroid glands as the cause of hypoparathyroidism.[39] Although two apparently unrelated kindreds from Missouri were reported with this disorder in 1960[40] and 1981,[41] more recent mitochondrial DNA analysis of some individuals has revealed that the kindreds are, in fact, related.[42] Linkage analysis has localized the underlying mutation to a 1.5 Mb region on Xq26-q27[43,44], and recent molecular studies have identified a deletion-insertion involving chromosomes Xq27 and 2p25 as the basis for the defect.[45] Further investigation is still needed to determine the genes involved in this deletion-insertion, and to delineate the role they play in the embryologic development of the parathyroids.

Disorders of Parathyroid Hormone Synthesis or Secretion

PTH Gene Mutations

Mutations in the PTH gene (located on 11p15.3-p15.1), while rare, can result in hypoparathyroidism through impaired synthesis of parathyroid hormone. The human PTH gene contains 3 exons which encode the prepro-PTH hormone. Two proteolytic cleavages are required to achieve the biologically active, 84-amino acid parathyroid hormone. The initial cleavage occurs as the nascent polypeptide traverses the translocation channel of the endoplasmic reticulum and is modified by a signal peptidase enzyme complex that catalyzes the endoproteolytic cleavage of the 25-amino acids signal sequence. Final processing of pro-PTH then takes place in the Golgi apparatus, where the 6-amino acid pro sequence is removed (Fig. 6).[46]

To date, there have been reports of only three kindreds with known mutations that affect PTH synthesis. In the first family, demonstrating an autosomal dominant form of familial isolated hypoparathyroidism, a point mutation (T $\rightarrow$ C) was idenitified in exon 2 resulting in a substitution of arginine for cystine of the signal sequence of preproPTH (Fig. 6), thus disrupting the normally hydrophobic core necessary for translocation into the endoplasmic reticulum.[47] Subsequent studies demonstrated that this mutation impairs translocation of the prepro-PTH precursor to the endoplasmic reticulum, where the mutant protein is inefficiently cleaved by signal peptidase to pro-PTH (Fig. 7).[48] The basis for the dominant negative phenotype of the mutant prepro-PTH molecule remains unknown, however, as competition experiments failed to demonstrate that the mutant protein could impair synthesis or secretion of wild-type PTH.[48] Another family, exhibiting an autosomal recessive inheritance pattern, possessed a nucleotide substitution (G $\rightarrow$ C) in intron 2 that resulted in the loss of exon 2, and consequently, impaired PTH secretion.[49] Most recently, another recessively inherited form of the disease was found to be due to a missense mutation in exon 2 of the prepro-PTH gene that replaces Ser with Pro at the −3 position of the signal peptide.[50] This mutation can potentially impair proteolytic cleavage of the prepro-hormone, resulting in degradation within the endoplasmic reticulum.[50]

Autosomal Dominant Hypocalcemic Hypercalciuria

Gain-of-function mutations in the gene encoding the calcium-sensing receptor (CaSR) are associated with autosomal dominant hypocalcemic hypercalciuria. The CaSR belongs to the superfamily of 7-transmembrane domain, G-protein coupled receptors. Binding by extracellular calcium activates the receptor and stimulates G-protein mediated activation of phospholipase C, leading to increased concentrations of intracellular calcium and inositol 1,4,5-trisphosphate (IP3). The elevated IP3 stimulates protein kinase C activity, which

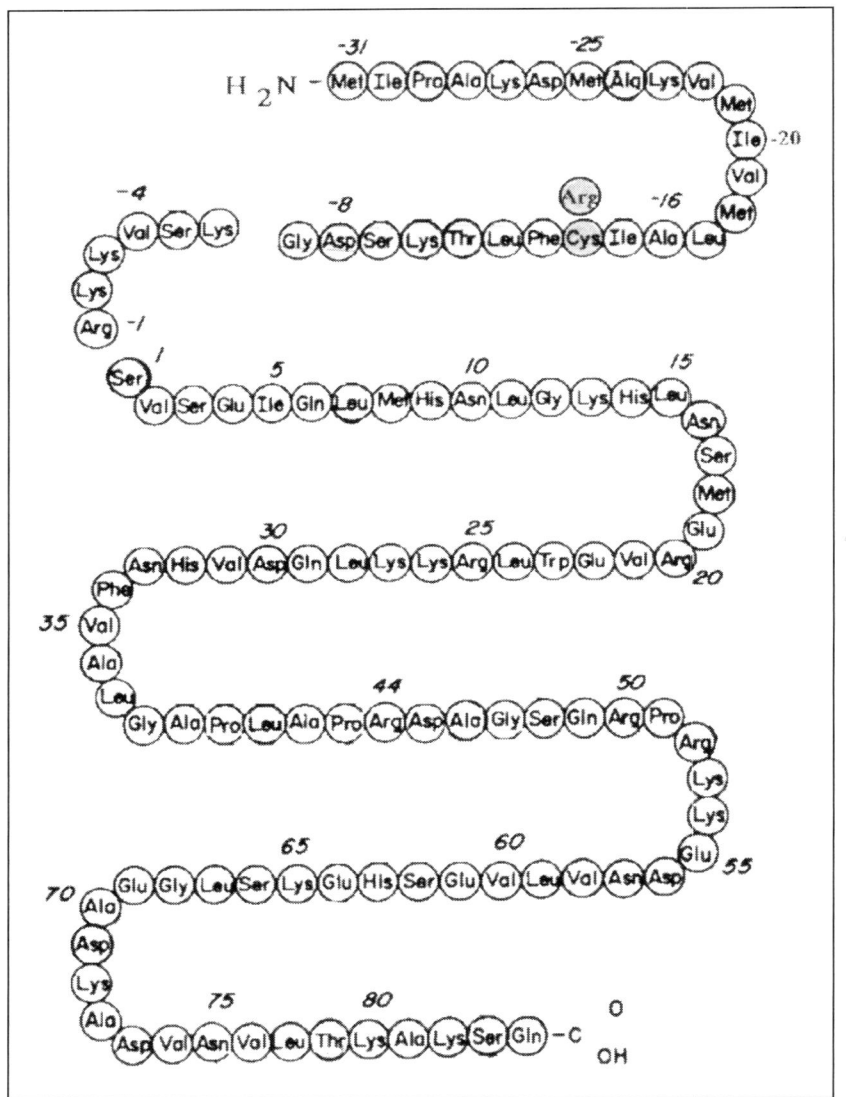

Figure 6. Synthesis and secretion of human parathyroid hormone. After translation in the nucleus, the preproPTH polypeptide enters the endoplasmic reticulum where it undergoes endoproteolytic cleavage of the signal sequence. Final processing of pro-PTH takes place in the Golgi apparatus, where the 6-amino acid pro sequence is removed. The biologically active PTH is then available for secretion.

ultimately decreases parathyroid hormone secretion.[51,52] Although the CaSR is present in many tissues, the clinically important features of this disorder result from abnormal expression in the renal tubules and parathyroid glands. In the parathyroid cells, constitutive activation of calcium-sensing receptors inhibits secretion of parathyroid hormone (Fig. 8). In the tubule cells of the thick ascending limb of the loop of Henle, activated CaSR's stimulate an inappropriate calciuresis despite low or low-normal serum calcium levels. Conversely,

Genetic Causes of Hypoparathyroidism

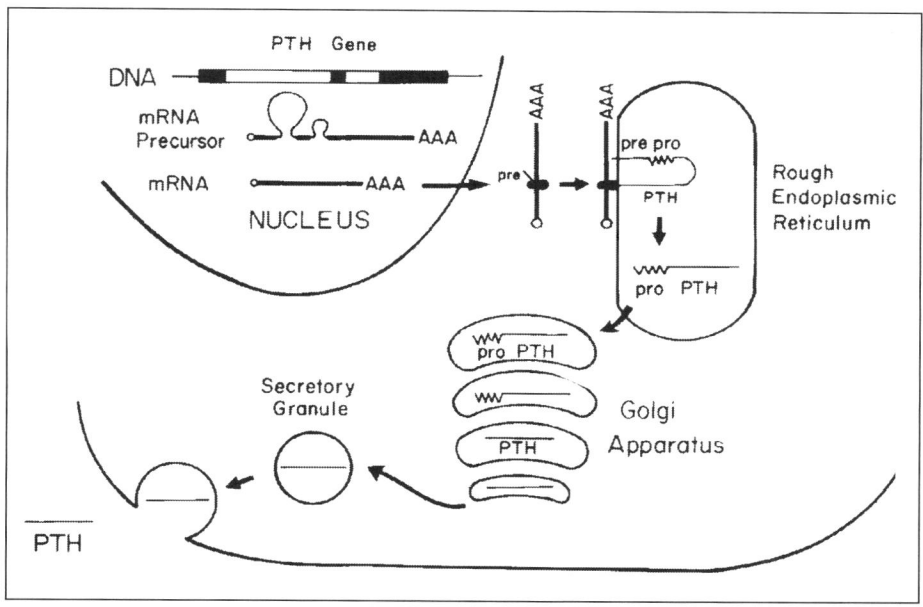

Figure 7. The codon for position 18 of the 31 amino acid prepro sequence of the preproparathyroid homone gene is changed from cysteine to arginine by a T → C point mutation, thus disrupting the hydrophobic core of the signal sequence. Reproduced with permission from Arnold et al.[47]

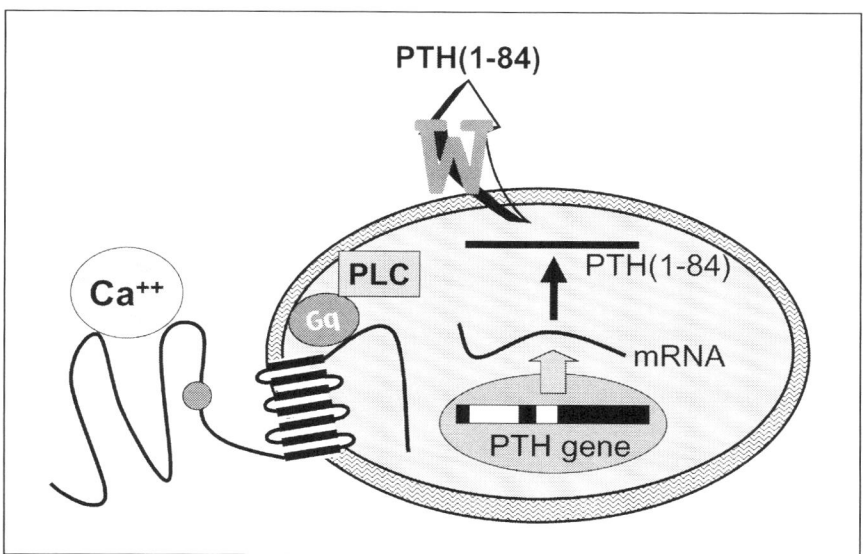

Figure 8. Impaired secretion of PTH due to activating mutations of the calcium-sensing receptor at the human parathyroid cell. Constitutive activation of the 7-transmembrane domain, G-protein coupled calcium-sensing receptor results in increased phospholipase C (PLC), intracellular calcium, and inositol 1,4,5 trisphosphate. The mechanism by which this signaling decreases PTH secretion is not well understood.

patients with inactivating mutations of the CaSR develop the syndrome of benign familial hypocalciuric hypercalcemia (heterozygotes) or, less frequently, neonatal severe hyperparathyroidism (most often homozygotes).[52]

The gene for the CaSR has been localized to chromosome 3q13.3-21. The initial report of a family with mild autosomal dominant hypoparathyroidism described a missense mutation in the aminoterminal, extracellular domain of the CaSR.[53] Subsequently, multiple other mutations have been identified in both the transmembrane and extracellular domains of the receptor with some mutations resulting in clinically severe hypoparathyroidism.[54-58] In addition to ADHH, de novo mutations have also been reported in cases of sporadic hypoparathyroidism.[59]

Parathyroid Gland Destruction

Autoimmune Polyendocrinopathy-Candidiasis-Ectodermal Dystrophy (APECED)

Autoimmune destruction of the parathyroids occurs most commonly in association with the complex of immune-mediated disorders that comprises the APECED (autoimmune polyendocrinopathy-candidiasis-ectodermal dystrophy) syndrome, sometimes referred to as APS I (Autoimmune Polyglandular Syndrome Type I). The syndrome's classic triad constitutes the HAM complex of hypoparathyroidism, adrenal insufficency, and mucocutaneous candidiasis. The temporal progression of features is surprisingly consistent, with the appearance of mucocutaneous candidiasis and hypoparathyroidism in the first decade of life, followed by primary adrenal insufficiency before 15 years of age (Fig. 9). While some patients do not manifest all three primary elements, other individuals may develop additional endocrinopathies such as hypogonadism, hypothyroidism, insulin-dependent diabetes, and hypophysitis. Nonendocrine components of the disorder that occur frequently include malabsorption, pernicious anemia, vitiligo, alopecia, nail and dental dystrophy, autoimmune hepatitis, and biliary cirrhosis (Fig. 10).[60,61]

APECED may be sporadic or familial; in the latter it demonstrates an autosomal recessive inheritance pattern. Although the disorder is seen worldwide, it is most prevalent in Finns, Sardinians, and Iranian Jews, and common gene mutations in some populations suggest significant founder effects.[62] The molecular basis for APECED appears to be inactivating mutations in the coding region of the *AIRE* (autoimmune regulator) gene, located on chromosome

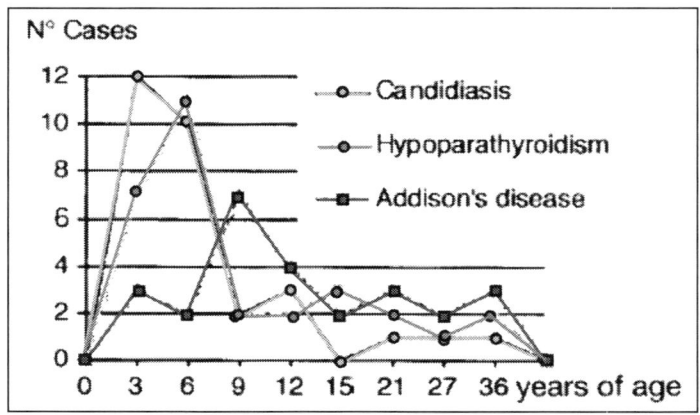

Figure 9. Timing of major clinical features in APECED. Adapted with permission from Betterle et al.[60]

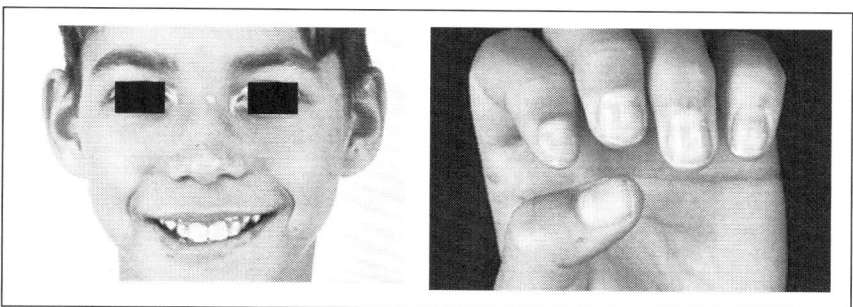

Figure 10. Photograph of a boy with APECED demonstrating vitiligo, dental abnormalities and nail dystrophy.

21q22.3.[62-65] *AIRE* encodes a 57.7-kDa protein that contains a nuclear localization signal and two PHD-type zinc-finger domains, features that are typical of proteins participating in chromatin-mediated transcriptional regulation.[66] The *AIRE* gene is expressed in tissues important for development and regulation of the immune system, such as the thymus and lymph nodes. Under normal circumstances, the protein product localizes to speckled domains in the nucleus while also colocalizing with cytoskeletal filaments.[67] The majority of *AIRE* mutations that have been identified in APECED patients are predicted to result in a truncated protein that lacks at least one of the PHD zinc-finger domains. While the exact function of *AIRE* is as yet unknown, expression in COS cells or fibroblasts of N-terminal *AIRE* protein fragments lacking the PHD domain demonstrates disrupted nuclear localization, suggesting that the effect of APECED mutations may occur primarily in the nucleus.[67]

Resistance to Parathyroid Hormone

In 1942, Albright, et al described patients with clinical hypoparathyroidism (hypocalcemia and hyperphosphatemia) and an unusual constellation of skeletal and developmental abnormalities who were resistant to the biological effects of endogenous PTH or administered bovine parathyroid extract.[68] Albright used the term pseudohypoparathyroidism (PHP) to characterize these initial patients, but the term now describes a heterogeneous group of disorders in which biochemical hypoparathyroidism results from target organ resistance to parathyroid hormone. In most cases of PHP, impaired responsiveness to PTH results from a defect in the expression or function of the α chain of G_s, the heterotrimeric G-protein that mediates PTH receptor stimulation of adenylyl cyclase.

Pseudohypoparathyroidism Type Ia (PHP Ia)

The term PHP Ia refers to patients with PTH resistance who also manifest the unusual developmental defects, now collectively referred to as Albright's Hereditary Osteodystrophy (AHO), which include brachydactaly, short stature, round facies, heterotopic subcutaneous ossification, and mental retardation (Fig. 11). Early studies indicated that patients with PHP Ia have defective excretion of nephrogenous cyclic AMP (cAMP) following PTH infusion[69] because of a deficiency of the α chain of G_s.[70,71]

The four subfamilies of the signal-transducing G proteins (G_s, G_i, G_q, and G_{12}) found coupled to over 1000 different receptors are comprised of three subunits ($\alpha\beta\gamma$). The α subunits are loosely associated with the tightly coupled $\beta\gamma$ dimers. The activity of the heterotrimeric protein is regulated by the binding and hydrolysis of GTP by the Gα subunit. Receptor activation results in release of GDP by the $\alpha\beta\gamma$ complex, binding of GTP to Gα, and dissociation of the

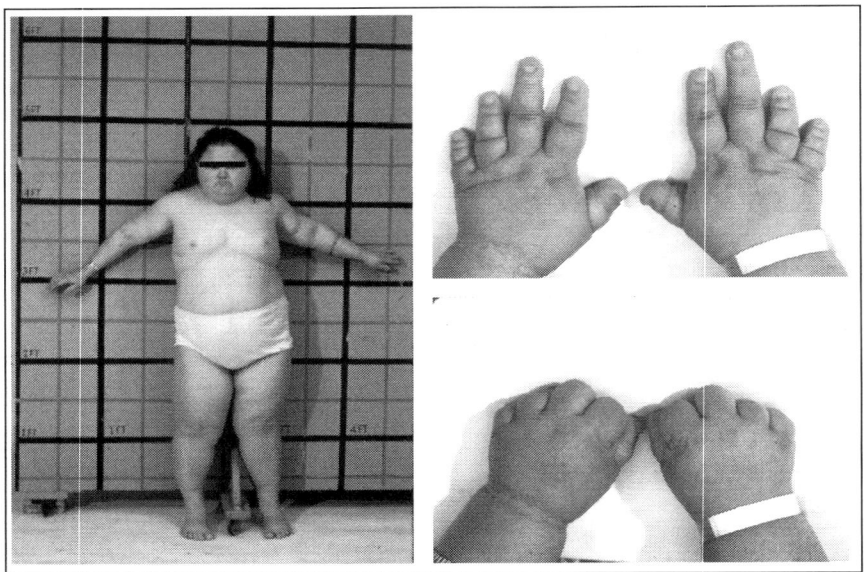

Figure 11. Photograph of a patient with the characteristic features of Albright's Hereditary Osteodystrophy including short stature, round facies, obesity, brachydactylay, and shortening of the 4th and 5th metacarpal bones. Reproduced with permission from Bilezikinn et al, The Parathyroids: Basic and Clinical Concepts, 2nd ed. San Diego: Academic Press, 2001:808.

Gα-GTP complex from the βγ dimer and receptor which lead to further regulation of downstream effectors. Members of the G_s family activate adenylyl cyclase, thus generating the second messenger cAMP necessary to mediate the actions of PTH in kidney and bone (Fig. 12).[72] Molecular analyses of patients with PHP Ia ultimately identified inactivating mutations of the *GNAS1* gene, located on chromosome 20q13.2-q13.3, which encodes $G_s\alpha$.[73] Gs activity is decreased by approximately 50% in many tissues in patients PHP Ia.[70] Because the G_s protein activity is essential for receptor function in a wide variety of tissues, some patients exhibit additional types of hormone resistance, such as hypothyroidism and hypogonadism.[74]

Pseudopseudohypoparathyroidism

Ten years after the initial description of PHP, Albright, et al reported an individual with AHO, the typical phenotype of PHP, without biochemical abnormalities.[75] This variant disorder, termed pseudopseudohypoparathyroidism (PPHP) or isolated AHO, was later found to have the same molecular defect as PHP Ia despite the absence of hypocalcemia or blunted cAMP excretion after PTH infusion. Review of family pedigrees revealed that when the mother was affected, the offspring developed PHP Ia while, if the father were affected, the children developed PPHP. In fact, patients with both PHP Ia and PPHP were often seen within the same kindred, leading researchers to believe that parathyroid hormone resistance is imprinted in a tissue specific manner.[76,77] Genomic imprinting occurs when specific regions on the maternal and paternal allele are differentially methylated, resulting in gene expression from only one allele.[77] To explain the clinical differences between patients with PHPIa and PPHP, this imprinting phenomenon would have to be tissue-specific such that $G_s\alpha$ is expressed only from the maternal allele in the proximal renal tubules while it is expressed biallelically in most other tissue. This model of tissue specific imprinting has been demonstrated in a $G_s\alpha$ knockout

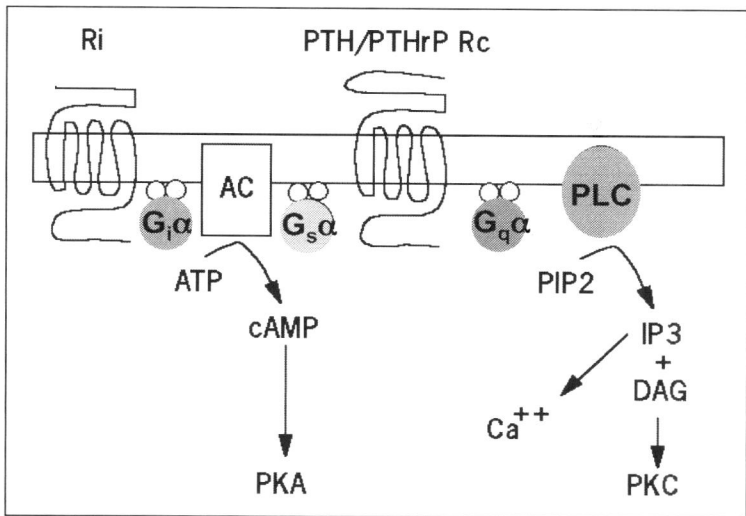

Figure 12. G-proteins involved in activation of the PTH/PTHrP receptor (Rc). Members of the stimulatory G_s family activate adenylyl cyclase (AC) to generate cAMP and protein kinase A (PKA). Other receptors, such as the Ca+-sensing receptor, bind to G_q proteins initiating the cascade resulting in production of protein kinase C (PKC). Inhibitory receptors (Ri) bind to members of the G_i family. ATP, adenosine triphosphate; DAG, diacylglycerol.

mouse.[78] Consequently, a paternally derived *GNAS1* mutation would not result in parathyroid hormone resistance but could yield abnormalities in other tissues (Fig. 13).

Pseudohypoparathyroidism Ib

Patients with pseudohypoparathyroidism Ib (PHP Ib), usually a sporadic disorder, exhibit PTH resistance similar to those with PHP Ia, yet they lack resistance to other hormones or an abnormal skeletal phenotype. In addition, $G_s\alpha$ activity is normal in the peripheral blood cells of affected individuals, virtually eliminating the possibility of inactivating mutation within the coding regions of *GNAS1*. Studies attempting to implicate a defect in the PTH receptor proved negative and mapping studies subsequently identified a paternally imprinted gene on 20q13.3,[79] in close proximity to the *GNAS1* gene responsible for PHPIa. Subsequent analysis revealed a region upsteam of the $G_s\alpha$ promoter, exon 1A, which is normally methylated on the maternal allele yet unmethylated on the paternal allele.[80] Evaluation of patients with both sporadic and familial forms of PHP Ib were found to have methylation defects in exon 1A resulting in a paternal-specific pattern on both alleles (Fig. 13).[80,81] Methylation defects have also been identified in other exons upstream of the known coding regions of *GNAS1*[82] as well as uniparental disomy resulting in loss of methylation in one patient with PHP Ib.[83] The exact mechanism by which these unmethylated regions cause decreased $G_s\alpha$ activity in the renal tubules versus other tissues is as yet unknown.

Pseudohypoparathyroidism 1C

There are rare reports of patients with patients with the biochemical and physical features of PHP Ia without measurable defect in *GNAS1* activity, classified as Pseudohypoparathyroidism 1C (PHP 1C). While the exact etiology is unknown, it may be related to a defect in expression of one or more isoforms of adenylyl cyclase.[84]

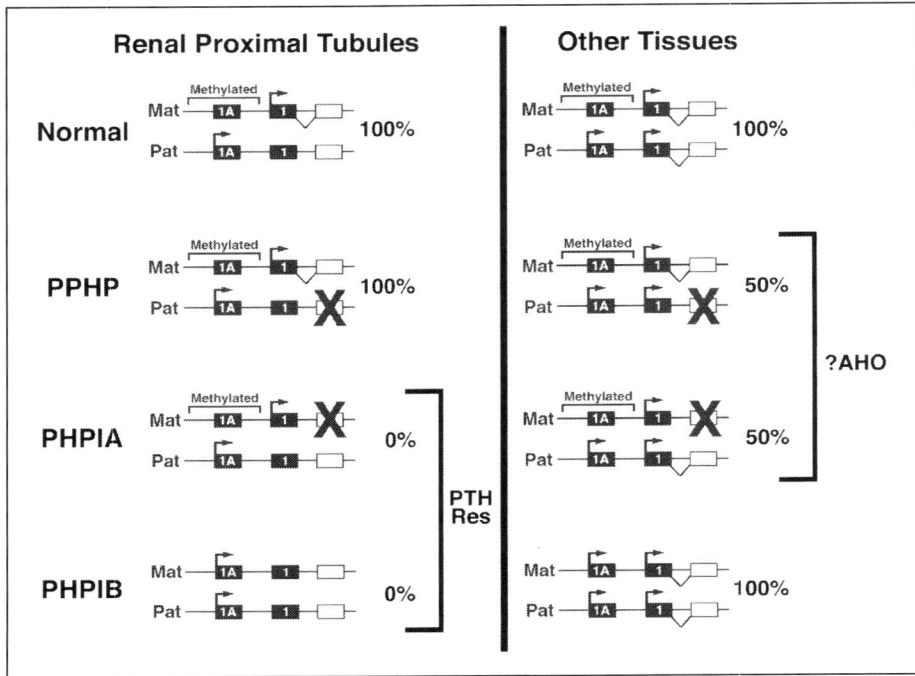

Figure 13. Mechanisms by which tissue-specific imprinting within the promoter regions may lead to variability in phenotype in *GNAS1* mutations. Exon 1A is located within a differentially methylated region (DMR) which is normally methylated on the maternal allele and unmethylated on the paternal allele. In most tissues, $G_s\alpha$ is expressed biallelically, independent of the exon 1A DMR. In the renal tubules, the unmethylated exon 1A DMR presumably suppresses $G_s\alpha$ expression by the paternal allele, resulting in maternal-specific expression in this tissue. In PHPIa and PPHP, inactivating mutations (X) on either allele results in a 50% reduction in $G_s\alpha$ in most tissues, seemingly responsible for the AHO phenotype. In the renal proximal tubules, however, a mutation in the already inactive paternal allele (PPHP) does not decrease $G_s\alpha$ activity and therefore does not result in the PTH resistance observed when the identical mutation is present on the maternal allele (PHPIa). In PHPIb, the defect is due to paternal-specific imprinting on both alleles, resulting in lack of $G_s\alpha$ expression in the renal tubules (i.e., PTH resistance), despite normal expression in other tissues. Reproduced with permission from Weinstein et al, Endocrine Reviews, 2001; 22:675-705.

Pseudohypoparathyroidism II

Patients with pseudohypoparathyroidism type II only manifest biochemical resistance to PTH. However, unlike patients with PHP Ib, they demonstrate a a blunted phosphaturic response to PTH infusion despite a normal increase in nephrogenous cAMP. Although these observations are consistent with a defect downstream of the second messenger cAMP, no specific defect has been elucidated.[85]

Conclusion

Hypoparathyroidism comprises a heterogeneous group of disorders that manifests at a variety of ages and in association with a wide spectrum of phenotypes. Innovative research using both humans and animals has brought us significantly closer to understanding the molecular mechanisms by which the parathyroid glands and parathyroid hormone regulate calcium homeostasis.

References

1. Marx SJ. Hyperparathyroid and hypoparathyroid disorders. N Engl J Med 2000; 343(25):1863-1875.
2. Sadler TW. Langman's Medical Embryology. 5th ed. Baltimore: Williams & Wilkins, 1985.
3. Kovacs CS, Manley NR, Moseley JM et al. Fetal parathyroids are not required to maintain placental calcium transport. J Clin Invest 2001; 107(8):1007-1015.
4. Kovacs CS, Lanske B, Hunzelman JL et al. Parathyroid hormone-related peptide (PTHrP) regulates fetal-placental calcium transport through a receptor distinct from the PTH/PTHrP receptor. Proc Natl Acad Sci USA 1996; 93(26):15233-15238.
5. Kovacs CS, Chafe LL, Fudge NJ et al. Pth regulates fetal blood calcium and skeletal mineralization independently of pthrp. Endocrinology 2001; 142(11):4983-4993.
6. Scambler PJ. The 22q11 deletion syndromes. Hum Mol Genet 2000; 9(16):2421-2426.
7. Shprintzen RJ. Velocardiofacial syndrome and DiGeorge sequence. J Med Genet 1994; 31(5):423-424.
8. Greig F, Paul E, DiMartino-Nardi J et al. Transient congenital hypoparathyroidism: resolution and recurrence in chromosome 22q11 deletion. J Pediatr 1996; 128(4):563-567.
9. Garcia-Garcia E, Camacho-Alonso J, Gomez-Rodriguez MJ et al. Transient congenital hypoparathyroidism and 22q11 deletion. J Pediatr Endocrinol Metab 2000; 13(6):659-661.
10. Cuneo BF, Driscoll DA, Gidding SS et al. Evolution of latent hypoparathyroidism in familial 22q11 deletion syndrome. Am J Med Genet 1997; 69(1):50-55.
11. Feller SM, Posern G, Voss J et al. Physiological signals and oncogenesis mediated through Crk family adapter proteins. J Cell Physiol 1998; 177(4):535-552.
12. Guris DL, Fantes J, Tara D et al. Mice lacking the homologue of the human 22q11.2 gene CRKL phenocopy neurocristopathies of DiGeorge syndrome. Nat Genet 2001; 27(3):293-298.
13. Lindsay EA, Vitelli F, Su H et al. Tbx1 haploinsufficieny in the DiGeorge syndrome region causes aortic arch defects in mice. Nature 2001; 410(6824):97-101.
14. Jerome LA, Papaioannou VE. DiGeorge syndrome phenotype in mice mutant for the T-box gene, Tbx1. Nat Genet 2001; 27(3):286-291.
15. Yagi H, Furutani Y, Hamada H et al. Role of TBX1 in human del22q11.2 syndrome. Lancet 2003; 362:1366-1373.
16. Baldini A. DiGeorge syndrome: an update. Curr Opin Cardiol 2004; 19:201-204.
17. Lindsay EA, Botta A, Jurecic V et al. Congenital heart disease in mice deficient for the DiGeorge syndrome region. Nature 1999; 401(6751):379-383.
18. Daw SC, Taylor C, Kraman M et al. A common region of 10p deleted in DiGeorge and velocardiofacial syndromes. Nat Genet 1996; 13(4):458-460.
19. Lichtner P, Konig R, Hasegawa T et al. An HDR (hypoparathyroidism, deafness, renal dysplasia) syndrome locus maps distal to the DiGeorge syndrome region on 10p13/14. J Med Genet 2000; 37(1):33-37.
20. Bilous RW, Murty G, Parkinson DB et al. Brief report: autosomal dominant familial hypoparathyroidism, sensorineural deafness, and renal dysplasia. N Engl J Med 1992; 327(15):1069-1074.
21. Van Esch H, Groenen P, Nesbit MA et al. GATA3 haplo-insufficiency causes human HDR syndrome. Nature 2000; 406(6794):419-422.
22. Hershkovitz E, Shalitin S, Levy J et al. The new syndrome of congenital hypoparathyroidism associated with dysmorphism, growth retardation, and developmental delay—a report of six patients. Isr J Med Sci 1995; 31(5):293-297.
23. Sanjad SA, Sakati NA, Abu-Osba YK et al. A new syndrome of congenital hypoparathyroidism, severe growth failure, and dysmorphic features. Arch Dis Child 1991; 66(2):193-196.
24. Parvari R, Hershkovitz E, Kanis A et al. Homozygosity and linkage-disequilibrium mapping of the syndrome of congenital hypoparathyroidism, growth and mental retardation, and dysmorphism to a 1-cM interval on chromosome 1q42-43. Am J Hum Genet 1998; 63(1):163-169.
25. Kelly TE, Blanton S, Saif R et al. Confirmation of the assignment of the Sanjad-Sakati (congenital hypoparathyroidism) syndrome (OMIM 241410) locus to chromosome 1q42-43. J Med Genet 2000; 37(1):63-64.
26. Diaz GA, Gelb BD, Ali F et al. Sanjad-Sakati and autosomal recessive Kenny-Caffey syndromes are allelic: evidence for an ancestral founder mutation and locus refinement. Am J Med Genet 1999; 85(1):48-52.

27. Parvari R, Hershkovitz E, Grossman N et al. Mutation of TBCE causes hypoparathyroidism-retardation-dysmorphism and autosomal recessive Kenny-Caffey syndrome. Nat Genet 2002; 32:448-452.
28. Tengan CH, Kiyomoto BH, Rocha MS et al. Mitochondrial encephalomyopathy and hypoparathyroidism associated with a duplication and a deletion of mitochondrial deoxyribonucleic acid. J Clin Endocrinol Metab 1998; 83(1):125-129.
29. Seneca S, De Meirleir L, De Schepper J et al. Pearson marrow pancreas syndrome: a molecular study and clinical management. Clin Genet 1997; 51(5):338-342.
30. Tanaka K, Takada Y, Matsunaka T et al. Diabetes mellitus, deafness, muscle weakness and hypocalcemia in a patient with an A3243G mutation of the mitochondrial DNA. Intern Med 2000; 39(3):249-252.
31. Morten KJ, Cooper JM, Brown GK et al. A new point mutation associated with mitochondrial encephalomyopathy. Hum Mol Genet 1993; 2(12):2081-2087.
32. Tyni T, Rapola J, Palotie A et al. Hypoparathyroidism in a patient with long-chain 3-hydroxyacyl-coenzyme A dehydrogenase deficiency caused by the G1528C mutation. J Pediatr 1997; 131(5):766-768.
33. Dionisi-Vici C, Garavaglia B, Burlina AB et al. Hypoparathyroidism in mitochondrial trifunctional protein deficiency. J Pediatr 1996; 129(1):159-162.
34. Hosoya T, Takizawa K, Nitta K et al. glial cells missing: a binary switch between neuronal and glial determination in Drosophila. Cell 1995; 82(6):1025-1036.
35. Gunther T, Chen ZF, Kim J et al. Genetic ablation of parathyroid glands reveals another source of parathyroid hormone. Nature 2000; 406(6792):199-203.
36. Kim J, Jones BW, Zock C et al. Isolation and characterization of mammalian homologs of the Drosophila gene glial cells missing. Proc Natl Acad Sci USA 1998; 95(21):12364-12369.
37. Kammerer M, Pirola B, Giglio S et al. GCMB, a second human homolog of the fly glide/gcm gene. Cytogenet Cell Genet 1999; 84(1-2):43-47.
38. Ding C, Buckingham B, Levine MA. Familial isolated hypoparathyroidism caused by a mutation in the gene for the transcription factor GCMB. J Clin Invest 2001; 108(8):1215-1220.
39. Whyte MP, Kim GS, Kosanovich M. Absence of parathyroid tissue in sex-linked recessive hypoparathyroidism. J Pediatr 1986; 109(5):915.
40. Peden VH. True idiopathic hypoparathyroidism as a sex-linked recessive trait. Am J Hum Genet 1960; 12:323-337.
41. Whyte MP, Weldon VV. Idiopathic hypoparathyroidism presenting with seizures during infancy: X-linked recessive inheritance in a large Missouri kindred. J Pediatr 1981; 99(4):608-611.
42. Mumm S, Whyte MP, Thakker RV et al. mtDNA analysis shows common ancestry in two kindreds with X-linked recessive hypoparathyroidism and reveals a heteroplasmic silent mutation. Am J Hum Genet 1997; 60(1):153-159.
43. Trump D, Dixon PH, Mumm S et al. Localisation of X linked recessive idiopathic hypoparathyroidism to a 1.5 Mb region on Xq26-q27. J Med Genet 1998; 35(11):905-909.
44. Thakker RV, Davies KE, Whyte MP et al. Mapping the gene causing X-linked recessive idiopathic hypoparathyroidism to Xq2-Xq27 by linkage studies. J Clin Invest 1990; 86(1):40-45.
45. Bowl MR, Nesbit MA, Harding B et al. X-linked recessive hypoparathyroidism is caused by a molecular deletion-insertion involving chromosomes Xq27 and 2p25. JBMR 2001; 16(Supp. 1): S152.
46. Habener JF, Rosenblatt M, Potts Jr JT. Parathyroid hormone: biochemical aspects of biosynthesis, secretion, action, and metabolism. Physiol Rev 1984; 64(3):985-1053.
47. Arnold A, Horst SA, Gardella TJ et al. Mutation of the signal peptide-encoding region of the preproparathyroid hormone gene in familial isolated hypoparathyroidism. J Clin Invest 1990; 86(4):1084-1087.
48. Karaplis AC, Lim SK, Baba H et al. Inefficient membrane targeting, translocation, and proteolytic processing by signal peptidase of a mutant preproparathyroid hormone protein. J Biol Chem 1995; 270(4):1629-1635.
49. Parkinson DB, Thakker RV. A donor splice site mutation in the parathyroid hormone gene is associated with autosomal recessive hypoparathyroidism. Nat Genet 1992; 1(2):149-152.
50. Sunthornthepvarakul T, Churesigaew S, Ngowngarmratana S. A novel mutation of the signal peptide of the preproparathyroid hormone gene associated with autosomal recessive familial isolated hypoparathyroidism. J Clin Endocrinol Metab 1999; 84(10):3792-3796.

51. Brown EM, MacLeod RJ. Extracellular calcium sensing and extracellular calcium signaling. Physiol Rev 2001; 81(1):239-297.
52. Mancilla EE, De Luca F, Baron J. Activating mutations of the Ca2+-sensing receptor. Mol Genet Metab 1998; 64(3):198-204.
53. Pollak MR, Brown EM, Estep HL et al. Autosomal dominant hypocalcaemia caused by a Ca(2+)-sensing receptor gene mutation. Nat Genet 1994; 8(3):303-307.
54. Baron J, Winer KK, Yanovski JA et al. Mutations in the Ca(2+)-sensing receptor gene cause autosomal dominant and sporadic hypoparathyroidism. Hum Mol Genet 1996; 5(5):601-606.
55. Pearce SH, Williamson C, Kifor O et al. A familial syndrome of hypocalcemia with hypercalciuria due to mutations in the calcium-sensing receptor. N Engl J Med 1996; 335(15):1115-1122.
56. Hirai H, Nakajima S, Miyauchi A et al. A novel activating mutation (C129S) in the calcium-sensing receptor gene in a Japanese family with autosomal dominant hypocalcemia. J Hum Genet 2001; 46(1):41-44.
57. Watanabe T, Bai M, Lane CR et al. Familial hypoparathyroidism: identification of a novel gain of function mutation in transmembrane domain 5 of the calcium-sensing receptor. J Clin Endocrinol Metab 1998; 83(7):2497-2502.
58. Hendy GN, D'Souza-Li L, Yang B et al. Mutations of the calcium-sensing receptor (CASR) in familial hypocalciuric hypercalcemia, neonatal severe hyperparathyroidism, and autosomal dominant hypocalcemia. Hum Mutat 2000; 16(4):281-296.
59. De Luca F, Ray K, Mancilla EE et al. Sporadic hypoparathyroidism caused by de Novo gain-of-function mutations of the Ca(2+)-sensing receptor. J Clin Endocrinol Metab 1997; 82(8):2710-2715.
60. Betterle C, Greggio NA, Volpato M. Autoimmune Polyglandular Syndrome Type 1. J Clin Endocrinol Metab 1998; 83(4):1049-1055.
61. Perheentupa J. Autoimmune polyendocrinopathy-candidiasis-ectodermal dystrophy (APECED). Horm Metab Res 1996; 28(7):353-356.
62. Pearce SH, Cheetham T, Imrie H et al. A common and recurrent 13-bp deletion in the autoimmune regulator gene in British kindreds with autoimmune polyendocrinopathy type 1. Am J Hum Genet 1998; 63(6):1675-1684.
63. Aaltonen J, Horelli-Kuitunen N, Fan JB et al. High-resolution physical and transcriptional mapping of the autoimmune polyendocrinopathy-candidiasis-ectodermal dystrophy locus on chromosome 21q22.3 by FISH. Genome Res 1997; 7(8):820-829.
64. Scott HS, Heino M, Peterson P et al. Common mutations in autoimmune polyendocrinopathy-candidiasis-ectodermal dystrophy patients of different origins. Mol Endocrinol 1998; 12(8):1112-1119.
65. Myhre AG, Halonen M, Eskelin P et al. Autoimmune polyendocrine syndrome type 1 (APS I) in Norway. Clin Endocrinol (Oxf) 2001; 54(2):211-217.
66. The Finnish-German APECED Consortium. An autoimmune disease, APECED, caused by mutations in a novel gene featuring two PHD-type zinc-finger domains. Nat Genet 1997; 17(4):399-403.
67. Rinderle C, Christensen HM, Schweiger S et al. AIRE encodes a nuclear protein colocalizing with cytoskeletal filaments: altered sub-cellular distribution of mutants lacking the PHD zinc fingers. Hum Mol Genet 1999; 8(2):277-290.
68. Albright F, Burnett CH, Smith PH et al. Pseudohypoparathyroidism - an example of "Seabright-Bantam syndrome". Endocrinology 1942; 3:922-932.
69. Chase LR, Melson GL, Aurbach GD. Pseudohypoparathyroidism: defective excretion of 3',5'-AMP in response to parathyroid hormone. J Clin Invest 1969; 48(10):1832-1844.
70. Levine MA, Ahn TG, Klupt SF et al. Genetic deficiency of the alpha subunit of the guanine nucleotide-binding protein Gs as the molecular basis for Albright hereditary osteodystrophy. Proc Natl Acad Sci USA 1988; 85(2):617-621.
71. Levine MA, Eil C, Downs Jr RW. et al. Deficient guanine nucleotide regulatory unit activity in cultured fibroblast membranes from patients with pseudohypoparathyroidism type I. a cause of impaired synthesis of 3',5'-cyclic AMP by intact and broken cells. J Clin Invest 1983; 72(1):316-324.
72. Farfel Z, Bourne HR, Iiri T. The expanding spectrum of G protein diseases. N Engl J Med 1999; 340(13):1012-1020.

73. Levine MA, Modi WS, O'Brien SJ. Mapping of the gene encoding the alpha subunit of the stimulatory G protein of adenylyl cyclase (GNAS1) to 20q13.2-q13.3 in human by in situ hybridization. Genomics 1991; 11(2):478-479.
74. Levine MA, Downs Jr RW. Moses AM et al. Resistance to multiple hormones in patients with pseudohypoparathyroidism. Association with deficient activity of guanine nucleotide regulatory protein. Am J Med 1983; 74(4):545-556.
75. Albright F, Forbes AP, Henneman PH. Pseudopseudohypoparathyroidism. Trans Assoc Am Physicians 1952; 65:337.
76. Davies SJ, Hughes HE. Imprinting in Albright's hereditary osteodystrophy. J Med Genet 1993; 30(2):101-103.
77. Weinstein LS, Yu S. The Role of Genomic Imprinting of Galpha in the Pathogenesis of Albright Hereditary Osteodystrophy. Trends Endocrinol Metab 1999; 10(3):81-85.
78. Yu S, Yu D, Lee E et al. Variable and tissue-specific hormone resistance in heterotrimeric Gs protein alpha-subunit (Gsalpha) knockout mice is due to tissue-specific imprinting of the gsalpha gene. Proc Natl Acad Sci USA 1998; 95(15):8715-8720.
79. Juppner H, Schipani E, Bastepe M et al. The gene responsible for pseudohypoparathyroidism type Ib is paternally imprinted and maps in four unrelated kindreds to chromosome 20q13.3. Proc Natl Acad Sci USA 1998; 95(20):11798-11803.
80. Liu J, Litman D, Rosenberg MJ et al. A GNAS1 imprinting defect in pseudohypoparathyroidism type IB. J Clin Invest 2000; 106(9):1167-1174.
81. Jan de Beur SM, Deng Z, Cho J et al. Loss of imprinting on the maternal GNAS1 allele in pseudohypoparathyroidism 1b. The 83rd Annual Meeting of the Endocrine Society 2001, 107 (abstract).
82. Bastepe M, Pincus JE, Sugimoto T et al. Positional dissociation between the genetic mutation responsible for pseudohypoparathyroidism type Ib and the associated methylation defect at exon A/B: evidence for a long-range regulatory element within the imprinted GNAS1 locus. Hum Mol Genet 2001; 10(12):1231-1241.
83. Bastepe M, Lane AH, Juppner H. Paternal uniparental isodisomy of chromosome 20q—and the resulting changes in GNAS1 methylation—as a plausible cause of pseudohypoparathyroidism. Am J Hum Genet 2001; 68(5):1283-1289.
84. Barrett D, Breslau NA, Wax MB et al. New form of pseudohypoparathyroidism with abnormal catalytic adenylate cyclase. Am J Physiol 1989; 257(2 Pt 1):E277-E283.
85. Drezner M, Neelon FA, Lebovitz HE. Pseudohypoparathyroidism type II: a possible defect in the reception of the cyclic AMP signal. N Engl J Med 1973; 289(20):1056-1060.

CHAPTER 13

Skeletal and Reproductive Abnormalities in *Pth*-Null Mice

Dengshun Miao, Bin He, Beate Lanske, Xiu-Ying Bai, Xin-Kang Tong, Geoffrey N. Hendy, David Goltzman and Andrew C. Karaplis

Abstract

We have examined the role of parathyroid hormone (PTH) in the postnatal state in a mouse model of PTH-deficiency generated by targeting the *Pth* gene in ES cells. Mice homozygous for the ablated allele, when maintained on a normal calcium intake, developed hypocalcemia, hyperphosphatemia, and low circulating 1,25-dihydroxyvitamin D_3 [1,25(OH)$_2$D$_3$] levels consistent with primary hypoparathyroidism. Fertility in mutant females was diminished due to abnormal ovarian function manifested in part by impaired angiogenesis in the developing corpus luteum. Even in the presence of ovarian dysfunction, bone turnover was reduced and trabecular and cortical bone volume were increased in PTH-deficient mice. When placed on a low calcium diet, fertility in female mice was completely abolished. Moreover, renal *25-hydroxyvitamin D 1 alpha-hydroxylase (Cyp27b1)* expression increased despite the absence of PTH, leading to a rise in circulating 1,25(OH)$_2$D$_3$ levels, marked osteoclastogenesis, and profound bone resorption. These studies demonstrate the dependence of the reproductive and skeletal phenotype in animals with genetically depleted PTH on the external environment as well as on internal hormonal and ionic circulatory factors. They point to the importance of calcium balance in reproduction and show that while PTH action is the first defense against hypocalcemia, 1,25(OH)$_2$D$_3$ can be mobilized, even in the absence of PTH, to guard against extreme calcium deficiency.

Introduction

Parathyroid hormone (PTH), the major peptide hormone regulator of calcium homeostasis, is produced almost exclusively by the parathyroid glands and is secreted in response to a decrease in extracellular calcium concentration. PTH enters the circulation and interacts with the type 1 PTH receptor (PTHR1) in target tissues, primarily bone and kidney.[1,2] This G protein-coupled cell surface receptor recognizes the 1-34 region of PTH, a sequence in which all the classic biological actions of the hormone (stimulation of bone resorption, increase in renal calcium reabsorption, phosphaturia, bicarbonaturia, 1α-hydroxylation of 25-hydroxyvitamin D, and cAMP production) reside.[3] PTHR1 possesses the unusual property of binding PTH as well as the paracrine factor PTH-related peptide (PTHrP) with nearly equal affinity. PTHrP was initially identified as the factor responsible for humoral hypercalcemia in patients with malignancy.[4] The capacity of PTHR1 to bind both PTH and PTHrP is based on sequence similarity in the N-terminal portion of these two ligands. Yet, PTHrP is

Molecular Biology of the Parathyroid, edited by Tally Naveh-Many. ©2005 Eurekah.com and Kluwer Academic / Plenum Publishers.

distinct from PTH in many structural features and certain biological effects, particularly in fetal development and physiology. Targeted disruption of either $Pthrp^5$ or $Pthr1^6$ in mice and defective PTHrP/PTHR1 signaling in man[7,8] lead to a form of lethal skeletal dysplasia characterized by decreased proliferation and accelerated differentiation of growth plate chondrocytes.

PTH also interacts with the type 2 PTH receptor (PTHR2), although the natural ligand for this receptor is likely the neuropeptide tuberoinfundibular peptide of 39 residues TIP39, rather than PTH itself.[9,10] Characterization of a third PTH receptor with specificity for the carboxyl-terminal region of PTH has also been reported in rat parathyroid cells, osteoblasts, and osteocytes that presumably exerts an antiresorptive effect on bone by impairing osteoclast differentiation.[11,12] It would seem, therefore, that several distinct properties could be attributed to PTH, likely mediated by a variety of receptors. Whether these non-classic biological effects of PTH have potential physiological relevance remains to be determined.

To better understand the physiologic actions of PTH on skeletal homeostasis and other biological systems, we have generated mice homozygous for a null *Pth* allele and examined the consequences associated with PTH deficiency in the post-natal state and the influence dietary calcium has in its absence.

Results

As a first step in generating PTH-null mice, we isolated mouse recombinant genomic DNA encompassing *Pth* and characterized the genomic organization and nucleotide sequence of the murine gene (Fig. 1A).[13] The strategy for ablating *Pth* focused on replacing the *Xho*I/*Xba*I genomic DNA fragment that encompasses part of exon 3 encoding the mature PTH peptide with the neomycin resistance (neo^r) selection cassette gene. Successfully targeted ES clones (Fig. 1B) were used to generate mice heterozygous for the mutation, which were then intercrossed to generate progeny homozygous for the disrupted *Pth* allele (Fig. 1C).

Pth-null mice were obtained at nearly the predicted Mendelian frequency and although lethality due to severe hypocalcemia was anticipated, the mice were viable. Yet, they did exhibit serum biochemical changes characteristic of primary hypoparathyroidism including moderate to severe hypocalcemia, hyperphosphatemia, and decreased serum 1,25-dihydroxyvitamin D_3 [$1,25(OH)_2D_3$] levels with complete absence of circulating PTH (Fig. 2A).

Histological examination of the post-natal parathyroids showed massive, diffuse enlargement of the glands consistent with the continuous stimulation by the prevailing hypocalcemia (Fig. 2B). PTH immunoreactivity was absent in the homozygous *Pth*-null glands thereby confirming the successful targeted disruption of the *Pth* gene. In contrast, expression of the calcium-sensing receptor (CaSR) protein was not markedly altered.

To delineate possible mechanism(s) that maintain calcemia to levels that sustain survival in the $Pth^{-/-}$ animals, we measured circulating serum levels of PTHrP. These were shown to be equivalent in wild-type and mutant litter mates (1.22 ± 0.06 and 1.12 ± 0.01 pmol/L, respectively). Moreover, thymectomy failed to reduce the survival of *Pth*-null mice compared to their normal litter mates (results not shown), indicating that the thymus does not serve as the tissue source of additional potential calcium-regulating factors other than PTH.[14]

Despite abnormalities in mineral homeostasis, mice grew normally and were fertile, although females were much less so than wild-type littermates (approximately 50-60% success rate). Intrigued by the decreased fertility of the *Pth*-null females, gross anatomic examination of the uterii and ovaries of 2-month-old mice was undertaken. The reproductive organs of these animals were similar in size to those from wild-type littermates (Fig. 3A,B). Histological assessment disclosed only very modest impairment in the development of the endometrium (data not shown) while in the ovaries, interstitial tissue was diffusely increased with rare corpora lutea (Fig. 3C). Associated with these alterations in ovarian morphology was the observation that serum levels of progesterone were significantly decreased in $Pth^{-/-}$ female mice relative

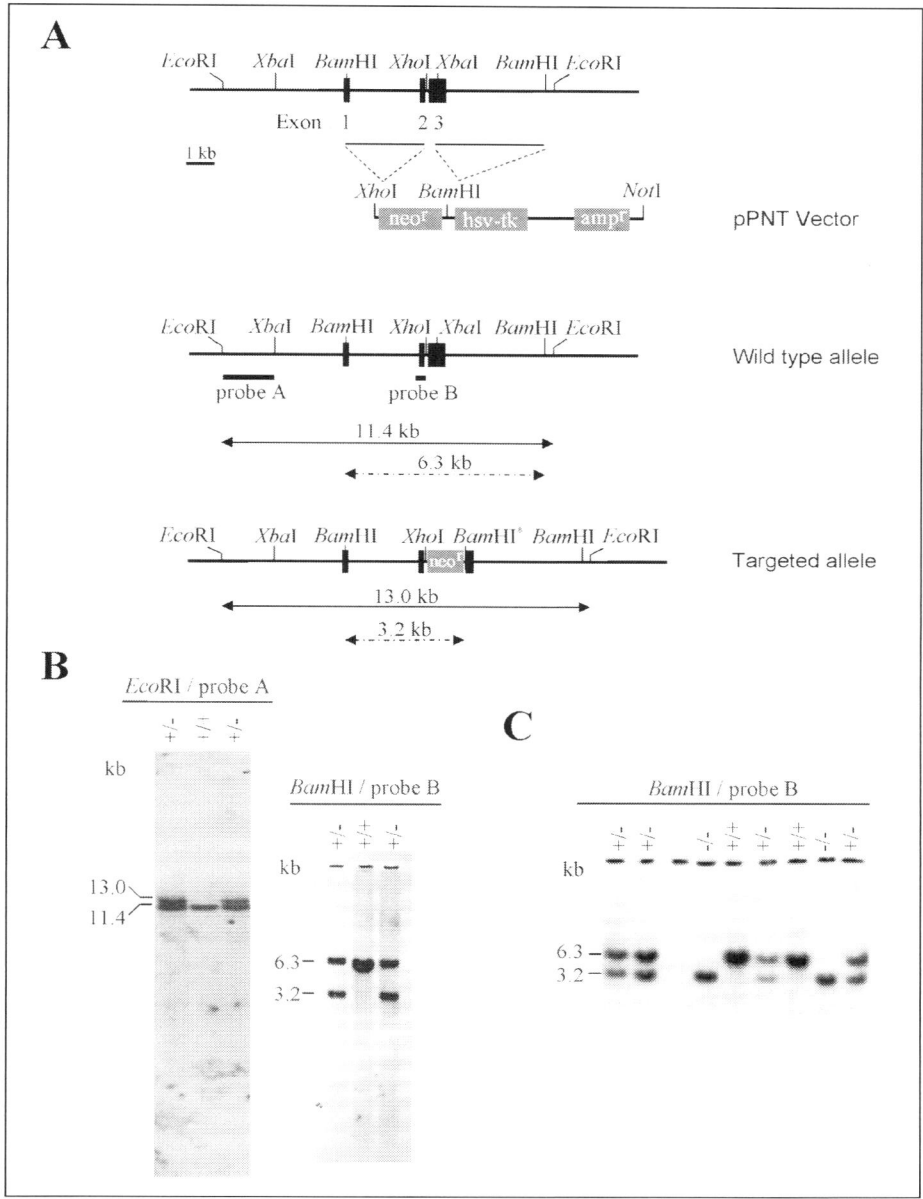

Figure 1. Targeted disruption of *Pth*. A) Genomic organization of *Pth* and generation of the targeting vector and targeted *Pth* allele. Asterix (*) indicates the *Bam*HI restriction site introduced in the genome by the targeting vector. B) Following digestion of genomic DNA from ES cell clones with *Eco*RI and hybridization with probe A (solid line) the targeted allele gives rise to a 13-kb fragment compared with an 11.4-kb fragment from the wild-type allele. A *Bam*HI digest and hybridization with probe B (dashed line) identifies a 3.2-kb band in the mutant allele compared with the 6.3-kb band from the wild-type allele. C) Identification of wild-type (+/+) mice and those heterozygous (+/-) and homozygous (-/-) for the disrupted *Pth* allele by Southern blot analysis of tail tip genomic DNA.

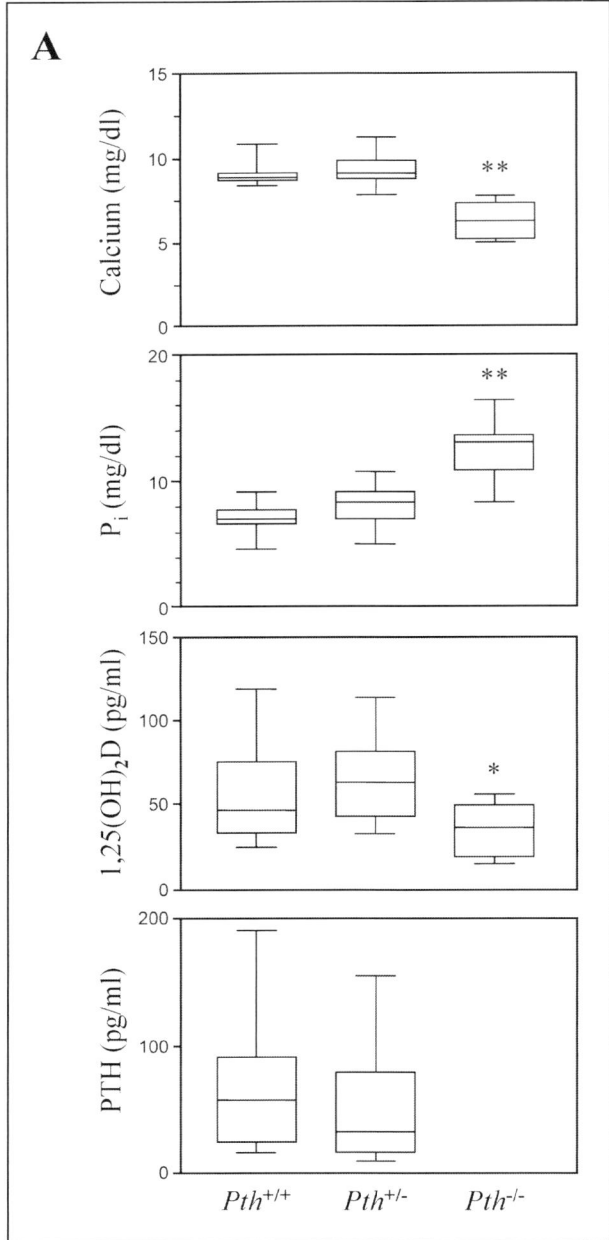

Figure 2. Hypocalcemia, hyperphosphatemia, reduced $1,25(OH)_2D_3$ levels, and parathyroid hyperplasia in *Pth*-null mice. A) Serum calcium, inorganic phosphate, $1,25(OH)_2D_3$, and PTH levels are shown as box and whiskers plots. The box extends from the 25th percentile to the 75th percentile, with the line at the 50th percentile (the median). The whiskers above and below the box show the highest and lowest values. * $P<0.05$; ** $P<0.01$ compared to wild-type mice (n=63 for +/+ and +/- groups; n=54 for -/- mice). Figure continued on next page.

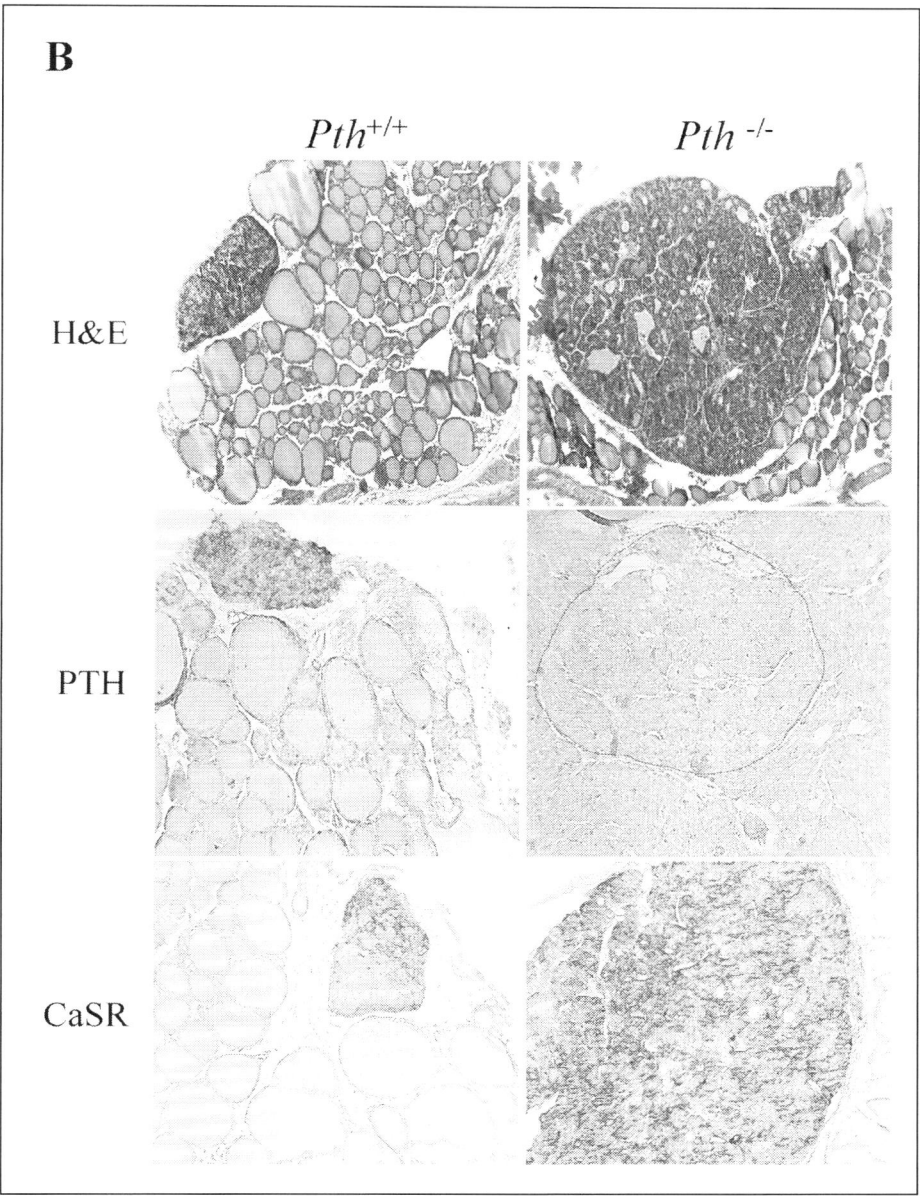

Figure 2, continued. B) Parathyroid hyperplasia in *Pth*-null mice. Top panels, H&E staining of parathyroid glands from wild-type and *Pth*-null mice. Diffuse parathyroid hyperplasia is noted. Middle panels, PTH immunostaining was evident in parathyroid tissue from wild-type animals but absent from mice with targeted disruption of *Pth*. Dashed circle demarcates the outline of the enlarged parathyroid gland. Bottom panels, Parathyroid tissue from both wild-type and mutant animals stained intensely for calcium-sensing receptor (CaSR) protein immunoreactivity. Results are representative of observations made from at least 4 animals in each group.

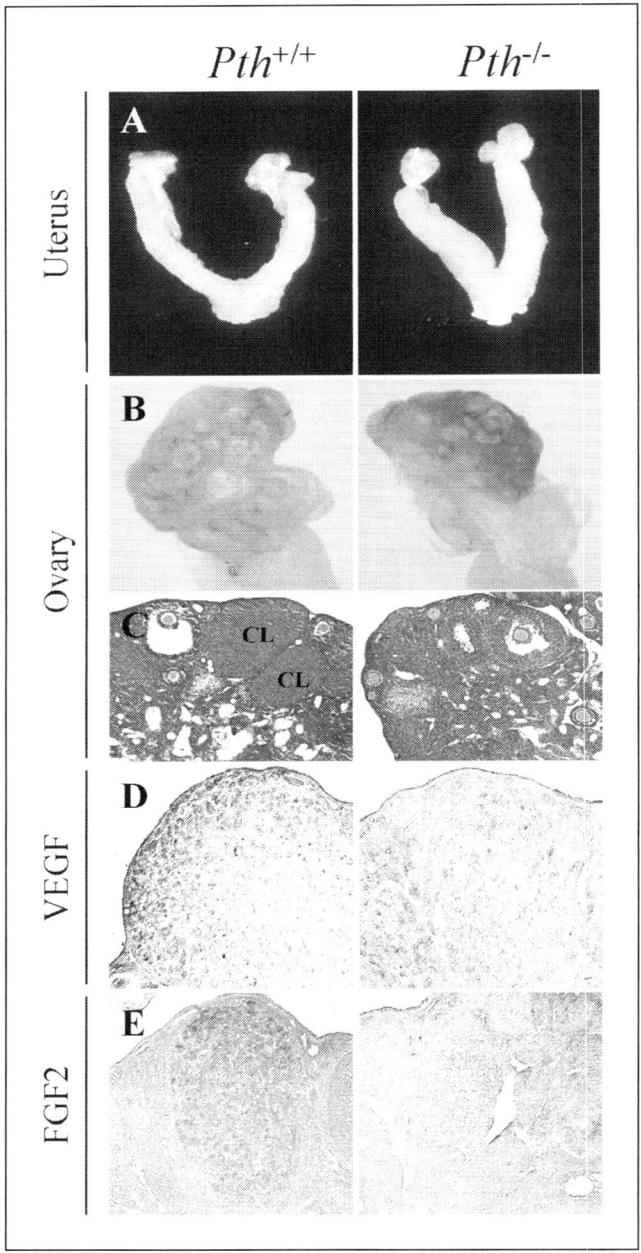

Figure 3. Abnormalities in $Pth^{-/-}$ female reproductive organs. A) Uteri of Pth-null animals were similar in size to those from wild-type litter mates. Histologically, the mutant uterine epithelium had slightly less prominent folds (not shown). B,C) Ovaries from Pth-null mice were of nearly normal size but the number of corpora lutea (CL) was markedly decreased. VEGF D) and FGF2 E) immunoreactivity was decreased in mutant corpora lutea compared to wild-type specimens. Results are representative of observations made from 16 animals in each group.

to wild-type littermates at the diestrus stage of the estrous cycle (8.2 ± 0.3 and 13.7 ± 3.7 nmol/L, respectively), while serum estradiol levels were comparable at the estrus stage (464.3 ± 26.6 and 437.1 ± 62.3 pmol/L, respectively). Male reproductive organs on the other hand were grossly and histologically normal in the mutant animals (data not shown).

Angiogenesis is very active during formation of the corpus luteum. We therefore examined the $Pth^{-/-}$ ovaries for expression of factors required for the development of this extensive neovascularization. Immunoreactivity for vascular endothelial growth factor (VEGF) and fibroblast growth factor-2 (FGF2) was diminished in the mutant tissue (Fig. 3D,E), implicating these angiogenic proteins as regulators of luteal angiogenesis whose expression is altered by the prevailing hypoparathyroid state.

Because PTH has major effects on bone remodeling, we next studied the skeletons of $Pth^{-/-}$ animals fed a normal calcium diet at 2, 4, 6 and 9 months of age. Histologically, cartilage development at the growth plate was normal at all times examined (Fig. 4A), with proper zone organization and adequate mineralization, suggesting that, in the postnatal state, PTH does not play a major role in chondrocyte biology. However, major bony alterations were observed in the mutant animals. Trabecular and cortical bone volume was consistently increased (1.8-fold) in these mice compared to sex-matched littermates (Fig. 4B). To define the cause of this increase in bone volume, dynamic histomorphometric analysis of bone was undertaken following double administration of calcein. Endosteal and trabecular Mean Apposition Rate (MAR), a parameter of bone formation, was profoundly decreased in the mutant mice compared with normal littermates and osteoblasts lining bone surfaces were also diminished (Fig. 5A,B). As well, the number of osteoclasts was reduced (50%) in the Pth-null animals (Fig. 5C), indicating that loss of PTH is associated with a generalized state of low bone turnover.

We next examined whether alterations in the calcium content of the diet consumed could alter the phenotype of the $Pth^{-/-}$ mice. For these studies, homozygous and heterozygous 2-month-old PTH deficient mice and wild-type litter mates were fed either regular or low calcium diets for 8 weeks at which time, samples of serum and bones were analyzed. Normal and heterozygous littermates were indistinguishable in all parameters examined (results not shown). In contrast, $Pth^{-/-}$ mice had persistent hypocalcemia and hyperphosphatemia that remained unaltered despite the dietary modification (Fig. 6A). Circulating levels of PTHrP also remained unchanged. In sharp contrast, serum $1,25(OH)_2D_3$ levels rose markedly in both groups of mice following institution of a low-calcium diet (4.9-fold in wild-type vs 6.0-fold in mutant mice), indicating that mechanisms had been set in motion in the Pth-null animals to overcome the absence of the stimulatory effect of PTH on vitamin D synthesis. This increase was reflected, in part, by a rise in renal proximal tubule *25-hydroxyvitamin D 1 alpha-hydroxylase* (*Cyp27b1*) transcript levels (Fig. 6B) and protein immunoreactivity (Fig. 6C), the former being more pronounced in wild-type animals likely due to the prevailing secondary hyperparathyroidism. Interestingly, under these dietary conditions, female $Pth^{-/-}$ mice lost completely their capacity to conceive, while the reproductive capability of male animals remained unaffected.

The source of calcium mobilized for maintenance of circulating calcium levels on the calcium-deficient diet became apparent when bones from these animals were examined (Fig. 7A). Trabecular bone volume in the wild-type lumbar vertebrae was reduced to 54% of the pretreatment content, while in the mutant specimens the corresponding reduction was to 29% of the basal level. Cortical bone thickness was unaffected in the wild-type mice but significantly decreased in the Pth-null animals (97% vs 56% of pretreatment thickness, respectively) (Fig. 7B). Because of the changes in bone volume that arose as a consequence of low dietary calcium, we then examined TRAP staining and quantified osteoclast number and size in these animals (Fig. 7C). In $Pth^{-/-}$ mice fed the normal calcium diet, these parameters were decreased compared to the wild-type littermates. However, when animals were moved to a low calcium diet for 3 days, TRAP staining and the number and size of osteoclasts increased, considerably

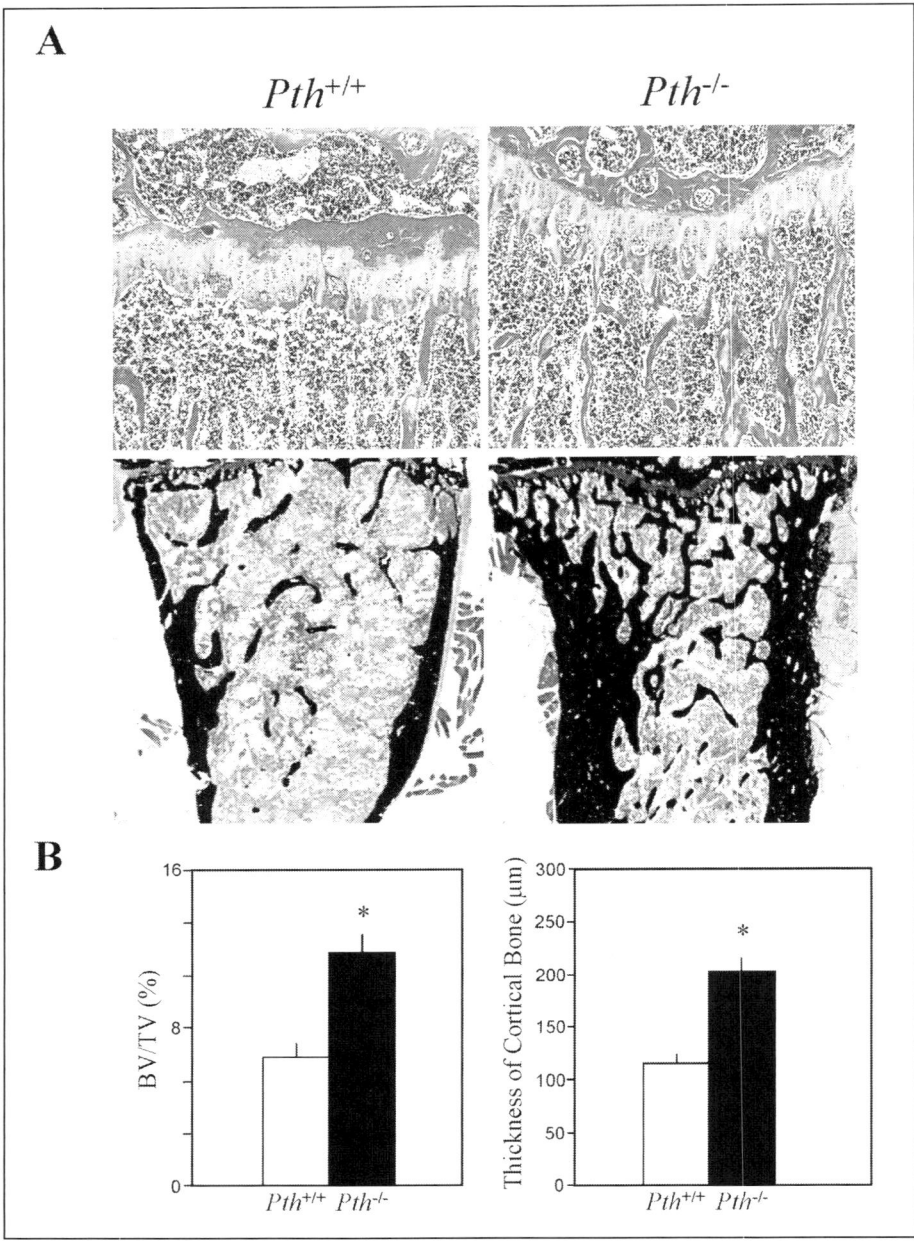

Figure 4. Altered bone development in *Pth*-null animals. A) Top panels: Histological analysis of the growth plate shows no abnormalities in the development of cartilage. Bottom panels: Von Kossa staining of metaphyseal region showing increased mineralized bone content in the mutant specimens. B) Quantitative assessment of histomorphometric measurements. Mice homozygous for targeted disruption of *Pth* demonstrated increased trabecular bone volume and cortical bone thickness compared to sex-matched littermates. BV/TV, Bone Volume/Total Volume. All data are mean ± SEM. * $P<0.05$ (n=15).

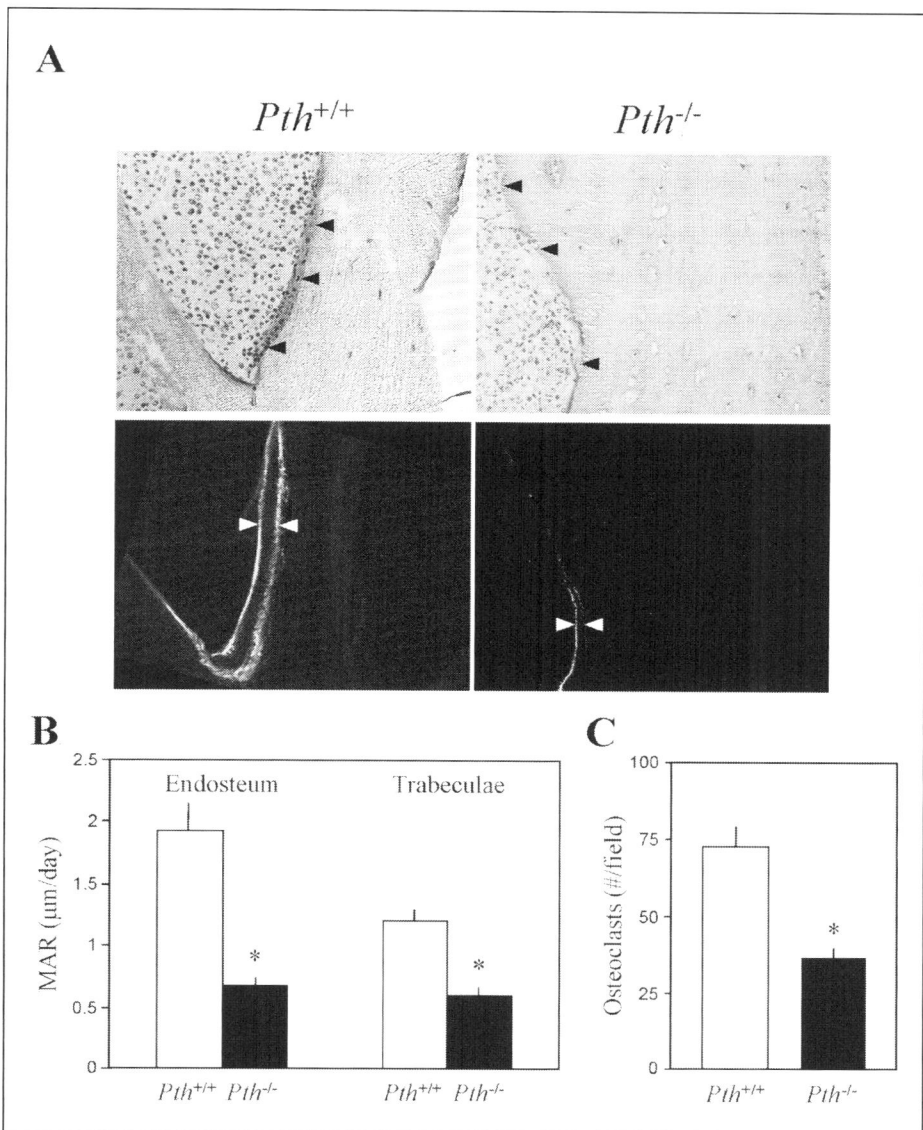

Figure 5. Increased bone content is related to decreased bone turnover in *Pth*-null mice. A) Top panels: Sections of cortical bone from wild-type and *Pth*-null mice. Black arrowheads outline the endosteal surface. Bottom panels: Double calcein-labeling as seen using fluorescence microscopy in the same cortical bone specimens. White arrowheads specify distance between labeled bone surfaces indicative of the amount of bone laid down over a period of 7 days. B) Mineral apposition rate (MAR), used as an indicator of bone formation rate, is significantly decreased in *Pth*-null cortical (endosteum) and trabecular bone. MAR is calculated by the thickness of bone between the two labels divided by the labeling interval (7 days). C) Osteoclast numbers are also diminished in the mutant mice. All data are reported as mean ± SEM * *P<0.05* (n=6).

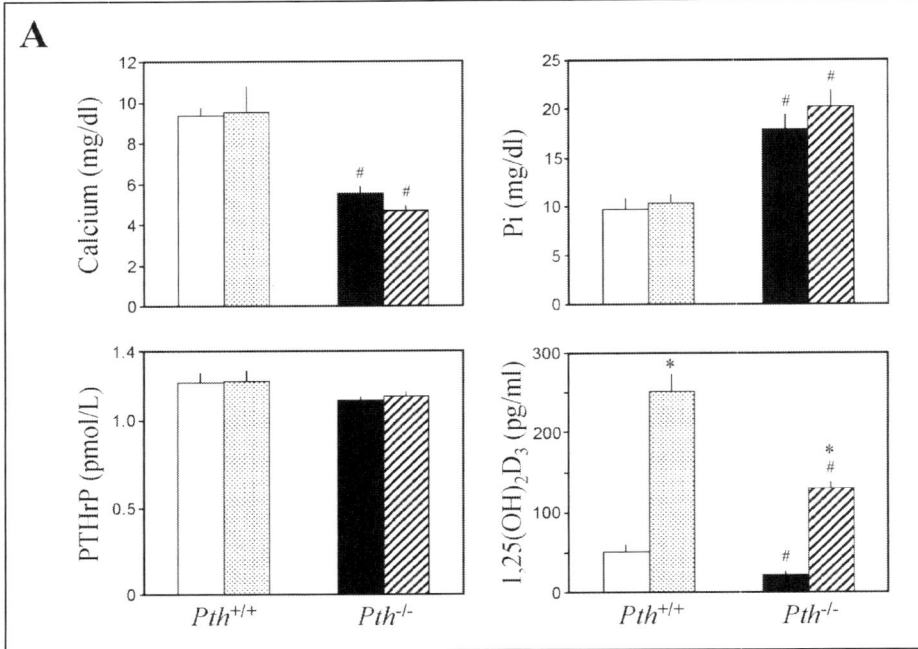

Figure 6. Effects of low-calcium diet on the biochemical and phenotype of wild-type and *Pth*-null mice. A) Serum levels of calcium, inorganic phosphate, and PTHrP remained unaltered in both wild-type and $Pth^{-/-}$ mice following administration of low calcium (stippled and striped bars, respectively) compared to a normal calcium (white and black solid bars, respectively) dietary intake, while circulating levels of $1,25(OH)_2D_3$ increased in both groups. All data are reported as mean ± SEM (n=6). *$P<0.05$ versus normal calcium diet; #$P<0.05$ versus wild-type mice. Figure continued on next page.

more so in the $Pth^{-/-}$ than in wild-type mice. These observations imply that a low calcium diet can potently increase $1,25(OH)_2D_3$ levels in vivo, even in the absence of PTH, and enhance bone resorption for maintenance of calcium homeostasis, but in the process can impact negatively on bone mass.

Discussion

Our findings in mice with targeted disruption of the *Pth* gene, demonstrate the profound manner in which developmental changes not only in the internal milieu but also in the external environment can modify the phenotype of animals with a single genetic alteration. Thus, our previous studies demonstrated that PTH-negative mice, in the protected intra-uterine environment of the fetus, show abnormalities at the chondro-osseous junction of the growth plate and in formation of the primary spongiosa and trabecular bone which point to an anabolic role for PTH at this stage of development.[15] Our current studies show that this anabolic effect of the hormone is transformed into a catabolic function in the post-natal state and this effect is modulated by both external factors as well as the ambient level of regulatory hormones and ions.

Furthermore, reproductive capacity is also influenced by these considerations. Thus, in the $Pth^{-/-}$ mice, reproductive organs appear to form normally in utero but in the post-natal state, defective ovarian function was noted with poor development of corpora lutea. The

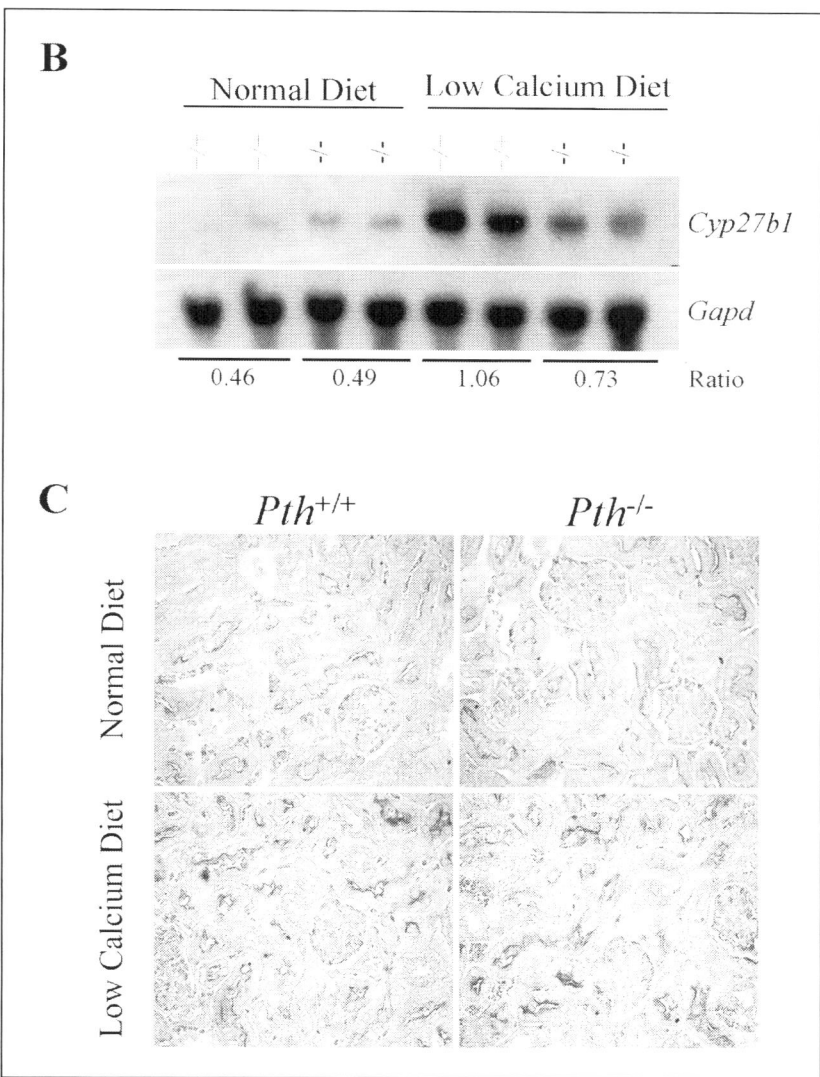

Figure 6, continued. B) Northern blot analysis for *Cyp27b1* transcript levels in total kidney RNA isolated from wild type and *Pth*$^{-/-}$ mice while on normal or low calcium diet. Similar analysis for *glyceraldehyde-3-phosphate dehydrogenase* (*Gapd*) mRNA was used to correct for differences in sample loading. C) Cyp27b1 immunoreactivity in renal tissues from wild-type and *Pth*-null mice on normal or low calcium diet. Results are representative of observations made from 4 animals in each group.

primary function of the corpus luteum is the secretion of progesterone, which is required for maintenance of normal pregnancy.[16] The corpus luteum develops from residual follicular granulosal and thecal cells after ovulation, as endothelial cells invade the ovulation site to form an extensive network of neovasculature that supports its rapid growth, which can exceed that of most rapidly growing tumors.[17] Hence, mediators of angiogenesis play a

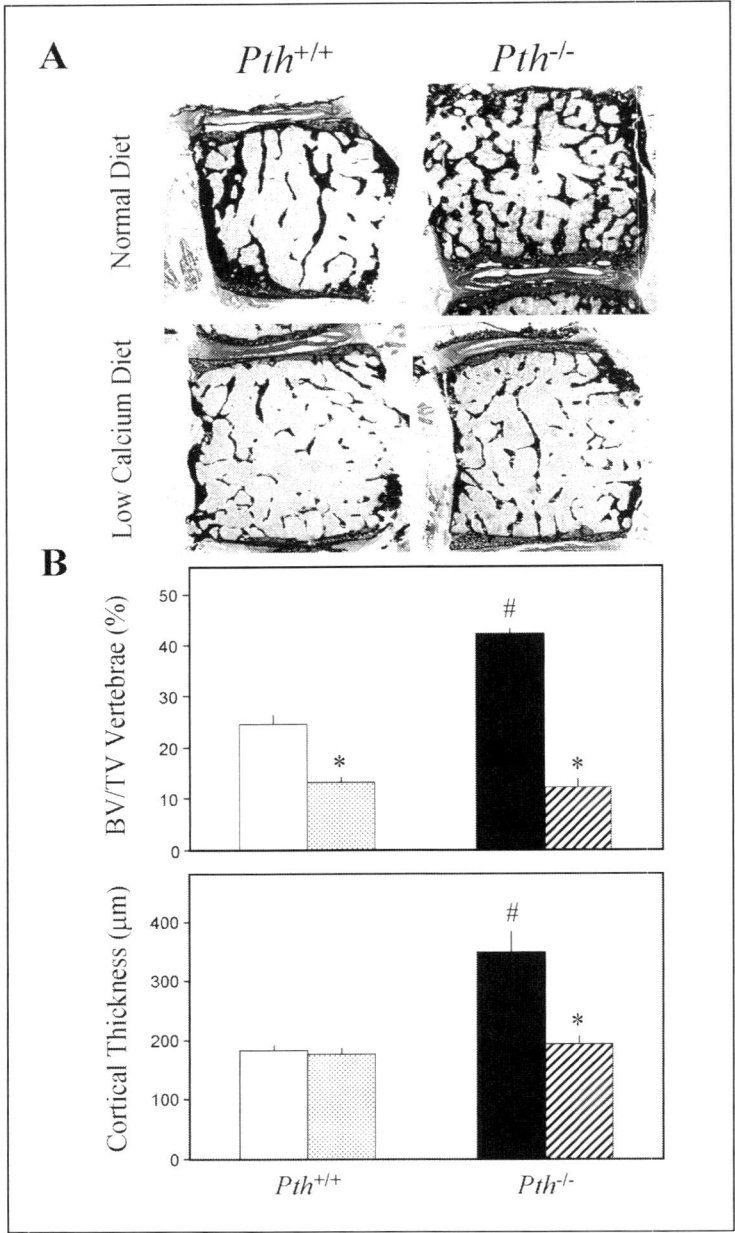

Figure 7. Calcium content of diet alters $Pth^{-/-}$ bone phenotype. A) Von Kossa stained sections of vertebral bodies from $Pth^{-/-}$ mice and wild type littermates on normal (top panels) and low (bottom panels) calcium diet. B) Quantification of changes in trabecular bone volume (BV/TV) and cortical bone thickness following intake of a low calcium diet (stippled and striped bars for wild-type and Pth-null mice, respectively). Figure continued on next page.

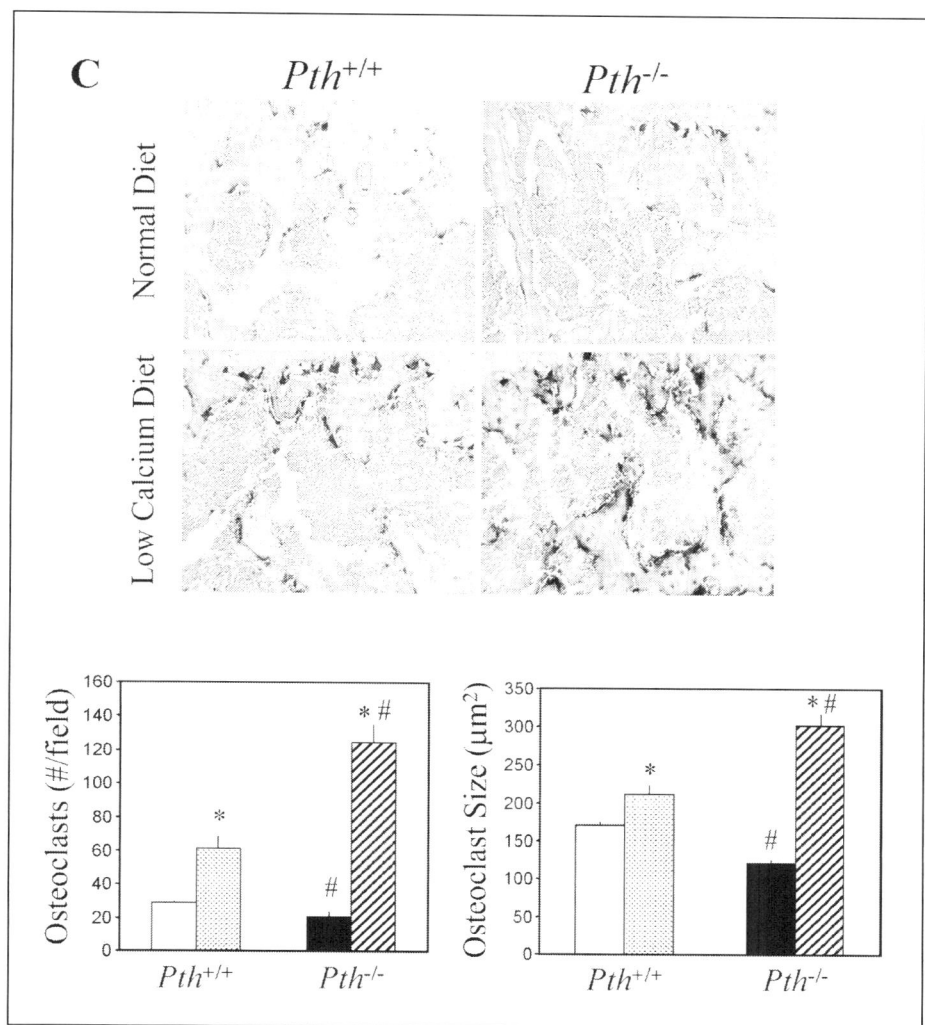

Figure 7, continued. C) Top: Decalcified paraffin embedded sections of femurs from the wild-type and PTH deficient mice fed with either normal (upper panels) or low (lower panels) calcium diet for three days were stained histochemically for TRAP. Below: Histomorphometric analysis showing that osteoclast number and size were increased in bones from mutant and wild-type litter mates following introduction of the low calcium diet. All data are reported as mean ± SEM (n=6). *$P<0.05$ versus normal calcium diet; #$P<0.05$ versus wild-type mice.

pivotal role in this process. VEGF, which is highly expressed during formation of the corpus luteum, and FGF2 have been implicated in luteal cell proliferation or turnover during early pregnancy, and may thereby contribute to the maintenance of luteal function, critical for the successful establishment of pregnancy. We show that VEGF and FGF2 expression was profoundly decreased in the ovaries of the *Pth* knockout mice, and these alterations negatively influenced neovasculature formation. Although it is unclear at present whether this diminished expression is a consequence of the lack of PTH per se, the prevailing hypocalcemia or of the decreased circulating $1,25(OH)_2D_3$ levels, the available experimental evidence would support hypocalcemia. For example, the reproductive dysfunction of the VDR-null mutant mice was corrected following initiation of a high calcium diet.[18] Moreover, here we demonstrate that the rise of endogenous $1,25(OH)_2D_3$ levels by low calcium dietary intake is insufficient to normalize the defective development of the corpus luteum, which in fact worsens, as fertility is completely lost. Lastly, the alterations in reproductive function observed in the *Pth*[-/-] mice have been described in humans with hypoparathyroidism and can be avoided by the early introduction of vitamin D and calcium replacement in this setting. This would suggest that PTH deficiency per se is not responsible for the impaired development of the corpus luteum.

The skeletal findings in the *Pth*-null mice are also of interest. First, they indicate that in the post-natal state PTH does not play a major role in cartilage development or in the development of the primary spongiosa, roles that may be subserved by locally produced PTHrP.[5,15] Second, in the presence of a normal calcium diet, the absence of PTH in the post-natal state was notable for markedly decreased bone turnover with resorption being particularly compromised. This leads to the marked increases in bone volume noted by our histomorphometric measurements. Observations in patients with hypoparathyroidism tend to support these findings. Increased bone mineral density (10-32%) has been reported in patients with chronic hypoparathyroidism following surgery for either thyroid cancer or hyperparathyroidism.[19] In addition, this condition provides protection against age-related bone loss in postmenopausal women, perhaps due to attenuation of the high turnover bone loss associated with menopause.[20] Although supplementation with vitamin D and calcium may have contributed to the increased bone mass in these patients, it would appear that high bone mineral density is a feature of hypoparathyroidism per se, as it has also been observed in untreated individuals with the disorder.[21,22] Consequently, in the post-natal state, where the maintenance of normal circulating calcium concentration in the organism is to a great extent dependent on more direct access to calcium in the external environment, the function of PTH appears to have evolved in order to primarily defend against decreases in the ambient calcium. This appears to involve stimulation of *Cyp27b1* expression to raise $1,25(OH)_2D_3$ levels, and induction of a catabolic action on bone to maintain normocalcemia.

Our findings in the post-natal *Pth*[-/-] mice and clinical observations in hypoparathyroid patients raise the possibility, therefore, that regulation of PTH secretion can provide a novel therapeutic avenue for the treatment of metabolic bone disease. Preliminary studies in animals tend to add credence to this hypothesis, as daily transient decreases in PTH levels following administration of the calcimimetic NPS R-568, a calcium-sensing receptor agonist, were shown to have an anabolic effect on uremic bones,[23,24] and to slow the rate of bone loss following ovariectomy.[25]

Finally, our studies of the *Pth*-null mice exposed to limiting amounts of calcium in the external environment point to additional mechanisms that can be mobilized to retain circulating calcium concentrations. Thus, in the presence of a low calcium diet, even in the absence of PTH, *Cyp27b1* expression in the kidney was increased, circulating $1,25(OH)_2D_3$ concentrations were augmented, bone resorption was enhanced, and the increased bone volume noted in the hypoparathyroid mice on a normal calcium diet was converted to an osteopenic state. Most likely, limiting amounts of dietary calcium resulted in transient further reduction of the hypocalcemia observed in the hypoparathyroid animals on normal calcium intake. This initiated *Cyp27b1* stimulation, increases in $1,25(OH)_2D_3$, augmented osteoclastogenesis in bone[26] and mobilization of calcium stores from bone. Hence, a new steady state was reached in which severe hypocalcemia was "re-set" to the moderate levels but at the expense of extreme osteopenia. This is consistent with previous reports suggesting that extracellular calcium concentrations can, independently of PTH, regulate *Cyp27b1* activity in vivo[27] and in vitro.[28] However, our studies suggest that in the presence of a normal calcium intake, the ensuing moderate hypocalcemia is less effective in enhancing *Cyp27b1* expression than in the presence of a reduced calcium intake where more extreme hypocalcemia may transiently exist. Consequently, the first line of defense in stimulating *Cyp27b1* transcription and maintaining a normal circulating calcium concentration appears to be augmentation of PTH levels whereas $1,25(OH)_2D_3$ may be directly mobilized, even in the absence of PTH, as hypocalcemia becomes more extreme.

An additional possibility is that intestinal epithelial cells directly play a pivotal role in defending against a further fall in calcium when dietary calcium is reduced. It is possible, that enterocytes have the capacity to sense the decreasing levels of dietary calcium intake and, in turn, release a putative signal, perhaps a circulating agent that acts at the level of the kidney to increase *Cyp27b1* expression. A concomitant effect of such a factor on the skeleton to directly promote bone resorption cannot be excluded.

Materials and Methods

Generation of Pth Knockout Mice

To clone the murine *Pth* gene, a radiolabeled cDNA encoding human PTH[29] was used as probe to screen a 129/Sv mouse genomic DNA library. Following isolation and characterization of the *Pth* gene,[13] the targeting vector was generated by introducing a 2.4 kb *Bam*HI/*Xho*I fragment corresponding to the 5' region of homology in the *Xho*I site of the pPNT vector,[30] while a 3.7 kb *Bam*HI fragment corresponding to the 3' region of homology was ligated in the unique *Bam*HI site of the vector. The strategy for ablating *Pth* focused on deleting PTH-encoding sequences from the mouse genome. In this construct, part of exon 3 encoding the entire sequence of the mature PTH form was replaced by the neomycin resistance (neo^r) selection gene cassette. The vector was linearized with *Not*I and electroporated into D3 ES cells maintained on mitotically-inactivated mouse primary embryonic fibroblasts resistant to G418. Following G418 (300 μg/ml) and 1-[2-deoxy,2-fluoro-β-D-arabinofuranosyl]-5-iodouracil (FIAU; 0.2 μM) selection, resistant colonies were isolated and the fidelity of the targeting event was verified by Southern blot analysis of genomic DNA. Appropriately targeted clones were then injected into 3.5-day *pc* C57B/L6 blastocysts and extensively chimeric male mice were mated to C57B/L6 females. Following germline transmission of the mutation, mice were bred to generate animals homozygous for the targeted *Pth* allele. Animals used in the present studies were obtained following at least six backcrossings into the C57B/L6 background.

Animal Experimentation

All animal experiments were reviewed and approved by the institutional animal care committee. The mice were housed in a 12-h light/12-h dark cycle. They were maintained in cages with wooden shavings and consumed water and either a normal calcium diet (0.95% calcium, 0.67% phosphorus and 4.5 IU/g vitamin D_3; PMI Feeds, Inc., St. Louis, MO) or a low calcium diet (0.001-0.005% calcium, 0.4% phosphorus and 2.4 IU/g vitamin D_3) ad libitum for the indicated time periods. Fertility in mice was defined as the number of successful pregnancies following visualization of a vaginal plug. The estrous cycle was staged by examining vaginal smears.

Serum Biochemistry

Serum concentrations of calcium and inorganic phosphorus were determined by routine methods using Sigma Diagnostics reagents (Sigma Diagnostics). Serum intact PTH was measured with an ELISA assay (Immutopics, Inc., San Clemente, CA) while serum PTHrP and

1,25$(OH)_2D_3$ determinations were performed using commercially available RIA kits (Nichols Institute Diagnostics, San Clemente, CA and Immunodiagnostic Systems, UK, respectively). Serum progesterone and estradiol levels were measured with an immunoassay from ADVIA Centaur Immunoassay System (Bayer Diagnostics, Tarrytown, NY).

Histology and Histochemistry

Ovaries, uterii, thyroparathyroidal tissue, femurs, tibiae, and vertebrae were removed from 6-week-old mice ($Pth^{+/+}$ and $Pth^{-/-}$ taken from the same litter) and fixed in PLP fixative (2% paraformaldehyde containing 0.075 M lysine and 0.01 M sodium periodate solution) overnight at 5°C prior to processing. Occasionally, bones were decalcified in ethylene-diamine tetra-acetic acid (EDTA) glycerol solution for 5-7 days at 5°C. Tissue samples were dehydrated and embedded in paraffin after which 5 µm sections were cut on a rotary microtome. The sections were stained with haematoxylin and eosin (H&E), for tartrate-resistant acid phosphatase (TRAP) or immunostained, as described below. Undecalcified bones were embedded in LR White acrylic resin (London Resin Company Ltd, U.K.). 1µm sections were cut on an ultramicrotome and stained for mineral with the von Kossa staining procedure using toluidine blue as counterstain.

Calcein labeling was performed by intra-peritoneal injection with 10 µg calcein/g bodyweight (C-0875, Sigma Chemical Co., St. Louis, MO) at 10 and 3 days before sacrifice. Bones were harvested and embedded in LR White acrylic resin. Serial sections were cut and the freshly cut surface of each section was imaged using fluorescence microscopy. The double calcein interlabel width in cortical and trabecular bone was measured using Northern Eclipse v6.0 (Empix Imaging Inc., Mississauga, ON) image software and the mineral apposition rate (MAR=interlabel width/labeling period) was calculated.

For immunohistochemistry, paraffin sections of thyroparathyroidal tissue were stained for PTH and calcium-sensing receptor (CaSR) immunoreactivity by the avidin-biotin-peroxidase complex (ABC) technique using goat serum against PTH 1-34 and mouse anti-CaSR monoclonal antibody, as described.[15] Ovaries were stained for vascular endothelial growth factor (VEGF) and fibroblast growth factor-2 (FGF2) immunoreactivity using goat antiserum against VEGF and rabbit antiserum against FGF2 (Santa Cruz Biotechnology Inc., Santa Cruz, CA). Kidney sections were immunostained for Cyp27b1 using purified rabbit antiserum.

Computer-Assisted Image Analysis

Computer-assisted image analysis was performed, as previously described.[31] For determining the area of the mineralized and unmineralized matrix, and the number and size of osteoclasts in stained bone sections, images of primary spongiosa and cortical bone were digitally recorded using a rectangular template and three different fields. In the primary spongiosa, each image was photographed from the edge of the metaphyseal border of the growth plate (i.e., at the level of the zone of vascular invasion). In cortical bone, images were taken from the diaphyseal bone close to the metaphysis. All digital images were captured with a Sony digital camera at a magnification of x200, producing a field area of 0.4 mm^2. The positive and negative areas staining in trabecular and cortical bone were measured by digital image analysis using Northern Eclipse v6.0 image software.

Northern Blot Analysis

A cDNA fragment corresponding to nucleotides 421-1474 of mouse 25-hydroxyvitamin D_3 1α-hydroxylase (*Cyp27b1*; Accession Number AB006034) was prepared by RT-PCR of mouse kidney RNA, subcloned, and sequenced. DNA probes for *Cyp27b1* and *Glyceraldehyde-3-phosphate dehydrogenase* (*Gapd*) were prepared by Random Primed DNA Labeling Kit (Roche) and [α-^{32}P]dCTP (800 Ci/mmol; NEN) Total RNA was isolated from kidney with Tripure Isolation Reagent (Roche), and 20 μg aliquots were fractionated by electrophoresis on a 1% formaldehyde agarose gel, transferred to nitrocellulose membranes and hybridized to the radiolabeled cDNA fragments (48% formamide, 10% dextran sulfate, 5xSSC, 1xDenhardt's and 100 μg/ml salmon sperm DNA) at 42°C overnight. The membranes were washed and autoradiograms were prepared using Kodak BioMax film at −80°C with intensifying screens. Quantification of signal intensity on autoradiograms was performed by Molecular Dynamics Personal Densitometer using ImageQuant software.

Statistical Analysis

Data from biochemical and image analyses are presented as means ± SEM. Statistical comparisons were made using Student's t test, with $P < 0.05$ being considered significant.

Acknowledgements

This work was supported by the Canadian Institutes of Health Research (CIHR) and the Canadian Arthritis Network (CAN). D.M. and A.C.K. are recipients of a CIHR Postdoctoral Fellowship and Scientist Award, respectively.

References

1. Juppner H, Abou-Samra AB, Freeman M et al. A G protein-linked receptor for parathyroid hormone and parathyroid hormone-related peptide. Science 1991; 254:1024-1026.
2. Schipani E, Karga H, Karaplis AC et al. Identical complementary deoxyribonucleic acids encode a human renal and bone parathyroid hormone (PTH)/PTH-related peptide receptor. Endocrinology 1993; 132:2157-2165.
3. Potts JT Jr, Kronenberg HM, Rosenblatt M. Parathyroid hormone: chemistry, biosynthesis, and mode of action. Adv Protein Chem 1982; 35:323-396.
4. Strewler GJ. The physiology of parathyroid hormone-related protein. N Engl J Med 2000; 342:177-185.
5. Karaplis AC, Luz A, Glowacki J et al. Lethal skeletal dysplasia from targeted disruption of the parathyroid hormone-related peptide gene. Genes Dev 1994; 8:277-289.
6. Lanske B, Karaplis AC, Lee K et al. PTH/PTHrP receptor in early development and Indian hedgehog-regulated bone growth. Science 1996; 273:663-666.
7. Jobert AS, Zhang P, Couvineau A et al. Absence of functional receptors for parathyroid hormone and parathyroid hormone-related peptide in Blomstrand chondrodysplasia. J Clin Invest 1998; 102:34-40.
8. Karaplis AC, He B, Nguyen MT et al. Inactivating mutation in the human parathyroid hormone receptor type 1 gene in Blomstrand chondrodysplasia. Endocrinology 1998; 139:5255-5258.
9. Usdin TB, Hilton J, Vertesi T et al. Distribution of the parathyroid hormone 2 receptor in rat: immunolocalization reveals expression by several endocrine cells. Endocrinology 1999; 140:3363-3371.
10. Usdin TB. The PTH2 receptor and TIP39: a new peptide-receptor system. Trends Pharmacol Sci 2000; 21:128-130.
11. Inomata N, Akiyama M, Kubota N et al. Characterization of a novel parathyroid hormone (PTH) receptor with specificity for the carboxyl-terminal region of PTH-(1-84). Endocrinology 1995; 136:4732-4740.
12. Divieti P, Inomata N, Chapin K et al. Receptors for the carboxyl-terminal region of PTH(1-84) are highly expressed in osteocytic cells. Endocrinology 2001; 142:916-925.
13. He B, Tong TK, Hiou-Tim FF et al. The murine gene encoding parathyroid hormone: genomic organization, nucleotide sequence and transcriptional regulation. J Mol Endocrinol 2002; 29:193-203.
14. Gunther T, Chen ZF, Kim J et al. Genetic ablation of parathyroid glands reveals another source of parathyroid hormone. Nature 2000; 406:199-203.

15. Miao D, He B, Karaplis AC et al. Parathyroid hormone is essential for normal fetal bone formation. J Clin Invest 2002; 109:1173-1182.
16. Jablonka-Shariff A, Grazul-Bilska AT, Redmer DA et al. Cellular proliferation and fibroblast growth factors in the corpus luteum during early pregnancy in ewes. Growth Factors 1997; 14:15-23.
17. Reynolds LP, Redmer DA. Expression of the angiogenic factors, basic fibroblast growth factor and vascular endothelial growth factor, in the ovary. J Anim Sci 1998; 76:1671-1681.
18. Johnson LE, DeLuca HF. Vitamin D receptor null mutant mice fed high levels of calcium are fertile. J Nutr 2001; 131:1787-1791.
19. Abugassa S, Nordenstrom J, Eriksson S et al. Bone mineral density in patients with chronic hypoparathyroidism. J Clin Endocrinol Metab 1993; 76:1617-1621.
20. Fujiyama K, Kiriyama T, Ito M et al. Attenuation of postmenopausal high turnover bone loss in patients with hypoparathyroidism. J Clin Endocrinol Metab 1995; 80:2135-2138.
21. Orr-Walker B, Harris R, Holdaway IM et al. High peripheral and axial bone densities in a postmenopausal woman with untreated hypoparathyroidism. Postgrad Med J 1990; 66:1061-1063.
22. Van Offel JF, De Gendt CM, De Clerck LS et al. High bone mass and hypocalcaemic myopathy in a patient with idiopathic hypoparathyroidism. Clin Rheumatol 2000; 19:64-66.
23. Ishii H, Wada M, Furuya Y et al. Daily intermittent decreases in serum levels of parathyroid hormone have an anabolic-like action on the bones of uremic rats with low-turnover bone and osteomalacia. Bone 2000; 26:175-182.
24. Olgaard K, and Lewin E. Prevention of uremic bone disease using calcimimetic compounds. Annu Rev Med 2001; 52:203-220.
25. Miller MA, Fox J. Daily transient decreases in plasma parathyroid hormone levels induced by the calcimimetic NPS R-568 slows the rate of bone loss but does not increase bone mass in ovariectomized rats. Bone 2000; 27:511-519.
26. Thomas GP, Baker SU, Eisman JA et al. Changing RANKL/OPG mRNA expression in differentiating murine primary osteoblasts. J Endocrinol 2001; 170:451-460.
27. Weisinger JR, Favus MJ, Langman CB et al. Regulation of 1,25-dihydroxyvitamin D3 by calcium in the parathyroidectomized, parathyroid hormone-replete rat. J Bone Miner Res 1989; 4:929-935.
28. Bland R, Walker EA, Hughes SV et al. Constitutive expression of 25-hydroxyvitamin D3-1alpha-hydroxylase in a transformed human proximal tubule cell line: evidence for direct regulation of vitamin D metabolism by calcium. Endocrinology 1999; 140:2027-2034.
29. Vasicek TJ, McDevitt BE, Freeman MW et al. Nucleotide sequence of the human parathyroid hormone gene. Proc Natl Acad Sci USA 1983; 80:2127-2131.
30. TybulewiczVL, Crawford CE, Jackson PK, et al. Neonatal lethality and lymphopenia in mice with a homozygous disruption of the c-abl proto-oncogene. Cell 1991; 65:1153-1163.
31. Bai X, Miao D, Panda D et al. Partial rescue of the Hyp phenotype by osteoblast-targeted PHEX (Phosphate-regulating gene with Homologies to Endopeptidases on the X chromosome) expression. Mol Endocrinol 2002; 16:2913-2925.

Index

A

1α(OH)D$_2$ 95, 102-104, 108, 109
Adenylate cyclase 44, 49
AU rich binding factor 1 (AUF1) 20, 57, 61, 65
AU rich element (ARE) 12, 20, 61
Autosomal dominant hypocalcemic hypercalciuria (ADHH) 167, 170
Autosomal recessive hypoparathyroidism 164, 166

B

Bone 1, 2, 26, 29, 37, 39, 40, 44, 45, 51, 57, 95-97, 99, 100, 102, 105-108, 113, 114, 130, 141, 145, 164, 172, 179, 180, 185-188, 190-194

C

Calcilytics 48, 50-52
Calcimimetics 48, 50-52, 91, 192
Calcium (Ca^{2+}) 1-3, 6, 16, 20, 25, 26, 29, 44-51, 57-61, 63-65, 68, 81, 84, 85, 87-91, 95-97, 99-108, 113, 123, 134, 141, 145-147, 149, 153, 159, 161, 167-169, 174, 179, 180, 182, 183, 185, 188-194
Calcium receptor 84, 89, 153
Calcium set point 134
Calreticulin 84, 89, 90
Catch-22 159, 160, 162-164
Cell cycle 49, 70, 100, 114, 115, 119, 120, 122, 130, 140, 143, 144, 150, 153
Chromosome 11q13 129, 131, 143, 144, 147-149, 152
Chronic renal failure 48, 51, 84, 89, 90, 92, 95, 96, 100, 109, 113
cis element 16, 19-23, 26, 63-65
Crystal structure 31, 33-37
Cyclic AMP response element (CRE) 24, 71, 80
Cyclin D1 49, 92, 114, 115, 119, 122, 129-131, 140, 142-147, 152, 153

D

DiGeorge syndrome (DGS) 1, 5, 6, 159, 160, 162-164
DiGeorge syndrome/Catch-22 159, 160
Direct repeat element 86

E

E2F 117, 118, 144
EGF receptor (EGFR) 119-124, 130
Embryonic day 11 (E11) 2, 3, 5
Extracellular calcium 1, 29, 44, 81, 134, 146, 167, 179, 192

F

Familial hypocalcuric hypercalcemia (FHH) 47-49, 134, 149

G

G-protein coupled receptor 61, 167, 169
Gcm2 1, 2, 3, 5, 166

H

HDR syndrome 6, 164, 165
Hoxa3 1, 4, 5
Hypercalcemia 29, 45, 47-49, 85, 90, 95-97, 99, 100, 102-104, 123, 124, 130, 131, 134, 140, 149, 151, 170, 179
Hyperparathyroidism 47-49, 52, 84, 87-92, 95-104, 109, 113-116, 120, 122, 128, 129, 132, 134, 140, 141, 145, 146, 149, 150, 152, 170, 185, 192
Hyperphosphatemia 51, 95, 96, 103, 113, 120, 122, 159, 171, 179, 180, 182, 185
Hypocalcemia 20, 48, 50, 57, 58, 65, 87, 88, 90, 91, 95, 96, 113, 145, 159, 162, 164, 166, 171, 172, 179, 180, 182, 185, 192, 193
Hypoparathyroidism 6, 48, 49, 159, 160, 162, 164, 166, 167, 170, 171, 174, 179, 180, 192
Hypophosphatemia 20, 57, 58, 65

I

Intracellular calcium 46, 49, 167, 169

J

JunD 148, 153

M

MAPK 44, 50, 52, 119
MENIN 131, 132
Multiple endocrine neoplasia type 1 (MEN1) 128, 129, 131, 140, 147-150, 152

N

19-nor-1,25(OH)$_2$ 95, 99, 106, 120-122
Nuclear magnetic resonance (NMR) 30, 31, 33-35, 37, 39, 40

O

1,25(OH)$_2$D$_3$ 44, 49, 58, 75, 84-92, 95-97, 99, 100, 102-109, 113, 116, 120-122, 179, 180, 182, 185, 188, 192-194
22-oxacalcitriol 95, 97, 104
Osteoblast 45, 87, 106, 108, 180, 185
Osteoclast 9, 106-108, 180, 185, 187, 191, 194
Osteoporosis 29, 40, 51, 52, 68

P

p21 92, 100, 116-124
Parathyroid adenoma 11, 115, 116, 128-130, 132-134, 140-146, 148-152
Parathyroid carcinoma 128, 129, 132, 133, 140, 141, 145, 151, 152
Parathyroid gland 1-5, 11, 21, 44, 45, 47, 49, 52, 57, 68, 85, 87-89, 91, 92, 95-97, 100, 104, 105, 108, 113-117, 119-124, 128, 140, 141, 145, 146, 152, 153, 159-162, 166-168, 170, 174, 179, 183
Parathyroid hyperplasia 47, 88, 96, 100, 113, 114, 120, 121, 123, 124, 129-131, 141, 182, 183
Parathyroid proliferation 57, 134, 150
Parathyroid tumorigenesis 128-134, 140, 144, 145, 150, 151
Pax1 1, 5
Pax9 1, 5
Phosphate 1, 10, 16, 20, 26, 49, 57-61, 63-65, 68, 84, 88, 89, 95, 96, 99-101, 103, 104, 113, 116, 120, 141, 159, 182, 188, 189, 194
Phosphate retention 95
Phospholipase C (PLC) 29, 38, 44, 46, 49, 50, 167, 169
Post-transcriptional regulation 20, 57, 58
PRAD1 114, 115, 129, 130
PreProPTH 8-10, 12, 16, 22, 25, 85, 167, 168
Protein kinase A (PKA) 46, 173
Protein kinase C (PKC) 46, 50, 167, 173
Protein modeling 65
Protein-RNA binding 63-65
Protein-RNA interaction 58
Protein structure 33
Pseudohypoparathyroidism (PHP) 161, 171-174
PTH (parathyroid hormone) 1-6, 8-12, 16-26, 29-31, 33-35, 37-40, 44-52, 57-65, 68, 70-81, 84-92, 95-97, 99-106, 109, 113-115, 121, 129, 130, 134, 140-147, 149, 152, 159, 161, 166-169, 171-174, 179, 180, 182, 183, 185, 188, 191-194
PTH 1-34 194
PTH mRNA 8-12, 16-21, 23-26, 57-65, 84, 85, 87, 88, 90, 91, 96, 97, 100, 113
Pth-null mice 179, 180, 182-184, 187-190, 192
PTH promoter 90
PTH/PTHrP receptor 29, 173
PTHrP (PTH-related protein) 3, 23, 29, 30, 33-35, 37-39, 50, 159, 173, 179, 180, 185, 188, 192, 193
PTH secretion 2, 44, 45, 50, 51, 57, 58, 84, 85, 90, 92, 95-97, 100, 134, 140, 146, 147, 167, 169, 192
PTH transcription 72, 85

Index

R

Rae28 1, 4
Regulatory element 68, 70, 71, 75, 80, 81, 143
Renal failure 48, 49, 51, 84, 89-92, 95-97, 100-103, 109, 113-116, 120-122, 124
RET 130, 149, 153
RNA stability 20, 61, 63, 65

S

Secondary hyperparathyroidism 48, 49, 84, 87-92, 95-104, 109, 113, 114, 120-122, 134, 141, 146, 152, 185
Signal transduction 46, 119, 129

T

Tbx1 5, 163
Tertiary hyperparathyroidism 89, 152
TGFα 92, 100, 117, 119-124
TGFβ 131, 132, 148
Thymus 2-6, 159, 162, 163, 166, 171, 180
TIP39 30, 180
trans factor 20, 57, 64, 65, 68, 130
Transcription factor (TF) 4-6, 68-70, 75-78, 80, 81, 86, 119, 129, 131, 144, 163, 166

U

3' untranslated region 11, 12, 20, 22, 23, 25, 113
Uremia 90-92, 113, 114, 116, 119-121, 123, 124, 134, 150, 152

V

Vitamin D 2, 24, 48, 49, 71, 84-92, 95, 97, 98, 100, 102, 104-109, 113-116, 118, 120-124, 134, 141, 146, 150, 153, 185, 192
Vitamin D analog 95, 97, 98, 100, 102, 104, 105, 108, 109, 121-123
Vitamin D receptor (VDR) 49, 70, 71, 75, 77, 78, 80, 81, 84-92, 97, 104-106, 108, 109, 113, 116, 120-123, 134, 146, 150, 153, 192
Vitamin D responsive element (VDRE) 24, 49, 70, 71, 75, 77-81, 84-90

X

X-linked hypoparathyroidism 167